AF875165

MICRONUTRIENTS AND MACRONUTRIENTS AS NUTRACEUTICALS

AAP Advances in Nutraceuticals

MICRONUTRIENTS AND MACRONUTRIENTS AS NUTRACEUTICALS

Edited by

Prakash Chandra Gupta, PhD

Sayan Bhattacharyya, MBBS, MD

Nisha Sharma, PhD, MPharm

Rajesh K. Kesharwani, PhD

Raj K. Keservani, PhD, MPharm

AAP | APPLE ACADEMIC PRESS

First edition published 2024

Apple Academic Press Inc.
1265 Goldenrod Circle, NE,
Palm Bay, FL 32905 USA

760 Laurentian Drive, Unit 19,
Burlington, ON L7N 0A4, CANADA

CRC Press
2385 NW Executive Center Drive,
Suite 320, Boca Raton FL 33431

4 Park Square, Milton Park,
Abingdon, Oxon, OX14 4RN UK

© 2024 by Apple Academic Press, Inc.

Apple Academic Press exclusively co-publishes with CRC Press, an imprint of Taylor & Francis Group, LLC

Reasonable efforts have been made to publish reliable data and information, but the authors, editors, and publisher cannot assume responsibility for the validity of all materials or the consequences of their use. The authors are solely responsible for all the chapter content, figures, tables, data etc. provided by them. The authors, editors, and publishers have attempted to trace the copyright holders of all material reproduced in this publication and apologize to copyright holders if permission to publish in this form has not been obtained. If any copyright material has not been acknowledged, please write and let us know so we may rectify in any future reprint.

Except as permitted under U.S. Copyright Law, no part of this book may be reprinted, reproduced, transmitted, or utilized in any form by any electronic, mechanical, or other means, now known or hereafter invented, including photocopying, microfilming, and recording, or in any information storage or retrieval system, without written permission from the publishers.

For permission to photocopy or use material electronically from this work, access www.copyright.com or contact the Copyright Clearance Center, Inc. (CCC), 222 Rosewood Drive, Danvers, MA 01923, 978-750-8400. For works that are not available on CCC please contact mpkbookspermissions@tandf.co.uk

Trademark notice: Product or corporate names may be trademarks or registered trademarks and are used only for identification and explanation without intent to infringe.

Library and Archives Canada Cataloguing in Publication

Title: Micronutrients and macronutrients as nutraceuticals / edited by Prakash Chandra Gupta, PhD, Sayan Bhattacharyya, MBBS, MD, Nisha Sharma, PhD, MPharm, Rajesh K. Kesharwani, PhD, Raj K. Keservani, PhD, MPharm.
Names: Gupta, Prakash Chandra (Lecturer in pharmaceutical sciences), editor. | Bhattacharyya, Sayan, editor. | Sharma, Nisha (Director of the School of Pharmaceutical Sciences), editor. | Kesharwani, Rajesh Kumar, 1978- editor. | Keservani, Raj K., 1981- editor.
Series: AAP advances in nutraceuticals.
Description: First edition. | Series statement: AAP advances in nutraceuticals | Includes bibliographical references and index.
Identifiers: Canadiana (print) 20230515258 | Canadiana (ebook) 20230515304 | ISBN 9781774912263 (hardcover) | ISBN 9781774913802 (softcover) | ISBN 9781003415381 (ebook)
Subjects: LCSH: Functional foods. | LCSH: Vitamins. | LCSH: Minerals in nutrition.
Classification: LCC QP144.F85 M53 2024 | DDC 613.2—dc23

Library of Congress Cataloging-in-Publication Data

CIP data on file with US Library of Congress

ISBN: 978-1-77491-226-3 (hbk)
ISBN: 978-1-77491-380-2 (pbk)
ISBN: 978-1-00341-538-1 (ebk)

ABOUT THE BOOK SERIES: AAP ADVANCES IN NUTRACEUTICALS

In the modern era, mankind has witnessed a paradigm shift with respect to fundamental eating behavior. The lack of physical workouts and busy schedules at offices and in households have promoted the consumption of junk foods, which eventually results in numerous diseases and disorders since the nutritional content from fast food is inadequate. Plumpness and obesity have become a global health threat. The leading causes of death from most developing nations are noncommunicable diseases such as cardiovascular diseases, cancer, arthritis, osteoporosis, and liver toxicity. Patients suffering from such lifestyle ailments are a bit apprehensive towards prolonged use of costly modern therapeutics, which encourages instead the use of alternative approaches for management of such diseases and disorders.

The emerging sector of the nutraceutical industry encompasses products derived from nature, dietary supplements, and functional foods. Nutraceuticals are used for treatment and prevention of a broad range of diseases, such as the common cold, arthritis, sleep-related disorders, cancers, cardiovascular complications, metabolic disorders, and others.

The research on nutraceuticals is increasing day by day, considering the beneficial effects of food or food supplements in the management of diverse diseases. The issue of paramount concern is standardization and establishment of clinical efficacy of nutraceuticals, which is in fact a challenge for researchers around the globe.

This book series aims at realizing the significance of the variety of nutraceuticals for human well-being. The books in the series will also emphasize the role of dieticians and nutritionists for the prescription of judicious eating habits. The key food components such as carbohydrates, proteins, and lipids as well as micronutrients (vitamins, minerals) are demonstrated to maintain good health, obviating the need for medicines. Thus, nutraceuticals that are indeed derived from food ingredients are believed to be potential alternative therapeutics.

Asymptomatic diseases/disorders necessitate proper diagnosis. Stressed circumstances have been reported to cause weight loss. Workplace stress is a key variable in the genesis and propagation of metabolic disorders.

The nuances of the underlying mechanisms are still to be deciphered. Poor lifestyle trends have been attributed along with the consumption of junk foods in several such instances. Sugar-rich carbonated beverages, the use of contraband drugs, and the consumption of liquor in excess are assigned as risk factors. This book series will cover advances and applications in addition to providing the basics of nutraceuticals.

The books in this series may be valuable resources for industry professionals to design and develop quality products for end users. In addition, the book series will be fortified by ideas of innovation, concept building, manufacturing aspects, quality control, and regulatory status of nutraceuticals.

Ostensibly, business has seen a rise in the nutraceutical consumption in recent years. Pure, safe, efficacious products are the need of the hour globally. There are a number of books on nutraceuticals in the market, yet these are only a few books presenting certain aspects of nutraceuticals therein. This book series aims to offer a holistic view of this promising strategy, which is picking up pace with time. This book series will be a unique endeavor to bring manufacturing, research and development, and marketing strategies under a single umbrella.

CURRENT BOOKS IN THE SERIES

- **Micronutrients and Macronutrients as Nutraceuticals**
  Editors: Prakash Chandra Gupta, PhD,
  Sayan Bhattacharyya, MBBS, MD, Nisha Sharma, PhD, MPharm,
  Rajesh K. Kesharwani, PhD, and Raj K. Keservani, PhD, MPharm
- **Formulations, Regulations, and Challenges of Nutraceuticals**
  Editors: Tingirikari Jagan Mohan Rao, PhD,
  Rajesh K. Kesharwani, PhD, Raj. K. Keservani, PhD, MPharm, and
  Anil K. Sharma, PhD
- **Food Supplements and Dietary Fibers in Health and Disease**
  Editors: Bhushan R. Rane, PhD, MPharm,
  Raj K. Keservani, PhD, MPharm, Durgesh Singh, DPhil,
  Nayan A. Gujarathi, PhD, and Ashish S. Jain, PhD

- **Nutraceuticals in Cancer Prevention, Management, and Treatment**
  Editors: Raj K. Keservani, PhD, MPharm, Bui Thanh Tung, PhD, Sippy Singh, DPhil, and Rajesh K. Kesharwani, PhD
- **Advances in Flavonoids for Human Health and Prevention of Diseases**
  Editors: Nisha Sharma, PhD, Deepika Saini, PhD, Rajesh K. Kesharwani, PhD, Prakash Chandra Gupta, and Raj. K. Keservani, PhD, MPharm
- **The Flavonoids: Extraction and Applications**
  Editors: Deepika Saini, PhD, Rajesh K. Kesharwani, PhD, and Raj K. Keservani, PhD, MPharm
- **Immune-Boosting Nutraceuticals for Better Human Health: Novel Applications**
  Editors: Urmila Jarouliya, PhD, Raj K. Keservani, PhD, MPharm, Rajesh K. Kesharwani, PhD, Virendra K. Patel, PhD, and Adi D. Bharti
- **Applications of Functional Foods in Disease Prevention**
  Editors: Raj K. Keservani, PhD, MPharm, and Eknath D. Ahire, MSPharm
- **Plant Metabolites and Vegetables as Nutraceuticals**
  Editors: Raj K. Keservani, PhD, MPharm, Bui Thanh Tung, PhD, Rajesh K. Kesharwani, PhD, MPharm, and Eknath D. Ahire, MSPharm
- **Preventive and Therapeutic Role of Vitamins as Nutraceuticals**
  Editors: Khemchand R. Surana, Eknath D. Ahire, Raj K. Keservani, PhD, MPharm, and Rajesh K. Kesharwani, PhD
- **Nutraceuticals and Bone Health**
  Editors: Deepak Sharma, PhD, Madan Mohan Gupta, PhD, Anil K. Sharma, PhD, MPharm, Raj K. Keservani, PhD, MPharm, and Rajesh K. Kesharwani, PhD
- **Flavonoids as Nutraceuticals**
  Editors: Rajesh K. Kesharwani, PhD, MTech, Deepika Saini, PhD, Raj K. Keservani, PhD, MPharm, and Anil Kumar Sharma, PhD, MPharm
- **Herbals as Nutraceuticals: Their Role in Healthcare**
  Editors: Raj K. Keservani, PhD, MPharm, Rajeshwar Kamal Kant Arya, PhD, MPharm, and Rajesh K. Kesharwani, PhD, MTech

- **Nutraceutical Fruits: Overview and Disease Prevention**
  Editor: Raj K. Keservani, PhD, MPharm
- **Nutrigenomics and Nutraceuticals**
  Editors: Raj K. Keservani, PhD, MPharm, Eknath D. Ahire, Shubham J. Khainar, PhD, Sanjay J. Kshirsagar, PhD, and Rajesh K. Kesharwani, PhD
- **The Nature of Nutraceuticals: History, Properties, Sources, and Nanotechnology**
  Editors: Rajesh K. Kesharwani, PhD, Prashant Kumar, PhD, and Raj K. Keservani, PhD, MPharm
- **Nutraceuticals for the Treatment and Prevention of Sexual Disorders**
  Editors: Raj K. Keservani, PhD, MPharm, Sharangouda J. Patil, PhD, and Ivan Aranha, PhD
- **Antioxidants as Nutraceuticals**
  Editors: Nayan A. Gujaratih, PhD, Raj K. Keservani, PhD, MPharm, Rajesh K. Kesharwani, PhD, Bhushan R. Rane, PhD, and Yogeeta Sameer Goyal, PhD
- **Nutraceuticals in Cardiac Health Management**
  Editors: Raj K. Keservani, PhD, Rajesh K. Kesharwani, PhD, Praveen Kumar Jain, PhD, Bhushan R. Rane, PhD, and Nayan A. Gujarathi, PhD
- **Nutraceuticals in Obesity Management and Control**
  Editors: Raj K. Keservani, PhD, and Alka Lohani, PhD
- **Herbals as Nutraceuticals: Their Role in Healthcare**
  Editors: Raj K. Keservani, PhD, Rajeshwar Kamal Kant Arya, PhD, MPharm, and Rajesh K. Kesharwani, PhD
- **Nutraceuticals in Insomnia and Sleep Problems**
  Editors: Raj K. Keservani, PhD, Sayan Bhattacharyya, PhD, and Rajesh K. Keservani, PhD
- **Nutraceuticals in Arthritis and Psoriasis**
  Editors: Meenakshi Jaiswal, PhD, Raj K. Keservani, PhD, Rajesh K. Kesharwani, PhD, and Swati G. Talele, PhD

ABOUT THE EDITORS

Prakash Chandra Gupta, PhD, MPharm
Assistant Professor, School of Pharmaceutical Sciences, Chhatrapati Shahu Ji MaharajUniversity, Kanpur, India

Prakash Chandra Gupta, PhD, is presently working as an Assistant Professor in the School of Pharmaceutical Sciences, Chhatrapati Shahu Ji Maharaj University, Kanpur, India. He has more than 15 years of academic and research experience. He has published many research and review articles in reputed national and international journals. He has received various research project grants from different funding agencies. Dr. Gupta received his BPharm degree in 2004 from the Department of Pharmacy, Kumaun University, Nainital, Uttarakhand, India, and MPharm degree (Pharmacognosy) in 2006 from the Institute of Pharmacy, Barkatullah University Bhopal, Madhya Pradesh, India, and PhD degree in 2014 from the Department of Pharmaceutical Science, Bhagwant University, Ajmer, Rajasthan, India. His research interests include the standardization of herbal medicine, pharmacological screening, and the development of herbal formulations.

Sayan Bhattacharyya, MBBS, MD
Associate Professor, Department of Microbiology, All India Institute of Hygiene & Public Health, Kolkata, India

Sayan Bhattacharyya, MD, MBBS, has more than 13 years of experience in teaching, patient care, laboratory work, and biomedical research in the field of medical microbiology. He has 75 publications in various peer-reviewed international medical journals. He pursued his MD in Microbiology at Chandigarh Postgraduate Institute of Medical Education and Research (PGIMER), Chandigarh, and MBBS at Medical College, Calcutta, India. He is now working as an Associate Professor in the Department of Microbiology, All India Institute of Hygiene and Public Health Kolkata. Before this, he had worked for more than four years in the Department of Microbiology, All India Institute of Medical Sciences Patna (AIIMS) as an Assistant Professor.

He has guided many students of different courses in their dissertation work. Dr. Bhattacharyya has won the second-best oral paper at the conferences STMIDI TROPICON 2018, Kolkata. He has also won the best English essay award in 'Rabies and One Health,' organized by Pashudhan Praharee in 2021. He is also engaged in whole genome sequencing from clinical bacterial isolates (project work) with the Central Research Laboratory, Kempegowda Institute of Medical Sciences (KIMS), Bengaluru. His fields of interest are medical mycology, nosocomial infections, antimicrobial stewardship, and public health.

Nisha Sharma, PhD, MPharm
Associate Professor, University Institute of Pharmacy, Chhatrapati Shahu Ji Maharaj University, Kanpur, India

Nisha Sharma, PhD, is presently working as Director of the School of Pharmaceutical Sciences, Chhatrapati Shahu Ji Maharaj University. University, Kanpur, India. With teaching experience of over 23 years, she has authored many research articles in journals and books. Dr. Sharma is a popular figure in the pharmaceutical field. She has chaired many scientific sessions at national and international conferences. She has served as a resource person for the All India Council for Technical Education faculty development program. Dr. Sharma received a BPharm from J. S. S. College of Pharmacy, Mysore, India, in 1995; MPharm from BITS Pilani, Rajasthan, India in 1997, and PhD From Bhagwant University, Ajmer, Rajasthan, India, in 2014. Her areas of research include phytochemistry, ethnopharmacology, and herbal formulations. She is associated with many professional bodies, such as IPGA, IPA, IHPA, APTI, and many others.

Rajesh K. Kesharwani, PhD
Associate Professor, Department of Computer Application, Nehru Gram Bharati (Deemed to be University), Prayagraj, India

Rajesh K. Kesharwani, PhD, MTech, is an Associate Professor, Department of Computer Application, Nehru Gram Bharati (Deemed to be University), Prayagraj, India. He has more than 11 years of research and nine years of teaching experience at various institutes of India, imparting bioinformatics and biotechnology education. He has received several awards, including the

NASI-Swarna Jayanti Puruskar by the National Academy of Sciences of India. He has supervised one PhD and more than 20 undergraduate and graduate students for their research work. Dr. Kesharwani has authored over 49 peer-reviewed articles, 20 book chapters, and 14 edited books with international publishers. He has been a member of many scientific communities as well as a reviewer for many international journals. He has presented many papers at various national and international conferences. Dr. Kesharwani received his PhD from the Indian Institute of Information Technology, Allahabad, and worked at NIT Warangal for two semesters. He has been a recipient of a Ministry of Human Resource Development (India) Fellowship and a Senior Research Fellowship from the Indian Council of Medical Research, India. His research fields of interest are medical informatics, protein structure and function prediction, computer-aided drug designing, structural biology, drug delivery, cancer biology, nanobiotechnology, and biomedical sciences.

Raj K. Keservani, PhD, MPharm
Associate Professor, Faculty of BPharm, CSM Group of Institutions, Prayagraj, India

Raj K. Keservani, PhD, MPharm, is working as an Associate Professor in the Faculty of B. Pharmacy at CSM Group of Institutions, Prayagraj, India. He has more than 12 years of academic (teaching) experience from various institutes of India in pharmaceutical education. He has supervised more than 25 UG students for their research work, has published 31 peer-reviewed papers in the field of pharmaceutical sciences in national and international journals, one patent, 50 book chapters, three co-authored books, and 40 edited books. He is also active as a reviewer for several international scientific journals. Dr. Keservani graduated with a pharmacy degree from the Department of Pharmacy, Kumaun University, Nainital (Uttarakhand), India. He received his Master of Pharmacy (MPharm) (specialization in pharmaceutics) from the School of Pharmaceutical Sciences, Rajiv Gandhi Proudyogiki Vishwavidyalaya, Bhopal, India. He received his PhD degree from IKG Punjab Technical University, Jalandhar, Punjab, India. Dr. Keservani is a life member of Society of Pharmaceutical Education and Research (SPER). His research interests include nutraceutical and functional foods, novel drug delivery systems (NDDS), transdermal drug delivery/drug delivery, health science, cancer biology, and neurobiology.

CONTENTS

CONTRIBUTORS

Shamshad Ahmed
Assistant Professor, Department of Community and Family Medicine, AIIMS, Patna, Bihar, India

Amit Banik
Associate Professor, Microbiology, All India Institute of Hygiene and Public Health (AIIH&PH), BN Campus, Kolkata, West Bengal, India

Mousumi Bhattacharya
Nutritionist, Kolkata, West Bengal, India; Test Coordinator, Tata Medical and Diagnostics Limited, Fort, Mumbai, Maharashtra, India

Sayan Bhattacharyya
Associate Professor, Microbiology, All India Institute of Hygiene and Public Health (AIIH&PH), Kolkata, West Bengal, India

Suddhasatwya Chatterjee
Sr. Consultant, Department of Internal Medicine and Rheumatology, Apollo Multispecialty Hospitals, Kolkata, West Bengal, India

Madhumita Dobe
Director/Professor (Public Health), All India Institute of Hygiene and Public Health (AIIH&PH), Kolkata, West Bengal, India

Meghna Dutta
Registrar, Critical Care Unit, Apollo Multispecialty Hospitals, Kolkata, West Bengal, India

Sumitra Gorain
Department of Biochemistry and Nutrition, All India Institute of Hygiene and Public Health, Kolkata, West Bengal, India

H. Haripriya
Junior Resident, Department of Community and Family Medicine, AIIMS, Phulwari Sharif, Patna, Bihar, India

Rashmi Jaiswal
Applied Nutrition Consultant, Nutritionist in HealthifyMe, Kolkata, West Bengal, India

Pragya Kumar
Additional Professor, Department of Community and Family Medicine, AIIMS, Phulwari Sharif, Patna, Bihar, India

K. G. Naveen
Junior Resident, Department of Community and Family Medicine, AIIMS, Patna, Bihar, India

Diksha Kumari Sharma
MSc in Applied Nutrition, All India Institute of Hygiene and Public Health, BN Campus, Saltlake, Kolkata, West Bengal, India

ABBREVIATIONS

AA	arachidonic acid
AAD	antibiotic-associated diarrhea
ABCA1	ATP-binding cassette A1
AD	Alzheimer's disease
ADME	absorption, distribution, metabolism, and elimination
AGE	advanced glycation end
AIDS	acquired immune deficiency syndrome
ALA	alpha-amino-levulinic acid
ALCAR	acetyl L-carnitine
AMP	antimicrobial peptide
AMPK	AMP related protein kinase
API	active pharmacological ingredient
ARMD	age-related macular degeneration
AT	active transport
BAT	brown adipose tissue
BCS	biopharmaceutical classification system
BDNF	brain-derived neurotrophic factor
BHA	butylated hydroxyanisole
BHT	butylated hydroxyl toluene
BL	permeability through bilayer
BMI	body-mass index
BR	bilirubin
CAT	catalase
CCS	chronic coronary syndrome
CD	celiac disease
CD	cyclodextrins
CETP	cholesteryl ester transfer protein
CFS	chronic fatigue syndrome
CHD	coronary heart disease
CKD	chronic kidney disease
CLA	conjugated linoleic acid
CN	cognitively normal
CO2	carbon dioxide
COPD	chronic obstructive pulmonary disease patient

COXs	cyclooxygenases
CPS	consumer product safety
CRN	council for responsible nutrition
CSF	cerebrospinal fluid
Cu	copper
CVD	cardiovascular disease
DCs	dendritic cells
DHA	docosahexaenoic acid
DHS	2,3-dehydrosilybin
DNA	deoxyribonucleic acid
DNMT	DNA N-methyl transferase
DSL	D-saccharic acid-1,4-lactone
DST	Department of Science and Technology
EAFE	European Academy of Food Engineering
EGCG	epigallocatechin-3-gallate
ENCODE	encyclopedia of DNA elements
eNOS	endothelial nitric oxide synthase
EPA	eicosapentaenoic acid
EPA	Environment Protection Agency
EPCs	endothelial progenitor cells
EPR	enhanced permeability and retention
ET	efflux transport
EU	European Union
FAD	factors ketone purine dinucleotide
FAO	Food and Agricultural Organization
FAS	fatty acid synthase
FDA	Food and Drug Administration
Fe2+	ferrous ions
FFA	free fatty acids
FODMAPs	fermentable oligosaccharides, disaccharides, monosaccharides, and polyols
FOS	fructooligosaccharides
GABA	gamma-aminobutyric acid
GAGs	glycosaminoglycans
GALT	gut-associated lymphoid tissues
GFR	glomerular filtration rate
GI	glycemic index
GIANT	genomic investigation of anthropometric traits
GIT	gastrointestinal tract

GlcUA	gluconic and glucuronic acids
GLS	glutaminase
GNPDA2	glucosamine-6-phosphate deaminase 2
GOS	galactooligosaccharides
GPx/GSHPX	glutathione peroxidase
GRAS	generally-regarded-as-safe
GS	glutamine metabolism
GSH	glutathione
GSIS	glucose-stimulated insulin secretion
GTP	green tea polyphenol
GWAS	genome-wide association studies
H2O2	hydrogen peroxide
HARPS	high-risk adenomatous polyps
HATs	histone acetyltransferases
HCA	hydroxycitric acid
HDAC	histone deacetylase
HDL	high-density lipoprotein
HDMs	histone demethylases
HFA	height/length-for-age
HFI	hypoxia-inducible factor
HMG-CoA	3-hydroxy-3-methyl-glutaryl-coenzyme A
HMOs	human milk oligosaccharides
HMTs	histone methyltransferases
HNE	4-hydroxy-2(E)-nonenal
Hs-CRP	high-sensitivity C-reactive protein
HSL	hormone-sensitive lipase
IBS	irritable bowel syndrome
IDX-1	islet/duodenum homeobox-1
IECs	intestinal epithelium of the host
IFA	iron folic acid
IHCP	Institute for Health and Consumer Protection
IL-10	interleukin-10
IL-2	interleukin-2
IL-6	interleukin-6
iNOS	inducible nitric oxide synthase
INSIG-2	insulin-induced gene-2
IQ	intelligent quotient
ISPEN	Indian Society for Parenteral and Enteral Nutrition
ITBNs	insecticide-treated bed nets

JNK	c-Jun N-terminal kinases
K/DOQI	kidney disease outcomes quality initiative
kDa	kiloDalton
LA	linoleic acid
LAB	lactic acid bacteria
LDL	low-density lipoprotein
LLINs	long-lasting insecticide-treated nets
LPL	lipoprotein lipase
LPO	lipid peroxidation
LXR	liver X receptor
MAM	moderate acute malnutrition
MAPK	mitogen-activated protein kinase
MBP	milk basic protein
MCI	mild cognitive impairment
MDD	minimum dietary diversity
ML	mucus layer
MMF	minimum meal frequency
MMKD	modified Mediterranean ketogenic diet
MNP	micronutrient powder
MoEF	Ministry of Environment and Forests
MoHFW	Ministry of Health and Family Welfare
MPO	myeloperoxidase
MRSA	methicillin-resistant Staphylococcus aureus
MS	multiple sclerosis
MSG	monosodium glutamate
MT	metallothionein
MTCH2	mitochondrial carrier homolog 2
MTs	metallothioneins
MUFA	monounsaturated fatty acids
NAC	N-acetyl cysteine
NAD	nicotinamide adenine dinucleotide
NAFLD	nonalcoholic fatty liver disease
NCDs	non-communicable diseases
NDOs	non-digestible oligosaccharides
NE	nanoemulsions
NEGR1	neuronal growth regulator 1
NHANES	National Health and Nutrition Examination Survey
NHGRI	National Human Genome Research Institute
NK	natural killer

NO	nitric oxide
NP	nanoparticles
NPC1L1	Niemann-PICK C1-like 1
NRotB	number of rotating bonds
NSAIDs	non-steroidal anti-inflammatory drugs
NSP	non-starchy polysaccharides
NSTI	nano-science and technology initiative
O/W	oil-in-water
O2	oxygen
OCD	obsessive-compulsive disorder
OGTT	oral glucose tolerance test
OH	hydroxyl radicals
OSHA	Occupational Health and Safety Administration
PCFT	proton-coupled folate transporter
PCOS	polycystic ovarian syndrome
PCSK9	proprotein convertase subtilisin/kexin type 9
PD	2,3-pentanedione
PDX-1	pancreas duodenum homebox-1
PEG	polyethylene glycol
PEM	protein energy malnutrition
PIC	phase inversion composition
PIT	phase inversion temperature
PLP	pyridoxal phosphate
PMP	pyridoxamine phosphate
PMS	premenstrual syndrome
PPAR	peroxisome proliferator-activated receptor
PPARγ	peroxisome proliferator-activated gamma
PTSD	post-traumatic stress disorder
PUFA	polyunsaturated fatty acids
QTL	quantitative trait locus
RA	rheumatoid arthritis
RANKL	receptor activator of nuclear factor kappa beta ligand
RBC	red blood cells
RDAs	recommended dietary allowance
RES	reticuloendothelial system
RNS	reactive nitrogen species
ROS	reactive oxygen species
RPE	retinal pigment epithelium
RUSF	ready-to-use supplementary foods

RYR	red yeast rice
SAM	severe acute malnutrition
SCF	supercritical fluid
SCFAs	short-chain fatty acids
SCOBY	symbiotic colony of acidic acid bacteria and yeast
Se	selenium
SEC16B	SEC16 homolog B
SH2B1	SH2B adaptor protein 1
sIgA	secretory IgA
SLE	systemic lupus erythematosus
SLN	solid-lipid NPs
SMVT	sodium-dependent vitamin pill transporter
SNPs	single nucleotide polymorphisms
SOD	superoxide dismutase
SREBP-1c	steroid regulating element binding protein 1c
TBHQ	tertiary butylhydroquinone
THF	tetrahydrofolate
TJ	tight junction
TLR2	toll-like receptor 2
TNBC	triple-negative breast cancer
TPN	total parenteral nutrition
TPP	tripolyphosphate
UA	uric acid
VRE	vancomycin-resistant enterococci
WBC	white blood cell
WFA	weight-for-age
WFH	weight-for-height/length
WHO	World Health Organization
Zn	zinc

PREFACE

Micronutrients, like vitamins and minerals, form an important and essential part of the diet of man and are required for our daily physiological needs. Many macronutrients are also needed, in appreciable amounts, for our homeostasis and maintenance of vital functions. Both can act as nutraceuticals or have some beneficial pharmacological effect. Some of these nutrients also have antioxidant-like properties. Sometimes microbes and their metabolic or other products also serve as excellent micronutrients and macronutrients and even have nutraceutical-like activity. Scientists, students, researchers, and even the common man need to know about these things since they are encountered in some form or the other in day-to-day life. Knowledge of food engineering and food processing is also needed for the packaging or purchase of these micronutrients and macronutrients.

The authors of this book belong to diverse fields and disciplines like nutrition, microbiology, general medicine, and preventive and social medicine. In this book, the authors have tried their best to delineate and describe in detail the various tenets and aspects of micronutrients, macronutrients, and their use as nutraceuticals, in the form of 12 chapters. Wherever possible, figures and tables have also been provided for a crisp and better understanding of the topics. I sincerely hope that the readers will find this book highly beneficial. The authors and the editors would deem their efforts fruitful if the book and its chapters are read and appreciated. Any feedback or constructive criticism in this context from the readers will be accepted wholeheartedly.

Functional foods and nutraceuticals play an important role in preventing various health problems and improving quality of life as discussed in Chapter 1 – "Functional Foods and Nutraceuticals," written by Sumitra Gorain.

Antioxidants that are obtained from outside the body are called exogenous. Our body has its personal antioxidant defenses to keep free radicals in check and its function is discussed in Chapter 2 – "Antioxidants Activity of Micronutrients and Macronutrients," written by Mousumi Bhattacharya.

Chapter 3 – "Vitamins as Nutraceuticals," written by Suddhasatwya Chatterjee and Meghna Dutta delves into the realm of vitamins and tries to explore their nutraceutical potential.

Nutraceuticals are nutrient molecules having some pharmacological properties and applications. Many microbes and microbial products or metabolites can be used as nutraceuticals and are described in Chapter 4 – "Microbes and Microbial Products as Nutraceuticals," written by Sayan Bhattacharyya.

Diksha Kumari Sharma reports the role of diet and, in particular, dietary carbohydrate, protein, and fat effective functioning of immune cells and how dietary intervention can be of benefit in order to improve intestinal health and immune status in Chapter 5 – "Immunity Boosting Macronutrients and Micronutrients."

Sayan Bhattacharyya summarizes micronutrients, macronutrients, and nutraceuticals that can be incorporated into the human diet for optimum health, good growth of beneficial bacteria in the gut, and inhibition of pathogens in Chapter 6 – "Gut Microbiota and Pathogens."

Chapter 7 – "Nutrient Delivery of Micronutrients and Macronutrients," written by Amit Banik discusses nutrition as derived from food materials that needs to be delivered efficiently to reap maximum benefits.

Disease-promoting habits often unfold in childhood, and for many, prenatally. Hence, primary and primordial levels of prevention will go a long.way in curbing the prevalence of these chronic diseases as discussed in Chapter 8 – "Micronutrients and Macronutrients in Nutrigenomics," written by Meghna Dutta and Suddhasatwya Chatterjee.

Chapter 9 – "Alternative Food and Therapeutic Food," written by Rashmi Jaiswal discusses alternative food as a therapeutic diet that bears high potentiality in preventing, treating, and curing many diseases.

Chapter 10 – "Micronutrients and Macronutrients in Food Engineering" written by H. Haripriya and Pragya Kumar tries to collate and present all scientific information on this interesting aspect of biomedical research.

K. G. Naveen and his associates discuss the available food processing techniques while delineating the effects of these techniques on food quality in Chapter 11 – "Food Processing of Micronutrients and Macronutrients."

Madhumita Dobe reports that nutraceuticals, micronutrients, and macronutrients are all essential for the sound health of the consumer in Chapter 12 – "Nutrition, Diet, and Health: Role of Macronutrients, Micronutrients, and Nutraceuticals."

CHAPTER 1

FUNCTIONAL FOODS AND NUTRACEUTICALS

SUMITRA GORAIN

Department of Biochemistry and Nutrition, All India Institute of Hygiene and Public Health, Kolkata, West Bengal, India

ABSTRACT

The concept of food has changed from only a means of satisfying hunger and nutritional needs in the past to means of preventing and reducing disease and improving overall health in recent years. This is due to Functional foods and nutraceuticals, which plays a role in preventing various health problems and improving quality of life.

1.1 INTRODUCTION

Food is vital for life. Food contains various dietary components that provide energy and other required nutrients to the body. When conventional food or food with added nutrients provides health benefits (health promoting or disease prevention) beyond nutritional properties is called functional food. Food components are sometimes extracted and supplemented in the form of drugs like pills, capsules, tablets, etc., for health promotion which are called nutraceuticals. The term nutraceutical, coined by DeFelice in 1989, is a combination of nutrition and pharmaceutical. Functional food and nutraceuticals are used interchangeably many times as they don't have many distinguished dissimilarities (Gul et al., 2016).

Micronutrients and Macronutrients as Nutraceuticals. Prakash Chandra Gupta, PhD, Sayan Bhattacharyya, MBBS, MD, Nisha Sharma, PhD, MPharm, Rajesh K. Kesharwani, PhD, Raj K. Keservani, PhD, MPharm (Eds.)
© 2024 Apple Academic Press, Inc. Co-published with CRC Press (Taylor & Francis)

Functional food can be grouped into three categories:

1. **Basic Foods:** Bioactive compounds are present naturally in the food or its product. Examples include turmeric and papaya with curcumin and beta-carotene present naturally, respectively (Wildman, 2007).
2. **Processed Foods with Added Bioactive Compounds:** This means that Bioactive compounds are added during processing. An example is orange juice with added Calcium and milk with added omega-3 fatty acids (Wildman, 2007).
3. **Foods Enhanced to have More of a Bioactive Component:** The level of bioactive elements present in food is modified or concentrated, for example by genetic engineering or breeding. Examples are tomatoes with increased lycopene content and eggs with increased levels of omega-3 fatty acids (Wildman, 2007).

1.2 CLASSIFICATION OF NUTRACEUTICALS

Nutraceuticals can be derived from plant, animal, or microbial sources. Different foods contain varying concentrations of nutraceuticals with different chemical natures and mechanisms of action. Hence can be classified in a different way based on different criteria as given in Table 1.1 (Wildman, 2007).

TABLE 1.1 Classification of Nutraceuticals on the Basis of Food Sources

• **Plants:** β-Glucan, Ascorbic acid, γ-Tocotrienol, Quercetin, Luteolin, Cellulose, Lutein, Gallic acid, Perillyl alcohol, Indole-3-carbonol, Pectin, Daidzein, Glutathione, Potassium, Allicin, δ-Limonene, Genestein, Lycopene, Hemicellulose, Lignin, Capsaicin, Geraniol, β-Ionone, α-Tocopherol, β-Carotene, Nordihydrocapsaicin, Selenium, Zeaxanthin, Minerals, MUFA.
• **Animal:** Conjugated linoleic acid (CLA), eicosapentaenoic acid (EPA), docosahexenoic acid (DHA), sphingolipids, choline, lecithin, calcium, coenzyme Q10, selenium, zinc, creatine, minerals.
• **Microbial:** *Saccharomyces boulardii* (yeast), *Bififidobacterium bififidus*, *B. longum*, *B. infantis*, *Lactobacillus acidophilus* (LC1), *Lactobacillus acidophilus* (NCFB 1748), and *Streptococcus salivarius* (subs. Thermophilus).

Source: Reprinted/Adapted from Wildman, 2007

Several nutraceuticals have various disease-preventing actions on the body. Different nutraceuticals (Keservani et al., 2010a) or groups of nutraceuticals exert their action in the body through different mechanisms and they are grouped accordingly as given in Table 1.2 (Wildman, 2007).

TABLE 1.2 Classification of Nutraceuticals based on Mechanism of Action

Anticancer	Positive Influence on Blood Lipid Profile	Antioxidant Activity	Anti-Inflammatory	Osteogenetic or Bone Protective
Capsaicin	β-Glucan	CLA	Linolenic acid	CLA
Genistein	γ-Tocotrienol	Ascorbic acid	EPA	Soy protein
Daidzein	δ-Tocotrienol	β-Carotene	DHA	Genistein
α-Tocotrienol	MUFA	Polyphenolics	GLA (gamma-linolenic acid)	Daidzein
γ-Tocotrienol	Quercetin	Tocopherols	Capsaicin	Calcium
CLA	ω-3 PUFAs	Tocotrienols	Quercetin	Casein phosphopeptides
Lactobacillus acidophilus	Resveratrol	Indole-3-carbonol	Curcumin	FOS (fructooligo-saccharides)
Sphingolipids	Tannins	α-Tocopherol		Inulin
Limonene	β-Sitosterol	Ellagic acid		
Diallyl sulfide	Saponins	Lycopene		
Ajoene	Guar	Lutein		
α-Tocopherol	Pectin	Glutathione		
Enterolactone		Hydroxytyrosol		
Glycyrrhizin		Luteolin		
Equol		Oleuropein		
Curcumin		Catechins		
Ellagic acid		Gingerol		
Lutein		Chlorogenic acid		
Carnosol		Tannins		
L. bulgaricus				

Many nutraceuticals are present in higher concentrations in specific food or food groups. Such examples are given in Table 1.3 (Wildman, 2007; Keservani et al., 2010b).

TABLE 1.3 Classification of Nutraceuticals According to Concentration Present in Different Foods

Nutraceutical (Substance/Family)	Foods of Remarkably High Content
Allyl sulfur compounds	Onions, garlic
Isoflavones (e.g., genestein, daidzein)	Soybeans and other legumes, Apios

TABLE 1.3 *(Continued)*

Nutraceutical (Substance/Family)	Foods of Remarkably High Content
Quercetin	Onion, red grapes, citrus fruit, broccoli, Italian yellow squash
Capsaicinoids	Pepper fruit
EPA and DHA	Fish oils
Lycopene	Tomatoes and tomato products
Isothiocyanates	Cruciferous vegetables
β-Glucan	Oat bran
CLA	Beef and dairy
Resveratrol	Grapes (skin), red wine
β-Carotene	Citrus fruit, carrots, squash, pumpkin
Carnosol	Rosemary
Catechins	tea, cocoa, apple, grapes
Adenosine	Garlic, onion
Indoles	Cabbage, broccoli, cauliflower, kale, brussels sprouts
Curcumin	Turmeric
Ellagic acid	Grapes, strawberries, raspberries, walnuts
Anthocyanates	Red wine
3-n-Butyl phthalide	Celery
Cellulose	Most plants (a component of cell walls)
Lutein, zeaxanthin	Kale, collards, spinach, corn, eggs, citrus
Psyllium	Psyllium husk
Inulin, Fructooligosaccharides (FOS)	Monounsaturated fatty acids, tree nuts, olive oil
Lactobacilli, Bifidobacteria	Whole grains, onions, garlic
Catechins Tea, cocoa, apples, grapes	Yogurt and other dairy
Lignans	Flax, rye

Source: Adapted from Wildman, 2007; Keservani et al., 2010b

Based on their chemical nature, nutraceuticals are grouped under seven categories as shown in Figure 1.1 (Wildman, 2007):

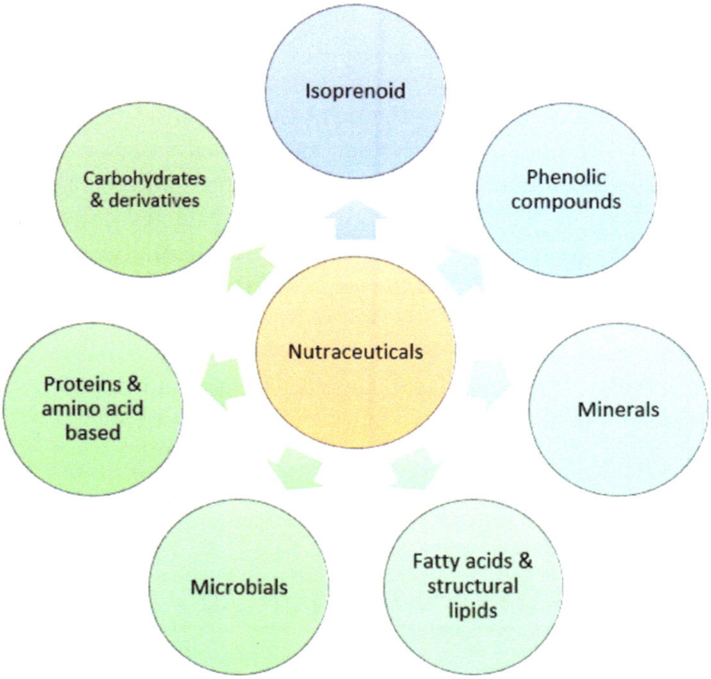

FIGURE 1.1 Classification of nutraceuticals based on chemical nature.
Source: Reprinted/Adapted from Wildman, 2007

1. **Isoprenoids (Terpenoids):** These are the largest groups of secondary metabolites of plants, so form the basis of many plant-derived nutraceuticals. Nutraceutical families such as carotenoids, tocopherols, tocotrienols, and saponins come under this. This group is also called isoprenoid derivatives because the main component is isoprene.
2. **Phenolic Compounds:** These are also secondary metabolites. The common nutraceuticals which come under this are flavonoids, anthocyanins, coumarin, tannin, lignin, isoflavones, flavonols, and flavones (Keservani & Sharma, 2014).
3. **Carbohydrates and Derivatives:** Nutraceuticals of this group include those substances that are carbohydrates or its derivative such as ascorbic acid, oligosaccharides, and non-starch polysaccharides.

4. **Fatty Acids and Structural Lipids:** Nutraceuticals like ω-3 PUFA, CLA, MUFA, sphingolipids, and lecithin fall under this group.
5. **Amino Acid Based:** This group includes amino acids, allyl sulfur compounds, indoles, folate, choline, isothiocyanates, and capsaicinoids.
6. **Microbials:** This group includes bacteria that are resistant to bile, acid conditions of the stomach, and digestive enzymes that are normally found in the human gastrointestinal tract (GIT). They are able to colonize in the human intestine and are safe for human consumption, have scientifically proven efficacy called probiotics. Examples of such bacteria having nutraceutical effects are *Lactobacillus acidophilus, L. plantarum, L. casei, Bifidobacterium bifidum, B. infantis*, and *Streptococcus salvarius subspecies thermophilus*. Some yeasts are also found like *Saccharomyces boulardii*.
7. **Minerals:** Minerals like Calcium, Potassium, Copper, Selenium, Manganese, and Zinc have nutraceutical properties.

The detailed role of each nutraceutical on human health is discussed later in this chapter.

1.3 BIOAVAILABILITY OF NUTRACEUTICALS

The bioavailability of nutraceuticals in food can be defined as the amount of nutraceuticals that reach the systemic circulation and are distributed to organs and tissues and show their biological activity (Dima et al., 2020). Factors affecting the bioavailability of healthcare products include the drug form of the drug or the type of food matrix, as well as the structure and physical and chemical properties of biologically active molecules, such as molecular weight, lipophilicity, and the ability to interact with receptors and transporters. The higher molecular weight of compounds affects bioavailability, while lipophilicity affects their solubility and permeability. High lipophilicity reduces the water solubility of molecules and increases permeability. In order for nutraceuticals to have a greater therapeutic effect, they must have structural elements that allow them to bind to transporters and receptors (such as hydrogen bonds).

Lipinski and his team found that if the oral bioavailability of nutritional supplements meets 3 of the following 4 rules, their oral bioavailability will be very good: $M \leq 500\ g/mol$; $log\ P \leq 5$; Hydrogen bond donor: HBD ≤ 5; Hydrogen bond acceptor: HBA ≤ 10. This Lipinski rule is called the Rule of Five (Ro5) (Lipinski et al., 1997). Subsequently, two factors were added, such as the area of the polar surface (PSA ≤ 140 Å2) and the number of rotating bonds (NRotB ≤ 10) (Palm et al., 1997; Veber et al., 2002). These rules are important for designing new drugs with high therapeutic effects.

1.4 FUNCTIONAL FOODS, NUTRACEUTICALS, AND HUMAN HEALTH

1.4.1 ANTIDIABETIC FUNCTIONAL FOODS WITH ANTIGLYCAN PROPERTIES

Diabetes mellitus is a metabolic disorder of high blood sugar levels due to insulin deficiency or insulin resistance (Harris et al., 2000). Chronic hyperglycemia in diabetic patients leads to unhealthy interaction between excess blood glucose and some proteins like collagen, hemoglobin, and lens crystallins which is called glycation (Harding & Ganea, 2006; Singh et al., 2017b). Some lists of foods with antiglycan properties are given in Table 1.4 (Kazeem & Bankole, 2018).

A possible mechanism by which these foods inhibit glycation and further formation of advanced glycation end (AGE) products may occur through one or more mechanisms like cleavage of the AGE-crosslink or also by blocking of the AGE receptor, trapping of reactive dicarbonyl species, as well as through antioxidant activities. Other mechanisms include the inhibition of aldose reductase and the reduction of blood glucose levels using hypoglycemic agents, such as acarbose and metformin (Kazeem & Bankole, 2018). The antiglycan roles of various foods are delineated in Table 1.4.

TABLE 1.4 Antiglycan Properties of Foods

Name	Part	Extract	Test	Method
Bael	Fruits	Methanol	*In vitro*	BSA-glucose
Apple	Fruits	Water	*In vitro*	BSA-fructose
Banana	Inflorescence	Methanol	*In vitro*	BSA-glucose
	Inflorescence	Ethanol	*In vitro*	BSA-fructose
	Flower	Diet	*In vivo*	STZ-diabetic rats
Black tea	Leaves	Water	*In vitro*	BSA-glucose
Buckwheat	Hull	Water	*In vitro*	BSA-glucose
	Seed	Ethanol	*In vitro*	BSA-carbonyl level

TABLE 1.4 *(Continued)*

Name	Part	Extract	Test	Method
Shaddock	Fruit pulp	Methanol	*In vitro*	BSA-fructose
Cornelian cherry	Fruits	Water	*In vitro*	BSA-fructose
	Fruits	Water	*In vivo*	STZ-diabetic rats
	Fruits	Glycosides	*In vivo*	STZ-diabetic rats
	Fruits	Morroniside	*In vivo*	STZ-diabetic rats
Cumin	Seeds	Water	*In vitro/vivo*	BSA/STZ-DM rat
	Seeds	Methanol	*In vitro/vivo*	BSA/STZ-DM rat
	Seeds	Cymene	*In vitro/vivo*	BSA/STZ-DM rat
Dendrobii	Whole plant	Polysaccharides	*In vitro*	BSA-glucose
Black nightshade	Leaf	Ethanol, water	*In vitro*	BSA-carbonyl level
Ginger	Rhizome	Ethylacetate	*In vitro*	BSA-glucose
	Rhizome	Polyphenol	*In vitro*	BSA-glucose
	Rhizome	Polyphenol	*In vivo*	STZ-diabetic rats
Guava	Leaf	Water	*In vitro*	BSA-fructosamine
	Leaf	Water	*In vitro*	Plasma-glycation
	Leaf	Methanol	*In vivo*	STZ-diabetic rats
	Leaf	Methanol	*In vivo*	STZ-diabetic rats
Yerba mate	Leaf	Water	*In vitro*	BSA-methylglyoxal
Common juniper	Fruits	Essential oil	*In vitro*	Insulin-glycation
Longan	Fruits	Polysaccharides	*In vitro*	BSA-glucose
Lotus	All parts	Methanol	*In vitro*	BSA-glucose
Saucer berry	Whole plant	Compounds	*In vitro*	BSA-glucose
Mung bean	Seed	Ethanol	*In vitro*	BSA-glucose
Passion fruit	Leaf	Water	*In vitro*	BSA-glucose
Pomegranate	Fruit	Polysaccharides	*In vitro*	BSA-glucose
Grapevine	Fruit	Water	*In vitro*	BSA-carbonyl level
Water apple	Leaf	Ethanol	*In vitro*	BSA-glucose
Turmeric	Rhizome	Ethylacetate	*In vitro*	BSA-glucose
Lingonberry	Fruit	Ethanol	*In vitro*	BSA-glucose
Bitter melon	Fruits	Capsule	*In vivo*	Diabetic subjects
Green tea	Leaf	Ethanol	*In vivo*	STZ-diabetic rats
Rambutan	Fruit	Ethanol	*In vitro*	BSA-glucose
	Fruit	Geraniin	*In vitro*	BSA-glucose

TABLE 1.4 *(Continued)*

Name	Part	Extract	Test	Method
Sickle senna	Seeds	Glucosides	*In vitro*	BSA-glucose
	Seeds	Anthraquinones	*In vitro*	BSA-glucose
Kokum	Fruits	Garcinol	*In vitro*	BSA-fructose
Luobuma tea	Leaves	Flavonoids	*In vitro*	BSA-glucose
Sangwhang	Fruiting body	Isolates	*In vitro*	HbA1c formation
Clove	Buds	Water	*In vitro*	BSA-fructose

Phenolic compounds from soybean have been found to have an inhibitory effect on α-amylase and α-glucosidase activity thus lowering post-prandial hyperglycemia. Black pepper and ginger also show anti-diabetic activity (Adefegha, 2018).

1.5 CARDIO PROTECTIVE FUNCTIONAL FOODS AND NUTRACEUTICALS

Cardiovascular disease (CVD) is one of the major causes of morbidity and mortality worldwide. It includes a wide range of health disorders of the heart and blood vessels. Dietary factors play an important role in contributing to cardiovascular risk, either directly, or indirectly via their effects on other cardiovascular risk factors including hypertension, dyslipidemia, and diabetes mellitus. Nutraceuticals have been shown to prevent or reduce the disease (Sosnowska et al., 2017). Some of them are discussed in subsections.

1.5.1 DIETARY FIBER

Dietary fiber reduces cholesterol by increasing bile acid fecal excretion and using cholesterol for the synthesis of hepatic bile acid. Another mechanism is by reducing insulin secretion due to its low glycemic effect on blood glucose which reduces cholesterol synthesis in the liver. There is a controversy over the explanation of fermentation of dietary fiber in the colon that produces short-chain fatty acids (SCFAs) (acetic, propionic, butyric acids) having an inhibitory effect on cholesterol biosynthesis in the liver. The overall efficacy of dietary fiber is moderate when oral intake is usual but when supplemented with different vegetable fibers including glucomannan, β-glucan, guar, pectin, psyllium, and hydroxypropyl methylcellulose, the efficacy is higher but mild reduction of LDL cholesterol (Bianconi et al., 2018; Zhen-Yu et al., 2008).

1.5.2 PHYTOSTEROLS

Plant sterols and stanols both reduce intestinal cholesterol by competing for sterol transporter Niemann-Pick C1-Like 1 (NPC1L1). They inhibit cholesterol absorption and decrease cholesterol synthesis by inhibiting HMG-CoA reductase gene expression. They reduce LDL cholesterol with little effect or without any effect on HDL or triglycerides. The mechanism by which they reduce LDL is by reducing the intestinal absorption of cholesterol, upregulating hepatic LDL receptors (consequently increasing hepatic cholesterol uptake) and reducing the production of endogenous cholesterol. It was found that a combination of statins and stanols/sterols was more effective in lowering the levels of total cholesterol and LDL-C than statins alone. HDL or triglyceride level is not altered by this combination (Sonowska et al., 2017; Bianconi et al., 2018; Zhen-Yu et al., 2008).

1.5.3 POLYPHENOLS

Polyphenols from green tea have a cholesterol-reducing effect. Green tea polyphenols (GTPs) also have antioxidant, anti-inflammatory, and anti-thrombotic properties. Cocoa polyphenols lower plasma total cholesterol and oxidized LDL without affecting HDL levels in persons with increased CVD risks. Resveratrol has many cholesterol-independent effects that benefit cardiovascular health, including antioxidant and anti-inflammatory effects, increased endothelial nitric oxide (NO) production and endothelial progenitor cell levels, and reduced platelet aggregation. Anthocyanins reduce cholesterol by inhibiting cholesteryl ester transfer protein (CETP) activity. It also improves arterial stiffness and endothelial function (Sosnowska et al., 2017; Bianconi et al., 2018; Zhen-Yu et al., 2008).

1.5.4 SOY PROTEIN

Soy is an important source of fibers, plant sterols, isoflavones, proteins, and additional bioactive compounds which exert a cholesterol-lowering effect by promoting liver LDL receptor expression and reducing intestinal cholesterol absorption. Some soy components also have antioxidant and anti-inflammatory effects which have positive cardiovascular health. Soy proteins also have anti-atherogenic properties (Bianconi et al., 2018; Zhen-Yu et al., 2008).

1.5.5 OTHER LEGUME PROTEINS

Other legume proteins like lupin, peas, beans, lentils, and butter beans have been shown to prevent hyperlipidemia. Lupin protein reduces LDL levels in hypercholesterolemic patients without affecting HDL levels and pea protein reduces plasma total cholesterol and triglyceride levels (Taylor et al., 2014).

1.5.6 RED YEAST RICE (RYR)

Red yeast rice (RYR) is obtained by fermentation of rice with red yeast (*Monascus purpureus*). It contains different cholesterol-lowering nutraceuticals like monacolins, plant sterols, fibers, and niacin. The cholesterol-lowering effect of RYR is mainly because of the monacolin K, which is analogous to lovastatin and thus inhibits 3-hydroxy-3-methyl-glutaryl-coenzyme A (HMG-CoA) reductase. Besides, it also has antioxidant and anti-inflammatory properties (Bianconi et al., 2018; Zhen-Yu et al., 2008).

1.5.7 BERBERINE

Berberine is a quaternary benzylisoquinoline alkaloid that is extracted from different species of plants including *Coptis, Hydrastis,* and *Berberis*. It shows a cholesterol-lowering effect by different mechanisms like upregulating the expression of LDL receptors through inhibition of proprotein convertase subtilisin/kexin type 9 (PCSK9) and the stabilization of the LDL receptor mRNA. It also lowers intestinal absorption of cholesterol and increases fecal excretion of bile acids. Besides this, it also has antioxidant and anti-inflammatory properties which have positive cardiovascular health (Bianconi et al., 2018).

1.5.8 CURCUMIN

Curcumin may inhibit intestinal absorption of cholesterol by downregulating the expression of NPC1L1 (Niemann-Pick-C1-Like), promote cholesterol efflux by upregulating the expression of ATP-binding cassette A1 (ABCA1), and enhance LDL particle liver uptake by downregulating PCSK9 expression. The effect of curcumin supplementation may be modest or absent depending on different target populations, duration of administration, and variable bioavailability of different oral formulations. Curcumin supplementation is also seen to have antioxidant and anti-inflammatory effects. It improves endothelial function and arterial stiffness by enhancing NO bioavailability, reducing inflammation and oxidative stress, and improving vascular remodeling (Bianconi et al., 2018).

1.5.9 BUCKWHEAT PROTEINS

Buckwheat herbs and seeds both are rich in rutin and also contain quercetin. It has hypocholesterolemic and hypotensive properties. Buckwheat has been found to reduce plasma total cholesterol and LDL cholesterol. It increases fecal excretion of bile acids as well as neutral sterols, thus exhibiting its action. Another mechanism may be, the insoluble fraction of buckwheat protein binds to cholesterol and prevent its solubility and uptake (Zhen-Yu et al., 2008; Taylor et al., 2014).

1.5.10 FERMENTED MILK

Fermented milk with *Lactobacillus*, *Bifidobacterium*, *Enterococcus*, and *Streptococcus* has a moderate cholesterol-lowering effect. The mechanism by which fermented milk may act on cholesterol levels are:

- The live bacteria alter the content and colonization of gut bacteria.
- These bacteria may ferment unabsorbed carbohydrates to produce SCFAs, lowering cholesterol by inhibiting HMG-CoA reductase.
- May bind to and absorb cholesterol, thus reducing cholesterol absorption in the intestine.
- Reduce reabsorption of bile acids through enterohepatic circulation by hydrolyzing and excreting conjugated bile acids (Zhen-Yu et al., 2008).

1.5.11 MARINE PROTEINS

Marine proteins from fish like Alaska Pollock, tuna, and scallops have been found to have hypocholesterolemic effects by increasing fecal excretion of bile acid and cholesterol or decreased reabsorption of bile acids from ileum through a decrease in IBAT (ileal bile acid transporter). The increased fecal cholesterol and bile acid excretion after eating these fish may be due to the digested products of fish protein which reduced micellar solubility of cholesterol and increased bile acid binding capacity. The freshwater clam (*Corbicula fluminea*) extract and hydrolysate protein fractions were also shown to reduce serum and hepatic cholesterol by increasing neutral sterols and bile acids excretion (Taylor et al., 2014).

1.5.12 OMEGA-3 POLY-UNSATURATED FATTY ACIDS (Ω-3 PUFAS)

DHA (docosahexaenoic acid, 22:6 ω-3), EPA (eicosapentaenoic acid, 20:5 ω-3), and ALA (α linolenic acid, 18:3 ω-3) are the main omega-3 PUFAs. ω-3 PUFAs are found to be beneficial against CVDs including hypotriglyceridemic, antithrombotic, anti-inflammatory, antiarrhythmic, and antiangiogenic effects. The mechanism by which ω-3 PUFAs exert their effect is many as given below:

- Upregulates gene expression involved in fatty acid oxidation through activating peroxisome proliferator-activated receptors (PPARα and PPARγ).
- Inhibits synthesis of hepatic fatty acid by suppressing sterol regulatory element binding protein (SREBP-1c) transcription and enhancing SREBP-1c protein degradation and mRNA decay.
- Suppresses insulin-induced SREBP-1c transcription via reducing the trans-activating capacity of LXRα.
- Increases fatty acid β-oxidation, ω-oxidation, and malic enzyme activities in the small intestine (Taylor et al., 2014).

Other single nutraceutical preparations that have cholesterol-lowering effects are:

- Allicin from garlic has cholesterol-lowering, antioxidant, anti-inflammatory, and anti-platelet properties. It reduces atherosclerosis progression and cardiovascular risk. It inhibits HMGR activity and/or CETP activity to improve lipid parameters in animals. However, the impact of garlic on human blood lipid parameters is controversial in human trials (Bianconi et al., 2018; Taylor et al., 2014).

- Artichoke extract has a lipid-lowering effect and reduces oxidative stress biomarkers and exerts beneficial effects on endothelial function (Bianconi et al., 2018).
- Brutieridin, melitidin, neoeriocitrin, and naringin from bergamot peel has a cholesterol-lowering effect, antioxidant, anti-inflammatory, and scavenging activity with a potential anti-atherogenic effect (Bianconi et al., 2018).
- Gamma oryzanol from rice bran oil also has a cholesterol-lowering effect. In addition, it has the downregulation of oxidative stress mediators, the reduction of the expression of adhesion molecules in vascular endothelial cells, and the inhibition of platelet aggregation (Bianconi et al., 2018; Zhen-Yu et al., 2008).
- Spirulina, a blue-green algae has lipid-lowering properties. It also possesses antioxidant, anti-atherogenic, and anti-inflammatory action which is good for cardiovascular health (Sosnowska et al., 2017; Bianconi et al., 2018).

Some foods too have certain cardio-protective effects as shown in Table 1.5 (Alissa & Ferns, 2012).

TABLE 1.5 Potential Cardiovascular Protective Effects of Some Functional Foods

Functional Foods	Bioactive Components	Potential Mechanism
Nuts	Tocopherols, omega-3 fatty acids	Lowering blood cholesterol
Legumes	Fiber and polyphenols	
Fruits and vegetables	Fiber (pectin)	
Margarine	Phytosterols	
Fish oil	Omega-3 fatty acids	
Whole grains	Fiber and phytochemicals	
Soy proteins	Genistein and daidzein	
Dark chocolate	Flavonoid	
Fish	Omega-3 fatty acids	Inhibition of LDL-C oxidation
Green leafy vegetables	Carotenoids	
Citrus fruits and vegetables	Vitamin C	
Tomato	Lycopene	
Extra virgin olive oil	Polyphenolics and oleic acids	
Green tea	Tea polyphenolics	
Soy protein	Genistein, daidzein, and glycitein	
Dark chocolate	Flavonoid	
Pomegranate	Polyphenols	

TABLE 1.5 *(Continued)*

Functional Foods	Bioactive Components	Potential Mechanism
Fish	Omega-3 fatty acids	Lowering blood triglycerides
Fish	Omega-3 fatty acids	Decreasing blood pressure
Legumes	Fiber	
Whole grains	Fiber and phytochemicals	
Citrus fruits	Ascorbic acid	
Ginseng	Ginsenosides	
Onion and garlic	Quercetin	
Green and black teas	Tea polyphenols	
Grapes and red wines	Grape polyphenols	
Dark chocolate	Flavonoid	
Fruits and vegetables	Folate and phytochemicals	Lowering blood homocysteine.
Whole grains	Fiber and phytochemical	
Citrus fruits and vegetables	Vitamin C	
Nuts, seeds, and oils	Vitamin E	
Tomatoes	Lycopene	Antioxidant action.
Green leafy vegetables, fruits	Carotenoids	
Vegetable oils	Tocopherol, tocotrienol	
Citrus fruits and vegetables	Vitamin C	
Soy protein	Genistein and daidzein	
Green and black teas	Tea polyphenols	
Grapes and red wines	Anthocyanins, catechins, cyanidins, flavonols, myricetin, and quercetin	
Nuts, seeds, and oils	Vitamin E	Anti-inflammatory action.
Fish	Omega-3 fatty acids	
Legumes	Polyphenols	
Grapes and red wines	Anthocyanins, catechins, cyanidins, flavonols, myricetin, and quercetin	
Fish	Omega-3 fatty acids	Endothelial function.
Nuts	Polyphenols	
Citrus fruits and vegetables	Vitamin C	
Grapes and red wines	Anthocyanins, catechins, cyanidins, flavonols, myricetin, and quercetin	
Dark chocolate	Flavonoids	
Grapes and red wines	Anthocyanins, catechins, cyanidins, flavonols, myricetin, and quercetin	Platelets aggregation.

Some nutraceuticals have a beneficial effect on chronic coronary syndrome (CCS) also called chronic ischemic heart disease which is summarized in Table 1.6 (Tognola et al., 2020).

TABLE 1.6 Principal Effect of Nutraceutical on CCS at Preclinical and Clinical Level

Nutraceutical	Preclinical Data	Translational Data
Vitamin C, A, E, B6	Effect on the cardiovascular system, free radicals scavenging, prevention of LDL oxidation. Supplementation lowers atherogenesis and improves HDL profile.	Lowering of SOD and mitochondrial factor in young athletes during endurance exercise. No significant effect on mortality or cardiovascular disease outcomes.
Vitamin D	Prevention of neointimal hyperplasia and restenosis after PTCA by inhibition of Cathepsin L.	Improves glycemic control, HDL, and CRP levels. There were no effects on triglycerides, total- and LDL-cholesterol levels. Combined calcium and vitamin D supplementation might increase the risk of stroke. No clear role in cardiovascular prevention.
Folic acid	–	Associated with lowering the risk for stroke.
Omega-3 PUFA	Effective in relieving IR injury: supplementation determines a reduction of the infarcted area in rats with induced myocardial ischemic damage.	Supplementation is related to reduced risk of cardiovascular events and mortality: beneficial effect on TG and lipids, improves endothelial function, anti-inflammatory, and antiplatelet effects, and plaque stabilization. Lowering of death from arrhythmic causes. No mortality and major CV events reduction in patients with AMI. The beneficial effect of flaxseed oil and fish oil on serum insulin, plasma total nitrite, and total antioxidant capacity among T2DM patients with CCS.

TABLE 1.6 *(Continued)*

Nutraceutical	Preclinical Data	Translational Data
Carnitine	Supplementation in the murine model depleted in fatty acids resulted in the restoration of the physiological oxidative cardiac metabolism and reduction in glucose utilization. Protective effect on IR injury. The possible effect of exogenous administration in the prevention of arrhythmias.	Supplementation in patients after CABG improves EF and lowers LVEDV. Carnitine treatment initiated early after AMI has a beneficial effect in LV reverse remodeling and reduced hospitalization for CHF.
Amino acids	Long-term supplementation improves myocyte survival by maintaining mitochondrial functional capacity during IR injury. Arginine administration in rabbits was superior to aspirin in the cardioprotective role in ischemic injury. Perinatal taurine depletion in rats on high sugar intake worsens cardiac damage and BP control after IR.	–
Flavonoids	Reduced progression of atherosclerosis, reduction of IR injury, and improved CV outcomes.	Decreased CVRFs, improved glycemic control (decreased fasting glucose and insulin resistance and significantly increased insulin sensitivity), HDL, and total cholesterol. Reduced CV events. Antioxidant effect, vasodilatation, improvement of endothelial function, suppression of platelet aggregation, and regulation of mitochondrial metabolic activity. An improved 6-minute walking test distance in patients after AMI.

TABLE 1.6 *(Continued)*

Nutraceutical	Preclinical Data	Translational Data
Ribose	Accelerates the regeneration of ATP, necessary in the way of pentose phosphates. IR damage and infarct area size reduction with an improved cardiac functioning parameter.	Improved cardio-pulmonary test performance in patients with chronic myocardial ischemia, reduced diastolic dysfunction, and improved quality of life in patients with ischemic heart failure.
Alpha lipoic acid	Clears ROS (reactive oxygen species), suppresses ROS generators and spares antioxidant compounds. Involved in cardiac energy metabolism. Improves glycemic control. Reduction of IR injury.	–
Magnesium	Reduced myocardial infarct size in animals by supplementation.	Supplementation in obese subjects reduces blood pressure, hyperinsulinism, and triglycerides. Supplementation reduces platelet aggregation. Intravenous supplementation in the first few hours after AMI showed a significant reduction in in-hospital mortality by reducing ventricular arrhythmias.
Selenium	Avoids IR injury	Plasma levels correlate with myocardial damage, defined as peak concentration of Troponin I, in subjects with AMI. Supplementation is mainly studied in subjects undergoing CABG that showed improvement in postoperative outcomes. Larger trials are needed to establish if low selenium concentrations could be a CVRF. Could have a beneficial effect on the inflammation regulation of CCS patients by modulating selenoprotein P and S genes expression in protein and mRNA levels.

TABLE 1.6 *(Continued)*

Nutraceutical	Preclinical Data	Translational Data
Zinc	Involved in the metabolism of free radicals and inflammatory processes, regulating atherosclerotic process: deficiency could be an RF for atherosclerosis. Supplementation in animal models has resulted in reduced myocardial infarct size through a protective effect against IR injury.	Inverse correlation with atherosclerosis. Low intake has an association with higher CIMT and HDL. Reduced during the first days after AMI. The inverse correlation was found with CK, CK-MB, and Troponin T.
Coenzyme Q10	Supplementation increases myocardial contractility, better endothelial function, and decreases the production of free radicals and other inflammatory molecules. Effective in reducing IR injury. Supplementation increases SOD and catalase (CAT) and reduces MDA and diene levels, with no effect on CRP, TNF-α, IL-6, and GPx levels among patients with chronic coronary syndrome.	Supplementation along with selenium reduces mortality from CV causes. Decrease total cholesterol and increase HDL. CoQ10 deficiency determined by statins has been identified as a possible cause of the known muscle side effect of these drugs and can be it can be used to reduce this side effect. Ameliorated statin-associated muscle symptoms.
Curcumin	Strong antioxidant activity and prevents cardiomyocyte damage in isoproterenol-induced myocardial infarction.	–
2,3-Dehydrosilybin (DHS)	Improves IR damage	–
Nicotinamide (Vitamin B3)	Reduced size of the infarct area and better cardiac function.	–
Red palm oil	Reduced size of the infarct area in animal models.	–
Beetroot juice	Improved heart function and reduced infarct size in mice.	–

1.5.13 FUNCTIONAL FOODS, NUTRACEUTICALS, AND CHRONIC KIDNEY DISEASE (CKD)

Chronic kidney disease (CKD) is a rapidly growing global health issue. In CKD, there is a progressive loss of kidney function leading to a decrease in the kidney's ability to excrete waste products leading to the accumulation of uremic toxins. CKD, according to K/DOQI (kidney disease outcomes quality initiative), is defined as kidney impairment that is confirmed by structural or functional abnormalities of the kidney or glomerular filtration rate (GFR) <60 mL/min/1.73 m^2 for 3 months or longer, regardless of cause (Coresh et al., 2007). A healthy balanced diet is important for maintaining good health status as it reduces the incidence of chronic diseases like CKD, diabetes mellitus, and hypertension. Functional molecules and nutrients like fatty acids, fiber, and nutraceuticals such as curcumin, steviol, glycosides, and resveratrol have beneficial effects on pro-inflammatory and anti-inflammatory pathways as well as on intestinal mucosa. Antioxidant-rich food consumption is associated with the reduction of oxidative stress in different health conditions (Rysz et al., 2021).

1.5.13.1 POLYPHENOLS

Polyphenols can reduce circulating uric acid (UA) and have a protective effect against vascular oxidative stress possibly inhibiting the enzyme xanthine oxidase required to produce UA or by acting as ROS scavengers. Gut dysbiosis was found to be a common feature in CKD so consuming prebiotics, probiotics, polyphenols, sugars, and proteins could be beneficial in altering the diversity of the gut microbiota and formation of uremic toxins (Rysz et al., 2021). The beneficial effects of some polyphenols and more nutraceuticals on CKD are discussed in subsections.

1.5.13.2 CURCUMIN

There are not enough beneficial effects on CKD patients. Short-term turmeric supplementation is shown to be able to attenuate proteinuria, TGF-β, and IL-8 in patients with type 2 diabetes nephropathy and a significant decrease of inflammatory markers TNFα, IL-6, and C-reactive protein is seen with *Curcuma longa* and *Boswellia serrata* in non-dialysis CKD patients.

Curcumin has also been shown to reduce lipid peroxidation (LPO) in patients with non-diabetic proteinuric CKD and boost antioxidant activity in those with diabetic proteinuric CKD (Rysz et al., 2021).

1.5.13.3 RESVERATROL

It has been found that resveratrol helps in improving vascular functions in CKD patients. According to a cross-sectional logistic regression analysis of the National Health and Nutrition Examination Survey (NHANES), light wine consumption (<1/day) significantly reduced the prevalence of CKD and decreased the rate of CVD in them when age, race, gender, diabetes mellitus, hypertension, and cholesterol levels were controlled. However, no major randomized, placebo-controlled, large clinical trials including CKD patients, have been conducted to support this concept. An *in vitro* study confirmed that resveratrol has anti-inflammatory and chondroprotective properties in human chondrocytes (Buhrmann et al., 2017; Csaki et al., 2009).

1.5.13.4 GREEN TEA AND COFFEE

It was found that daily supplementation of catechin from decaffeinated green tea is effective in reducing ROS production, decreasing hemodialysis-induced plasma hypochlorous acid activity and pre-dialysis plasma hydrogen peroxide (H_2O_2) activity, and lowering phosphatidylcholine hydroperoxide, C-reactive protein, and pro-inflammatory cytokine concentrations (Hsu et al., 2007). Since oxidized pro-atherosclerotic compounds such as oxidized LDL and phosphatidylcholine hydroperoxide are not removed mechanically during hemodialysis so the use of potent antioxidants that limit ROS production and protect against oxidative damage appears to be beneficial in this group of patients as it slows down the progression of atherosclerotic vascular disease (Cosola et al., 2018). The consumption of green tea by patients with CKD was associated with significantly improved FMD in the catechin group. However, in dialysis dependent CKD patients, no significant improvement was seen in clinical features, oxidative stress, inflammatory markers, or the quantity of circulating endothelial progenitor cells (EPCs) (Park et al., 2010).

1.5.13.5 CAFFEINE

Caffeine consumption was shown to reduce the risk of development of CKD in healthy consumers compared with non-drinkers. However, data supporting this is sparse (Rysz et al., 2021).

1.5.13.6 VITAMIN D

Vitamin D supplementation reduces the mortality risk from CVD in CKD patients. In patients with non-diabetic CKD stage 1–4, oral cholecalciferol markedly enhances endothelium-dependent brachial artery flow-mediated dilation that improves vascular function, and the improvements are linked with 25(OH)D levels (Rysz et al., 2021; Singh et al., 2017).

1.5.13.7 POLYUNSATURATED FATTY ACIDS (PUFA)

An inverse relation was found with higher consumption of dietary eicosapentaenoic (EPA) acid and docosahexaenoic acid (DHA) to the severity of albuminuria in type 1 diabetes but not to the incidence of albuminuria (Lee et al., 2010). However, due to the poor quality of the papers included in the meta-analysis, larger investigations are needed to corroborate the findings. Such findings are encouraging because even a small decrease in urinary albumin excretion has been linked to a lower risk of renal failure and CVD (Matsushita et al., 2010). In comparison to corn oil (an omega-6 fatty-acid source), adding fish oil to the diet of hemodialysis patients was related to a significant rise in HDL levels, but also with an increase in LDL concentrations (Bowden et al., 2009). Administration of (45% EPA and 37.5% DHA) in hemodialysis patients resulted in a substantial reduction in serum triglycerides in the n–3 PUFA group compared with olive oil (control supplement) (Svensson et al., 2008). Total cholesterol, high-density lipoprotein (HDL) cholesterol, low-density lipoprotein (LDL) cholesterol, Lipoprotein (a), and apoB, on the other hand, showed no significant effect. In another trial, omega-3 supplementation (2 g/day) did not affect lipid levels when compared to a similar amount of olive oil, but it did lower systolic, diastolic, and mean blood pressure (Vernaglione et al., 2008). The findings of polyunsaturated fatty acid research are inconsistent. However, it appears that consuming omega-3 fatty acids can help HD patients reduce their risk of CVD by influencing risk factors such as blood pressure, LDL, HDL, and triglyceride levels (Ajay et al., 2019).

1.5.13.8 CONJUGATED LINOLENIC ACID (CLA)

It has been shown in several animal studies that CLA has anti-inflammatory, anti-atherosclerotic, and anti-diabetic qualities, as well as lowers renal PGE2 production and has considerable renal anti-inflammatory and anti-fibrotic effects. As a result, CLA appears to be a promising molecule for the dietary treatment of renal illness (Rysz et al., 2021; Ajay et al., 2019; Ogborn et al., 2003).

1.6 FUNCTIONAL FOODS, NUTRACEUTICALS, AND OBESITY

Obesity is a complex chronic disease or syndrome characterized primarily by energy expenditure and intake imbalance. Obesity, along with its comorbidities, is the fifth largest cause of death worldwide. Several epidemiological studies have found that consuming functional foods/nutraceuticals reduces the incidence of a variety of chronic disorders, including obesity. They are discussed in Table 1.7 (Venkatakrishnan et al., 2019; Jarouliya & Keservani, 2019).

1.6.1 FUNCTIONAL FOODS, NUTRACEUTICALS, AND OSTEOPOROSIS

Osteoporosis is a very common bone disorder among the aged population and post-menopausal women. It occurs due to defects in the bone remodeling process (Rajput et al., 2018). There are various nutraceuticals that are helpful in osteoporotic patients for bone health which are classified as:

1. **Minerals**: Minerals like Calcium, Magnesium, Potassium, Boron, and Copper are found good for osteoporosis management. Calcium is helpful in attaining peak bone density. Potassium helps to maintain an alkaline environment which reduces the requirement of bone calcium for regulating pH. Copper helps in crosslinking of collagen and removes free radicals from our bodies. Boron prolongs Vitamin D and estrogen half-lives. It also prevents bone mineralization. Magnesium acts as a buffer against acidosis. PTH levels in the body are maintained by this mineral (Rajput et al., 2018).
2. **Herbs and Phytochemicals:** These includes alfa-alfa, red clover, horsetail, citrus fruits, onion, green tea, *Herba epimedii*, berries,

TABLE 1.7 Functional Foods and Nutraceuticals Mechanism in Obesity Management

Functional Food	Bioactive Component	Anti-Obesity Mechanism
Garcinia cambogia plant	Hydroxycitric acid (HCA), garcinol	Indirectly inhibit the production of acetyl CoA and mitochondrial enzyme citrate lyase and thereby halt the lipogenesis process (Vasques et al., 2014).
		Increase the availability of serotonin and thereby suppress the appetite as well as Regulate adipocyte differentiation via the mammalian target of the rapamycin (mTOR) pathway (Kim et al., 2004).
		Regulate both carbohydrate (glycolysis and glycogenesis) and lipid metabolism (altering lipid metabolizing enzymes) both oxidation and synthesis via upregulating AMP-related protein kinase (AMPK) and serine protein kinase B/phosphatidylinositol-3-kinase (Akt/PI3K) signaling pathway (Saito et al., 2005).
Bitter melon	Sterols and triterpenes such as charantin, vicine, cucurbitane, momordicin II, Kuguaglycoside G, conjugated fatty acids like α-eleostearic acid and bioactive proteins like polypeptide p.	It decreases adipocytokines release, inhibits pre-adipocyte differentiation, and positively regulates the expression of peroxisome proliferator-activated receptor (PPAR) α and γ (via the AMPK pathway) (Bai et al., 2016).
		Protect the islet of Langerhans, thus improving insulin production as well as lower adipocytokines and thereby improving insulin sensitivity as well as enhancing the enzyme expression involved in glycolysis (Zhu et al., 2016).
		It increases β-oxidation and decreases fatty acid synthesis via modulating Akt/PI3K pathway (Alam et al., 2015).
Capsicum	Capsaicin/Capsaicinoid	Exhibit thermogenic or improved energy expenditure via SNS in brown adipose tissue (BAT) was exerted by the positive impact on various mitochondrial uncoupling proteins (UCP2 and 3) (Masuda et al., 2003).
		Induce pre-adipocyte apoptosis, inhibit adipogenesis, and regulate adipocyte differentiation through altering c-Jun N-terminal kinases (JNK), extracellular signal-regulated kinases (ERK½), AMPK and NF-κB signaling pathway (Lu et al., 2018).
		Increase adiponectin synthesis (via upregulating PPARα) and upregulate the mRNA expression of hormone-sensitive lipase (HSL) (Hsu & Yen, 2007; Kang et al., 2007).

TABLE 1.7 *(Continued)*

Functional Food	Bioactive Component	Anti-Obesity Mechanism
Polyunsaturated fatty acids (PUFA)/ omega-3-fatty acids	ALA, CLA, DHA, and EPA	Modulate the expression of adipocytokines and thereby regulate the proliferation and differentiation of adipocytes (via Akt/PI3K and AMPK pathway), activate PPAR-α, decrease leptin expression, trigger thermogenesis in the BAT (Balaji et al., 2016).
		Enhance fat oxidation (thermogenesis) as well as inhibit adipogenesis/lipogenesis via downregulating the JNK signaling pathway (Sung et al., 2018).
		TG (triglyceride) synthesis is inhibited DHA and EPA by decreasing the circulating FFA and enhancing β oxidation (Flachs et al., 2005).
Probiotics	*Bifidobacterium* and *Lactobacillus* spp.	Probiotics can absorb cholesterol on their own and possess bile acid hydrolase activity and thus aids in the binding of cholesterol with deconjugated bile and excreted. Hence, two-way probiotics can regulate cholesterol levels either by excretion or utilization and thus indirectly attenuate obesity and CVD risk (Hunter & Hegele, 2017).
		SCFA produced from probiotics binds to PPAR as well as alters the activity of a lipolytic enzyme called lipoprotein lipase (LPL) and thereby halts lipogenesis (Dahiya et al., 2017).
		SCFA upregulates the production of GLP-2, PPY, and CCK by triggering POMC and results in satiety as well as improving insulin sensitivity (Baboota et al., 2013).
Psyllium husk	Dietary fiber	Increase gastric transient time (delay gastric emptying) and favors satiety (alter Gut-Brain Axis) by upregulating GLP-1/2 and CCK hormones as well as inhibiting ghrelin secretion (Jane et al., 2018).
		Possess a lesser glycemic index and hence revert insulin resistance and enhance the thermogenesis process to maintain energy homeostasis (Khossousi et al., 2008).
		SCFA produced by fermentation act as a prebiotic hence inhibiting PPAR to exhibit lipolysis by upregulating the activity of LPL as well as improving insulin sensitivity (Baboota et al., 2013).

TABLE 1.7 *(Continued)*

Functional Food	Bioactive Component	Anti-Obesity Mechanism
Oats bran	Soluble fiber (β-glucan)	Enhance gastric emptying as well as increase bile acid excretion and thus regulate cholesterol homeostasis (HMG-CoA reductase) (Liao et al., 2019).
		Elevate post-prandial CCK, and PYY hormone and lower insulin response and thereby suppress appetite (Beck et al., 2009).
		Similar to psyllium it can also produce SCFA by fermentation as they indirectly act as prebiotics and bind to PPAR to favor lipogenesis by enhancing LPL activity (Dahiya et al., 2017).
Green tea	Catechin	Increase GLP-1 level as well as improve insulin resistance, a pancreatic lipase inhibitor, increase thermogenesis, increase satiety (enhance serotonin/dopamine uptake by altering gut-brain axis) downregulate the protein expression of insulin resistance-related adipokine resistin via extracellular signal-related kinase (ERK½) signaling pathway and NF-κB pathway (Beck et al., 2009).
		It also enhances fat oxidation (β oxidation), inhibits intestinal glucose absorption, enhances glucose uptake, and upregulates enzymes involved in glycolysis/glycogenesis via the AMPK/PPAR pathway and increases fecal lipid excretion (Chen et al., 2016).
Quercetin	Flavonoid	Antioxidant (lower oxidative stress via Nuclear factor erythroid derive 2 like 2; Nrf2 pathway), anti-inflammatory (downregulate NF-κB pathway), hypolipidemic, thermo-genesis (upregulating UPC-1 and PPAR gamma coactivator 1 alpha: PGC1α) activities, increase insulin sensitivity by improving adiponectin level and reduce nitrate level (iNOS), lipid metabolizing enzymes like fatty acid synthase (FAS), acyl-CoA carboxylase (ACC) as well as inhibit adipogenesis by suppressing the mRNA expression of PPARγ, CCAAT-enhancer binding protein alpha (C/EBPα), LPL (Lu et al., 2018; Nabavi et al., 2015).
		Induce apoptosis (mature adipocytes) and carbohydrate metabolism (glycolysis) via AMPK (positively regulate) and mitogen-activated protein kinase (MAPK; negative regulate) signaling pathway (Rivera et al., 2008; Moon et al., 2013).

TABLE 1.7 *(Continued)*

Functional Food	Bioactive Component	Anti-Obesity Mechanism
Resveratrol	Polyphenol	Enhance GLP-1 level and thereby delay gastric emptying (alter gut-brain axis) and food intake as well as lower blood glucose level (Park et al., 2012).
		Inhibition of adipogenesis and trigger apoptosis in matured adipocytes along with quercetin (Yang et al., 2008).
		Inhibit the expression of various signaling molecules like steroid regulating element binding protein 1c (SREBP-1c), liver X receptor (LXR), and PPAR to downregulate the expression of respective genes (related to insulin resistance) of leptin, resistin, FAS, ACC through downregulating toll-like receptor 2 (TLR2) (Kim et al., 2011).
		Increase the thermogenesis process by activating the AMPK signaling pathway and increase leptin sensitivity by inhibiting the NF-κB pathway (Lu et al., 2018; Wang et al., 2017).
Turmeric	Curcumin	It exhibits potent antioxidant (upregulating Nrf2 signaling pathway), anti-inflammatory, anti-hyperlipidemic (downregulate SREBP, LXR), and thermogenic (by upregulating UPC1, PGC-1α) activities (Lu et al., 2018).
		It suppresses the conversion of pre-adipocytes to adult or mature adipocytes (differentiation-adipogenesis) as well as inhibits the proliferation of pre-adipocytes via the AMPK-PPAR signaling pathway (Bradford, 2013).
		It stimulates lipolysis (lower FFA circulation), suppresses lipogenesis, and enhances the apoptosis of adipocytes via MAPK and ERK ½ pathway (Lee et al., 2008).
Ginger	Zingerone, zingiberene, gingerols, 6-shogaol, gingerenone A and volatile oils.	Ginger (gingerols, 6-shogaol) is reported to modulate PPAR (α/γ) signaling pathway (C/EBPα) and thus suppress lipogenesis (adipogenesis) as well as increase β oxidation (lipolysis) (Lu et al., 2018).
		Improve insulin sensitivity by upregulating UCP-1 and PGC-1α, increasing serotonin/dopamine uptake and satiety, inhibiting intestinal fat/cholesterol absorption (pancreatic lipase), inhibiting adipogenesis, and altering carbohydrate metabolism (glucose uptake and utilization) via MAPK, Akt/PI3K and NF-κB signaling pathways (Attari et al., 2018; Misawa et al., 2015).

plums, resveratrol, black cohosh. The active compound involved in bone protection, their function, and their mechanism of action is described in Table 1.8 (Rajput et al., 2018).

TABLE 1.8 Mechanism of Actions of Herbs/Phytochemicals for Improving Bone Health

Herb/ Phytoconstituent	Active Constituent	Mechanism of Action	Effect on Bone
Alfalfa	Ipriflavone	Increases alkaline phosphatase, type 1 collagen. Decreases secretion of cathepsin K, cathepsin L, and β-glucuronidase.	Enhances osteoblastogenesis. Inhibits osteoclastogenesis.
Red clover	Genistein and daidzein.	Inhibition of topoisomerase II and tyrosine kinase.	Inhibits osteoclast-mediated bone resorption.
Horsetail	Silica	Formation of collagen.	Bone mineralization.
	Quercetin and kaempferol	Activates anti-inflammatory pathway.	Inhibits osteoclast differentiation.
	Luteolin	Decrease bone marrow mononuclear cell differentiation.	Decrease bone resorption.
Citrus fruits	Limonoids and flavonoids.	Reduce oxidative stress.	Decrease bone resorption.
Onion	Quercetin, rutin, and organo-sulfur compounds.	Reduce oxidative stress.	Decrease bone resorption.
Green tea	Epigallocatechin-3-gallate.	Activation of caspase-3. Suppression of matrix metalloproteinase-9. Suppression of osteoclast formation. Inhibition of IL-6, down-regulation of RANKL. Increases alkaline phosphatase suppresses TNF-α.	Osteoclast cell death. Suppression of osteoclast formation. Suppression of osteoclast differentiation. Enhances osteoblastogenesis.

TABLE 1.8 *(Continued)*

Herb/ Phytoconstituent	Active Constituent	Mechanism of Action	Effect on Bone
Herba epimedii	Icariin	Inhibits TNF-α, IL-6, cyclo-oxygenase type-2, and prostaglandin E2.	Prevents inflammatory bone loss.
Berries	Anthocyanidine	Inhibits receptor activator of nuclear factor kappa beta ligand (RANKL) receptors.	Inhibits osteoclastogenesis
Plums	Polyphenols	Inhibits RANKL receptors.	Inhibits osteoclastogenesis.
Resveratrol	Resveratrol	Modulate osteoblast and osteoclast gene expression.	Enhances osteoblastogenesis, and inhibits osteoclastogenesis.
		Reduces oxidative stress, and inflammatory cytokines.	Antioxidant and anti-inflammatory.
Black cohosh	25-acetylcimigenol xylopyranoside	Inhibits RANKL and TNF-α	Inhibits osteoclastogenesis

Source: Adapted/Reprinted from Rajput et al, 2018

3. **Dairy Products:** These includes milk, cheese, yogurt. Calcium is found in abundance in milk and milk products, which is necessary for the prevention of osteoporosis. They are also good sources of proteins, lipids, potassium, sodium, zinc, phosphate, and vitamins A and B2. The functional compounds in milk which are beneficial for bone health are casein phosphopeptide, milk basic protein (MBP), and lactoferrin. MBP increases the proliferation and differentiation of osteoblasts and prevents osteoclast generation and bone resorption. MBP contains Cystatin C which inhibits cysteine protease causing the release of calcium from bones under the influence of thrombin, interleukin-1, and prostaglandin E2, and also digests bone collagen. It also elevates serum levels of osteocalcin which is a chief non-collagenous protein present in the bone. Lactoferrin has anabolic effects on bone due to the low-density lipoprotein receptor-related protein that stimulates mitogenic-activated protein kinases to confer rapid cell mitogenesis (Rajput et al., 2018).

1.6.2 PROBIOTIC

Probiotics are bacterial components found in normal human gut flora that are capable of producing short-chain fatty acid and lactate metabolic end products. Production of vitamin D, C, K, and folate, which are associated with calcium metabolism and are essential for bone formation, decreases in PTH by producing SCFAs thereby increasing mineral absorption, produces antioxidant status thereby decreasing osteoclast differentiation, and finally reduces intestinal inflammation and cytokine levels while increasing calcium uptake in the bones are all possible mechanisms by which probiotics provide good bone health. Lactobacillus and Bifidobacterium are two bacteria that have been studied for bone (Rajput et al., 2018).

1.6.3 PREBIOTICS

Prebiotics are non-digestible food ingredients that are fermented in the colon by intestinal microflora and support the growth of other beneficial bacteria called probiotics. Non-digestible oligosaccharides (NDOs) are prebiotics that increases bone mineralization by increasing the absorption of minerals. Supplementing with galacto-oligosaccharides (GOS) increased magnesium absorption, calcium uptake into the femur, calcium, and magnesium retention, and femur and tibia breaking strength, demonstrating that GOS has positive effects on bone (Rajput et al., 2018).

PUFAs are divided into two groups: n–6 (omega 6) and n–3 (omega 3) fatty acids. Vegetable oils are the source of n–6 fatty acids like linoleic acid (LA) and fish oil contain n–3 fatty acids like eicosapentaenoic acid (EPA) and docosahexaenoic acid (DHA). n–3 fatty acids regulate bone remodeling in three important ways: calcium absorption, inflammatory cytokines, and peroxisome proliferator-activated receptor gamma. Long-chain PUFAs enhance Calcium ATPase, an enzyme that regulates calcium absorption in the intestine. The n–6 fatty acids in the cell membrane are considerably replaced by dietary consumption of long-chain n–3 fatty acids, resulting in a rise in the omega 3/omega-6 fatty acid ratio, and a decrease in the levels of IL-1, IL-6, and TNF-alpha as a result of the change in membrane function. PPAR-gamma is a regulatory mechanism for controlling the differentiation of stem cells into osteoblasts or bone marrow (Rajput et al., 2018; Keservani et al., 2010a; Rane & Keservani, 2020; Keservani & Sharma, 2020).

Melatonin – Pineal gland produces a hormone called melatonin which is synthesized and released at night causing diurnal fluctuation. Melatonin raises the concentration of protein, such as procollagen type I c-peptide which is incorporated in the bone matrix. It also raises the osteoblastic protein, osteoprotegerin which prevents differentiation of osteoclast (Rajput et al., 2018).

1.6.4 FUNCTIONAL FOODS, NUTRACEUTICALS, AND THEIR ANTI-CANCER ROLE

Cancer occurs over a few decades and is characterized by a complex interaction between multiple genes within the cell and its neighboring tissues, leading to the gradual conversion of healthy cells into cancerous cells (Upadhyay et al., 2009; Kesharwani & Misra, 2011; Manidhar et al., 2013). It is initiated by DNA damage as a result of genetic, epigenetic, and post-translational events involved in carcinogenesis (Upadhyay et al., 2009; Kesharwani & Misra, 2011; Kesharwani, et al., 2015; Manidhar et al., 2013). Less than 10% of all kinds of cancer are caused by germline mutation, whereas 90% are attributable to somatic mutations, lifestyle factors (including dietary habits), and different types of environmental carcinogens (Kapinova et al., 2017; Aggarwal et al., 2009; Grivennikov et al., 2010). Moreover, current research shows that errors associated with DNA replication also play a significant part in the progression of cancer (Kapinova et al., 2017; Tomasetti et al., 2017). The best way to prevent cancer is through rational dietary habits and behaviors, and the consumption of sufficient amounts of antioxidants and bioactive plant-derived compounds that demonstrate protective effects against carcinogenesis (Kapinova et al., 2017; Liu, 2013; Wilett, 1995, 2002). The most common cancer is breast cancer, which is most commonly occurring, and it is now the first leading cause of cancer-associated deaths of women globally. According to WHO figures from 2020, 2.3 million women worldwide were diagnosed with breast cancer, with 6,85,000 deaths (WHO, 2021; Augimeri et al., 2021).

Several epidemiological and clinical studies have shown that long-term and continuous consumption of nutraceuticals can reduce breast cancer risk. Fruits, vegetables, or legumes significantly reduced the risk of breast cancer, particularly in the triple-negative breast cancer (TNBC) subtype (Castello et al., 2014). Frequent consumption of extra virgin oil is good for preventing breast cancer in high-risk women (Toledo et al., 2015). An increased intake of energy from phytochemical-rich diets is linked with a lower incidence of breast cancer and blueberries and peaches showed a reduction in the incidence of ER breast cancer in postmenopausal women (Fung et al., 2013).

Intake of folate decreases breast cancer risk, especially for women with higher alcohol intake levels (Chen et al., 2014). Breast cancer is reduced in postmenopausal women who consume flavonols and flavones (Hui et al., 2013). Administration of resveratrol to women at an increased risk for breast cancer decreased methylation of four cancer-related genes (p16, RASSF-1α, APC, and CCND2) on mammary tissue biopsies after treatment (Zhu et al., 2012). According to a study, EGCG can help conventional breast cancer anticancer therapy work better. After radiation, EGCG ingestion lowers serum levels of VEGFs, hepatocyte growth factor, MMP-2, and MMP-9 (Zhang et al., 2012). Though there are not enough data supporting the failure or harmful effects of isolated compounds in breast cancer there are studies supporting that folic acid treatment is not associated with the reduction of total cancer incidence in a variety of cancers including breast cancer (Qin et al., 2013, 2015). Retinoids, either alone or in combination with PPAR agonists, are cell differentiation agents that affect cell proliferation and metabolism (Augimeri et al., 2021). The mechanism by which these functional foods and nutraceuticals help in preventing breast cancer is illustrated in Figure 1.2 (Kapinova et al., 2017).

The anti-cancer effects of different functional foods and nutraceuticals are summarized in Table 1.9 (Kapinova et al., 2017; Augimeri et al., 2021).

TABLE 1.9 Anticancer Effects of Functional Foods and Nutraceuticals

Functional Foods/Nutraceuticals	Anti-Cancer Effects
Phytochemical-rich foods like fruits, vegetables, whole grains, legumes, and nuts.	Reduction of breast cancer risk, especially in the case of the TNBC subtype.
Extra virgin oil	Chemo-preventive effect in high-risk women.
Blueberries and peaches	Reduction of the incidence of ER-breast cancer in post-menopausal women.
A mixture of green tea catechins.	Anti-proliferative effect.
Folate	Reduction of breast cancer risk, especially for women with higher alcohol intake levels.
Flavonols and flavones	Reduction of breast cancer in postmenopausal women.
Resveratrol	Decrease in methylation of cancer-related genes (p16, RASSF-1α, APC, CCND2).
Epigallocatechin-3-gallate (EGCG) (isolated from green tea, Taiyo Green Power Co.)	Increasing the effect of radiotherapy, anti-metastatic effect.
Retinoids	Control cell proliferation and metabolism.

Source: Adapted from Kapinova et al., 2017; Augimeri et al., 2021

1.7 FUNCTIONAL FOODS, NUTRACEUTICALS, AND AUTOIMMUNE DISORDER

ROS (reactive oxygen species) are a type of chemical that causes DNA damage. They are formed as a result of cellular metabolism in reaction to harmful substances or when exogenous events intervene. The ROS family includes chemical products such as superoxide anion radicals (O_2), H_2O_2, hydroxyl radicals (OH), and singlet oxygen (O_2) (Casciaro et al., 2017). Several autoimmune illnesses are linked to oxidative stress and autoimmunity with genetic susceptibility, but the mechanism of integrating these two pathways is not understood (Wang et al., 2019; Ruggeri et al., 2016). Rheumatoid arthritis (RA), systemic lupus erythematosus (SLE), multiple sclerosis (MS), and diabetes mellitus are some of the most common autoimmune disorders (DM) Type 1 patient have a different ratio of Type 1 helper (Th1) to Type 2 helper (Th2) activity, as well as increased Type 17 helper (Th17) responses and serum levels of Type 0 helper (Th0) and Th1 cytokines (Aryaeian et al., 2019). ROS effects are counterbalanced by antioxidant mechanisms including enzymes such as superoxide dismutase (SOD), catalase (CAT), glutathione peroxidase (GPx), and myeloperoxidase (MPO). Non-enzymatic and exogenous antioxidants can also help to counteract oxidative damage (Vaccaro et al., 2017; Cannavo et al., 2019). These antioxidants are obtained from diet which includes Vitamin C, E, carotenoids, and flavonoids (Carr & Maggini, 2017). These functional foods and nutraceuticals having antioxidant effects and their mechanism in preventing autoimmune disorders are discussed below (Mannucci et al., 2021):

1. **Vitamins:** Vitamin E is a good antioxidant working as a peroxyl radical scavenger. The study has shown that vitamin E helps in ameliorating oxidative stress in DM-1 subjects (Gupta et al., 2011). Vitamin C acts along with vitamin E. It removes excess radicals and replenishes vitamin E. Carotenoids protect against singlet oxygen. Vitamin D contains anti-inflammatory, antioxidant, and anti-fibrotic properties. Low levels of vitamin D can induce autoimmune diseases (Mannucci et al., 2021).
2. **Flavonoids:** These exert their effect by targeting multiple cell systems. Some of their targets are NO and ROS which improve antioxidant enzymes such as inducible nitric oxide synthase (iNOS) and cyclooxygenases (COXs) (Mannucci et al., 2021).
3. **Coenzyme Q:** It also acts as an antioxidant and defends against ROS. The electron carrier coenzyme Q10 (ubiquinone) also serves as a protein translocator. After transforming to ubiquinol it regenerates vitamin E and performs its antioxidant role (Mannucci et al., 2021).

4. **Probiotics:** These were shown to have positive effects on several immune-related disorders which include RA, ulcerative colitis, MS, and hepatic encephalopathy. *Lactobacilli* and *Bifidobacteria* were found effective in balancing between Treg and Th17 action. Also, in some lung diseases like asthma and COPD probiotics helped in ameliorating oxidative stress status (Mannucci et al., 2021).
5. **Medicinal Plants:** *Allium Sativum* known as garlic have several beneficial effects but at high doses can be detrimental to the stomach. Its extracts and oils might have an improved antioxidant effect than fresh garlic. *Origanum Vulgare* contains phenolic components like flavonoids and phenolic acids, which have potent antioxidant properties. Some findings have shown RA, MS, psoriasis, and inflammatory bowel illness all benefit from curcumin. This action of curcumin may be due to immune cell regulation of inflammatory cytokines and related signaling cascades (Bright, 2007).
6. **Lichen Species:** Depsides, depsidones, dibenzofurans, and pulvinic acid derivatives are lichen secondary metabolites that serve as antioxidants by increasing endogenous enzymatic defenses and non-enzymatic mechanisms (Mannucci et al., 2021).
7. **Melatonin:** These show antioxidant activity by acting as a scavenger and inducing genes related to antioxidant system activity (Mannucci et al., 2021).
8. **Arginine:** Arginine ameliorate oxidative stress tolerance and balance immunity. The exact antioxidant mechanism is unclear, but some hypothesized on the basis of its ability to boost endogenous antioxidant defense (Mannucci et al., 2021).
9. **Omega-3 Fatty Acid:** ω-3 PUFAs and their metabolites have been demonstrated to play a role in the prevention and treatment of autoimmune diseases such as RA, SLE, type 1 diabetes, and MS in some clinical studies and data (Mannucci et al., 2021; Li et al., 2019).
10. **Minerals:** Selenium supplementation has been shown to lower circulating thyroid autoantibodies in people with chronic autoimmune thyroiditis. Zinc helps to regulate the immune system. It has been established that patients with autoimmune disorders have a zinc deficit (Mannucci et al., 2021; Sanna et al., 2018). Some experimental animal studies have brought the hypothesis that zinc supplementation can aid in the management of the course and effects of diabetes mellitus. Zinc induces metallothionein (MT) gene expression that boosts antioxidant performance (Ozcelik et al., 2012).

1.7.1 FUNCTIONAL FOODS AND NEURODEGENERATIVE DISORDERS

The activation and production of proinflammatory cytokines, as well as related pathways, are inhibited by nutraceuticals. Bacosides A and B, brahmine, quercetin, kaempferol, withanine, somniferine, asiatic acid, and bhilavanol A and B are only a few examples (Keservani et al., 2016; Makkar et al., 2020).

Increased oxidative stress and free radicals are linked to Alzheimer's disease (AD). The majority of nutraceuticals that are antioxidant in nature are used to treat this condition. Flavonoids (fruits, vegetables, tea, wine, and coffee); carotenoids (lutein, zeaxanthin, lycopene, and carotenes); anthocyanidins (cyanidin); flavones (luteolin, apigenin) (Makkar et al., 2020; Keservani & Kesharwani, 2015).

Uninhibited oxidative stress and free radicals, in combination with abnormally misfolded proteins, neuroinflammation, and dysfunctional mitochondria, cause cellular metabolism and energy to be compromised in Parkinson's disease, affecting brain function, and leading to neurodegenerative disorders such as PD (Keservani et al., 2015). Vitamin A, Omega-3 fatty acids, lycopene, vincamine, gallic acid, curcumin, and Mito Q are just a few examples (Makkar et al., 2020).

Nutraceuticals that limit monoamine re-uptake have anti-inflammatory and antioxidant characteristics, making them ideal for treating depression. Omega-3 fatty acids, folic acid, S-adenosyl methionine, zinc, N-acetyl cysteine (NAC), L-tryptophan/5-HTP, and Vitamin D are other examples (Makkar et al., 2020).

In the treatment of psychosis, nutraceuticals that increase neurotransmission in dopaminergic serotoninergic neurons can be used. These mostly consist of several vitamins and omega-3 fatty acids (Makkar et al., 2020).

1.8 FOOD WASTE AS NUTRACEUTICAL

Food waste is a by-product of various industrial, agricultural, domestic, and other activities in the food sector, and these activities continue to be on the rise due to the increase in such activities. Various studies have shown that different types of food waste obtained from fruits, vegetables, grains, and other food processing industries can be used as potential sources of bioactive and nutraceutical compounds and have important applications in the treatment of various diseases. Various extraction methods like solvent extraction, supercritical fluid (SCF) extraction, subcritical water extraction, microwave-assisted extraction using ultrasound, enzyme assisted extraction have been used to extract different secondary metabolites, minerals, and vitamins from food waste. Over the next years, these methods can provide an innovative way to increase the production of the specific compound, which can be used as an ingredient in the nutraceutical or functional food design. The nutraceuticals thus obtained can provide health benefits in various ailments like AD, and cancer, hypertension, inflammation, CVD (Keservani et al., 2016a), Parkinson's disease (Keservani & Kesharwani, 2015; Keservani et al., 2016b), leukemia, cataract, hyperlipidemia. Some of them are good anti-oxidants and also possess anti-microbial activity (Kumar et al., 2017). Nutraceuticals from different food waste residues are tabulated in Table 1.10 (Kumar et al., 2017).

TABLE 1.10 Bioactive Components in Different Industrial Food Ingredients

Food Industry	Source	Residue	Bioactive Components
	Apple	Peel and pomace	Epicatechin, catechins, anthocyanins, quercitin glycosides, chlorogenic acid, hydroxycinnamates, phloretin glycosides, procyanidins
	Avocado	Peel and seeds	Epicatechin, catechin, gallic acid, chlorogenic acid, cyanidin 3-glucoside, homogentisic acid.
	Banana	Peel	Gallocatechin, anthocyanins, delphinidin, cyaniding, catecholamine
Fruits	Citrus fruits	Peel	Hesperidin, naringin, eriocitrin, narirutin
	Grapes	Seeds and skin	Coumaric acid, cafeic acid, ferulic acid, chlorogenic acid, cinnamic acid, neochlorogenic acid, *p*-hydroxybenzoic acid, protocatechuic acid, vanillic acid, gallic acid, proanthocyanidins, quercetin 3-*o*-glucuronide, quercetin, resveratrol.
	Guava	Seeds and skin	Catechin, cyanidin 3-glucoside, galangin, gallic acid, homogentisic acid, kaempferol.
	Litchi	Pericarp, seeds	Cyanidin-3-glucoside, cyanidin-3-rutonoside, malvidin-3-glucoside, gallic acid, epicatechin-3-gallate.
	Mango	Kernel	Gallic acid, ellagic acid, gallates, gallotannins, condensed tannins.

TABLE 1.10 *(Continued)*

Food Industry	Source	Residue	Bioactive Components
	Palm	Byproducts of palm oil milling.	Tocopherols, tocotrienols, sterols, and squalene, are phenolic antioxidants.
	Pomegranate	Peel and pericarp	Gallic acid, cyanidin-3,5-diglucoside, cyanidin-3-diglucoside, delphinidin-3,5-diglucoside.
Vegetables	Carrot	Peel	Phenols, beta-carotene
	Cucumber	Peel	Chlorophyll, pheophytin, phellandrene, caryophyllene.
	Potato	Peel	Gallicacid, cafeic acid vanillic acid
	Tomato	Skin and pomace	Carotenoids
Cereal crops	Barley	Bran	β-Glucan
	Rice	Bran	γ-Oryzanol, bran oil
	Wheat	Bran and germ	Phenolic acids and antioxidants

Source: Reprinted/Adapted from Kumar et al., 2017

KEYWORDS

- **bioactive**
- **bioavailability**
- **dietary fiber**
- **functional foods**
- **nutraceuticals**
- **poly unsaturated fatty acids**
- **polyphenols**

REFERENCES

Adefegha, S. A., (2018). Functional foods and nutraceuticals as dietary intervention in chronic diseases; novel perspectives for health promotion and disease prevention chronic diseases; novel perspectives for health promotion and. *J. Diet. Suppl. [Internet], 15*(6), 977–1009. Available from: https://doi.org/10.1080/19390211.2017.1401573.

Aggarwal, B. B., Van, K. M. E., Iyer, L. H., Harikumar, K. B., & Sung, B., (2009). Molecular targets of nutraceuticals derived from dietary spices: Potential role in suppression of inflammation and tumorigenesis. *Exp. Biol. Med. (Maywood), 234*, 825–849.

Aggarwal, B. B., Vijayakshmi, R. V., & Sung, B., (2009). Targeting inflammatory pathways for prevention and therapy of cancer: Short-term friend, long-term foe. *Clin. Cancer Res., 15*, 425–430.

Ajay, A. K., Vig, S., & Sabbisetti, V. S., (2019). Mechanism of action of functional lipids and metabolites for patients with chronic kidney disease. *Funct. Foods Health Dis., 9*(6), 412.

Alam, M. A., Uddin, R., Subhan, N., Rahman, M. M., Jain, P., & Reza, H. M., (2015). Beneficial role of bitter melon supplementation in obesity and related complications in metabolic syndrome. *J. Lipids,* 496169.

Alissa, E. M., & Ferns, G. A., (2012). Functional Foods and Nutraceuticals in the Primary Prevention of Cardiovascular Diseases. Volume 2012, Article ID 569486, https://doi.org/10.1155/2012/569486.

Aryaeian, N., Shahram, F., Mahmoudi, M., Tavakoli, H., Yousefi, B., Arablou, T., et al., (2019). The effect of ginger supplementation on some immunity and inflammation intermediate genes expression in patients with active rheumatoid arthritis. *Gene, 698*, 179–185.

Attari, V. E., Mahdavi, A. M., Javadivala, Z., Mahluji, S., Vahed, S. Z., & Ostadrahimi, A., (2018). A systematic review of the anti-obesity and weight lowering effect of ginger (*Zingiberofficinale* Roscoe) and its mechanisms of action. *Phytother. Res., 32*, 577–585.

Augimeri, G., Montalto, F. I., Giordano, C., Barone, I., Lanzino, M., & Catalano, S., (2021). *Nutraceuticals in the Mediterranean Diet: Potential Avenues for Breast Cancer Treatment,* 1–24.

Baboota, R. K., Bishnoi, M., Ambalam, P., Kondepudi, K. K., Sarma, S. M., Boparai, R. K., et al., (2013). Functional food ingredients for the management of obesity and associated comorbidities: A review. *J. Funct. Foods, 5*, 997–1012.

Bai, J., Zhu, Y., & Dong, Y., (2016). Response of gut microbiota and inflammatory status to bitter melon (*Momordica charantia* L.) in high fat diet induced obese rats. *J. Ethnopharmacol., 194*, 717–726.

Balaji, M., Ganjayi, M. S., Hanuma, K. G. E. N., Parim, B. N., Mopuri, R., & Dasari, S., (2016). A review on possible therapeutic targets to contain obesity: The role of phytochemicals. *Obes. Res. Clin. Pract., 10*(4), 363–380.

Beck, E. J., Tosh, S. M., Batterham, M. J., Tapsell, L. C., & Huang, X. F., (2009). Oat β-glucan increases postprandial cholecystokinin levels, decreases insulin response and extends subjective satiety in overweight subjects. *Mol. Nutr. Food Res., 53*, 1343–1351.

Bianconi, V., Mannarino, M. R., Sahebkar, A., Cosentino, T., Pirro, M., & Cosentino, T., (2018). Cholesterol-lowering nutraceuticals affecting vascular function and cardiovascular disease risk. *Curr. Cardiol. Rep., 20*(7), 53. doi: 10.1007/s11886-018-0994-7.

Bowden, R. G., Jitomir, J., Wilson, R. L., & Gentile, M., (2009). Effects of omega-3 fatty acid supplementation on lipid levels in end-stage renal disease patients. *J. Ren. Nutr., 19*, 259–266.

Bradford, P. G., (2013). Curcumin and obesity. *Bio Factors, 39*(1), 78–87.

Bright, J. J., (2007). Curcumin and autoimmune disease. *Adv. Exp. Med. Biol., 595*, 425–451.

Buhrmann, C., Popper, B., Aggarwal, B. B., & Shakibaei, M., (2017). Resveratrol downregulates inflammatory pathway activated by lymphotoxin α (TNF-β) in articular chondrocytes: Comparison with TNF-α. *PLoS One, 12*(11), e0186993.

Cannavò, S. P., Riso, G., Casciaro, M., Di Salvo, E., & Gangemi, S., (2019). Oxidative stress involvement in psoriasis: A systematic review. *Free Radic. Res., 53*(8), 829–840.

Cao, Y., Xiao, J., Song, M., & Ho, C. T., (2018). Molecular mechanisms of the anti-obesity effect of bioactive ingredients in common spices: A review. *Food Funct, 9*(9), 4569–4581.

Carr, A., & Maggini, S., (2017). Vitamin C and immune function. *Nutrients, 9*(11), 1211.

Casciaro, M., Di Salvo, E., Pace, E., Ventura-Spagnolo, E., Navarra, M., & Gangemi, S., (2017). Chlorinative stress in age-related diseases: A literature review. *Immun. Ageing, 14*, 21.

Castello, A., Pollán, M., Buijsse, B., Ruiz, A., Casas, A. M., Baena-Cañada, J. M., et al., (2014). Spanish Mediterranean diet and other dietary patterns and breast cancer risk: Case-control EpiGEICAM study. *Br. J. Cancer, 111*, 1454–1462.

Chen, I. J., Liu, C. Y., Chiu, J. P., & Hsu, C. H., (2016). Therapeutic effect of high-dose green tea extract on weight reduction: A randomized, double-blind, placebo-controlled clinical trial. *Clin. Nutr., 35*(3), 592–599.

Chen, P., Li, C., Li, X., Li, J., Chu, R., & Wang, H., (2014). Higher dietary folate intake reduces the breast cancer risk: A systematic review and meta-analysis. *Br. J. Cancer, 110*(9), 2327–2338.

Chronic Kidney Disease Prognosis Consortium, Matsushita, K., Van, D. V. M., Astor, B. C., Woodward, M., Levey, A. S., et al., (2010). Association of estimated glomerular filtration rate and albuminuria with all-cause and cardiovascular mortality in general population cohorts: A collaborative meta-analysis. *Lancet, 375*(9731), 2073–2081.

Coresh, J., Selvin, E., Stevens, L. A., Manzi, J., Kusek, J. W., Eggers, P., et al., (2007). Prevalence of chronic kidney disease in the United States. *JAMA, 298*(17), 2038–2047.

Cosola, C., Sabatino, A., Di Bari, I., Fiaccadori, E., & Gesualdo, L., (2018). Nutrients, nutraceuticals, and xenobiotics affecting renal health. *Nutrients [Internet], 10*(7), 808. http://dx.doi.org/10.3390/nu10070808.

Csaki, C., Mobasheri, A., & Shakibaei, M., (2009). Synergistic chondroprotective effects of curcumin and resveratrol in human articular chondrocytes: Inhibition of IL-1beta-induced NF-kappa B-mediated inflammation and apoptosis. *Arthritis Res. Ther., 11*(6), R165.

Dahiya, D. K., Renuka, Puniya, M., Shandilya, U. K., Dhewa, T., Kumar, N., et al., (2017). Gut microbiota modulation and its relationship with obesity using prebiotic fibers and probiotics: A review. *Front Microbiol., 8*, 563.

Dima, C., Assadpour, E., Dima, S., & Jafari, S. M., (2020). Bioavailability and bioaccessibility of food bioactive compounds; overview and assessment by in vitro methods. *Compr. Rev. Food Sci. Food Saf., 19*(6), 2862–2884.

Flachs, P., Horakova, O., Brauner, P., Rossmeisl, M., Pecina, P., & Hal, N. F., (2005). Polyunsaturated fatty acids of marine origin up-regulate mitochondrial biogenesis and induce b-oxidation in white fat. *Diabetologia, 48*, 2365–2375.

Fung, T. T., Chiuve, S. E., Willett, W. C., Hankinson, S. E., Hu, F. B., & Holmes, M. D., (2013). Intake of specific fruits and vegetables in relation to risk of estrogen receptor-negative breast cancer among postmenopausal women. *Breast Cancer Res. Treat., 138*, 925–930.

Grivennikov, S. I., Greten, F. R., & Karin, M., (2010). Immunity, inflammation, and cancer. *Cell, 140*(6), 883–899.

Gul, K., Singh, A. K., & Jabeen, R., (2016). Nutraceuticals and functional foods: The foods for the future world nutraceuticals and functional foods: The foods for the future world. *Crit. Rev. Food Sci. Nutr., 56*(16), 2617–2627. doi: 10.1080/10408398.2014.903384.

Gupta, S., Sharma, T. K., Kaushik, G. G., & Shekhawat, V. P. S., (2011). Vitamin E supplementation may ameliorate oxidative stress in type 1 diabetes mellitus patients. *Clin. Lab, 57*(5, 6), 379–386.

Harding, J., & Ganea, E., (2006). Protection against glycation and similar post-translational modifications of proteins. *Biochimicaet Biophysica Acta (BBA)-Proteins and Proteomics, 1764*(9), 1436–1446.

Harris, I. M., & Eastman, R. C., (2000). Early detection of undiagnosed diabetes mellitus: A US perspective. *Diabetes/Metabolism Res. Rev., 16*(4), 230–236.

Hsu, C. L., & Yen, G. C., (2007). Effects of capsaicin on induction of apoptosis and inhibition of adipogenesis in 3T3-L1 cells. *J. Agric. Food Chem., 55*(5), 1730–1736.

Hsu, S. P., Wu, M. S., Yang, C. C., Huang, K. C., Liou, S. Y., Hsu, S. M., et al., (2007). Chronic green tea extract supplementation reduces hemodialysis-enhanced production of hydrogen peroxide and hypochlorous acid, atherosclerotic factors, and proinflammatory cytokines. *Am. J. Clin. Nutr., 86*(5), 1539–1547.

Hui, C., Qi, X., Qianyong, Z., Xiaoli, P., Jundong, Z., & Mantian, M., (2013). Flavonoids, flavonoid subclasses and breast cancer risk: A meta-analysis of epidemiologic studies. *PLoS One, 8*(1), e54318.

Hunter, P. M., & Hegele, R. A. (2017). Functional foods and dietary supplements for the management of dyslipidaemia. *Nature Reviews Endocrinology. 13*(5), 278–288.

Jane, M., McKay, J., & Pal, S., (2018). Effects of daily consumption of psyllium, oat bran and PolyGlycopleX® on obesity-related disease risk factors. *Nutr., 57*, 84–91.

Jarouliya, U., & Keservani, R. K., (2019). Pathways leading to child obesity. In: Debasis, B., (ed.), *Global Perspectives on Childhood Obesity: Current Status, Consequences and Prevention* (2nd edn., pp. 137–146). Chapter 12, Academic Press, Elsevier. ISBN: 9780128128404.

Jespersen, T., Kruse, N., Mehta, T., Kuwabara, M., Noureddine, L., & Jalal, D., (2018). Light wine consumption is associated with a lower odd for cardiovascular disease in chronic kidney disease. *Nutr. Metab. Cardiovasc. Dis., 28*(11), 1133–1139.

Kang, J. H., Kim, C. S., Han, I. S., Kawada, T., & Yu, R., (2007). Capsaicin, a spicy component of hot peppers, modulates adipokine gene expression and protein release from obese-mouse adipose tissues and isolated adipocytes, and suppresses the inflammatory responses of adipose tissue macrophages. *FEBS Lett., 581*(23), 4389–4396.

Kapinova, A., Stefanicka, P., Kubatka, P., Zubor, P., Uramova, S., Kello, M., et al., (2017). Are plant-based functional foods better choice against cancer than single phytochemicals? A critical review of current breast cancer research. *Biomed. Pharmacother, 96*, 1465–1477.

Kazeem, M. I., & Bankole, H. A., (2018). Antidiabetic functional foods with antiglycation properties antidiabetic functional foods with antiglycation properties. In: *Book: Bioactive Molecules in Food.* doi: 10.1007/978-3-319-54528-8_16-1.

Keservani, R. K., & Sharma, A. K., (2014). Flavonoids: Emerging trends and potential health benefits. *Journal of Chinese Pharmaceutical Sciences, 23*(12), 815.

Keservani, R. K., Kesharwani, R. K., Sharma, A. K., & Ahmad, M. F., (2015). Stress, Parkinson's, Alzheimer's disease: Role of dietary supplements. In: Debasis, B., Harry, G. P., & Anand, S., (eds.), *Nutraceutical and Functional Foods in Human Life and Disease Prevention* (pp. 241–253). Chapter 14, Boca Raton, Fl. CRC, Press, Taylor and Francis. ISBN: 9781482237214.

Keservani, R. K., Kesharwani, R. K., Sharma, A. K., Vyas, N., & Chadoker, A., (2010b). Nutritional supplements: An overview. *International Journal of Current Pharmaceutical Review and Research, 1*(1), 59–75.

Keservani, R. K., Kesharwani, R. K., Vyas, N., Jain, S., Raghuvanshi, R., & Sharma, A. K., (2010a). Nutraceutical and functional food as future food: A review. *Der Pharmacia Lettre, 2*(1), 106–116.

Keservani, R. K., Sharma, A. K., & Kesharwani, R. K., (2016a). Nutraceutical and functional foods for cardiovascular health. In: *Food Process Engineering* (pp. 257–278, 291–312). Apple Academic Press, CRC Press. ISBN: 9781771884020.

Keservani, R. K., Sharma, A. K., & Kesharwani, R. K., (2016b). Medicinal effect of nutraceutical fruits for the cognition and brain health. *Scientifica,* 10. Article ID 3109254. doi: 10.1155/2016/3109254.

Keservani, R. K., Sharma, A. K., & Kesharwani, R. K., (2020). *Nutraceuticals and Dietary Supplements: Applications in Health Improvement and Disease Management.* CRC Press. ISBN: 9781771888738.

Keservani, R. K., Singh, S., Singh, V., Kesharwani, R. K., & Sharma, A. K., (2015). Nutraceuticals and functional foods in the prevention of mental disorder. In: Debasis, B., Harry, G. P., & Anand, S., (eds.), *Nutraceutical and Functional Foods in Human Life and Disease Prevention* (pp. 255–268). Chapter 15, Boca Raton, Fl. CRC Press, Taylor and Francis. ISBN: 9781482237214.

Kesharwani, R. K., & Misra, K. (2011). Prediction of binding site for curcuminoids at human topoisomerase II α protein; an in silico approach. *Current Science. 25*, 1060–1065.

Kesharwani, R. K., Srivastava, V., Singh, P. et al. (2015). A Novel Approach for Overcoming Drug Resistance in Breast Cancer Chemotherapy by Targeting new Synthetic Curcumin Analogues Against Aldehyde Dehydrogenase 1 (ALDH1A1) and Glycogen Synthase Kinase-3 β (GSK-3β). *Appl Biochem Biotechnol 176*, 1996–2017.

Khossousi, A., Binns, C. W., Dhaliwal, S. S., & Pal, S., (2008). The acute effects of psyllium on postprandial lipaemia and thermogenesis in overweight and obese men. *Br. J. Nutr., 99*(5), 1068–1075.

Kim, M. S., Kim, J. K., Kwon, D. Y., & Park, R., (2004). Anti-adipogenic effects of garcinia extract on the lipid droplet accumulation and the expression of transcription factor. *Biofactors, 22*, 193–196.

Kim, S., Jin, Y., Choi, Y., & Park, T., (2011). Resveratrol exerts anti-obesity effects via mechanisms involving down-regulation of adipogenic and inflammatory processes in mice. *Biochem. Pharmacol., 81*(11), 1343–1351.

Kumar, K., Yadav, A. N., Kumar, V., Vyas, P., & Dhaliwal, H. S., (2017). Food waste: A potential bioresource for extraction of nutraceuticals and bioactive compounds. *Bioresour. Bioprocess, 4*, 18.

Lee, C. C., Sharp, S. J., Wexler, D. J., & Adler, A. I., (2010). Dietary intake of eicosapentaenoic and docosahexaenoic acid and diabetic nephropathy: Cohort analysis of the diabetes control and complications trial. *Diabetes Care, 33*(7), 1454–1456.

Lee, Y. K., Lee, W. S., Hwang, J. T., Kwon, D. Y., Surh, Y. J., & Park, O. J., (2008). *J Agric. Food Chem., 57*, 305–310.

Li, X., Bi, X., Wang, S., Zhang, Z., Li, F., & Zhao, A. Z., (2019). Therapeutic potential of ω-3 polyunsaturated fatty acids in human autoimmune diseases. *Front Immunol., 10*, 2241.

Liao, M. Y., Shen, Y. C., Chiu, H. F., Ten, S. M., Lu, Y. Y., Han, Y. C., et al., (2019). Down-regulation of partial substitution for staple food by oat noodles on blood lipid levels: A randomized, double-blind, clinical trial. *J. Food Drug Anal., 27*(1), 93–100.

Lipinski, C. A., Lombardo, F., Dominy, B. W., & Feeney, P. J., (1997). Experimental and computational approaches to estimate solubility and permeability in drug discovery and development settings. *Advanced Drug Delivery Rev., 23*, 3–25. https://doi.org/10.1016/S0169-409X(00)00129-0.

Liu, R. H., (2013). Dietary bioactive compounds and their health implications: Dietary bioactive compounds and health. *J. Food Sci., 78*(s1), A18–25.

Liu, R. H., (2013). Health-promoting components of fruits and vegetables in the diet. *Adv. Nutr., 4*, 384–392.

Makkar, R., Behl, T., Bungau, S., Zengin, G., & Mehta, V., (2020). Nutraceuticals in neurological disorders. *Int. J. Mol. Sci., 21*(12), 4424.

Manidhar, D. M., Kesharwani, R. K., Reddy, N. B., et al. (2013). Designing, synthesis, and characterization of some novel coumarin derivatives as probable anticancer drugs. *Med Chem Res 22*, 4146–4157.

Mannucci, C., Casciaro, M., Sorbara, E. E., Calapai, F., Salvo, E. D., Pioggia, G., et al., (2021). Nutraceuticals against oxidative stress in autoimmune disorders. *Antioxidants (Basel), 10*(2), 261. doi: 10.3390/antiox10020261.

Mao, Y., Hao, J., Jin, Z. Q., Niu, Y. Y., Yang, X., Liu, D., et al., (2017). Network pharmacology-based and clinically relevant prediction of the active ingredients and potential targets of Chinese herbs in metastatic breast cancer patients. *Oncotarget,* 27007–27021.

Masuda, Y., Haramizu, S., Oki, K., Ohnuki, K., Watanabe, T., & Yazawa, S., (2003). Upregulation of uncoupling proteins by oral administration of capsiate, a nonpungent capsaicin analog. *J. Appl. Physiol., 95*, 2408–2415.

Misawa, K., Hashizume, K., Yamamoto, M., Minegishi, Y., Hase, T., & Shimotoyodome, A., (2015). Ginger extract prevents high-fat diet-induced obesity in mice via activation of the peroxisome proliferator-activated receptor δ pathway. *J. Nutr. Biochem., 26*(10), 1058–1067.

Moon, J., Do, H. J., Kim, O. Y., & Shin, M. J., (2013). Anti-obesity effects of quercetin-rich onion peel extract on the differentiation of 3T3-L1 preadipocytes and the adipogenesis in high fat-fed rats. *Food Chem. Toxicol., 58*, 347–354.

Nabavi, S. F., Russo, G. L., Daglia, M., & Nabavi, S. M., (2015). Role of quercetin as an alternative for obesity treatment: You are what you eat! *Food Chem., 179*, 305–310.

Ogborn, M. R., Nitschmann, E., Bankovic-Calic, N., Weiler, H. A., Fitzpatrick-Wong, S., & Aukema, H. M., (2003). Dietary conjugated linoleic acid reduces PGE2 release and interstitial injury in rat polycystic kidney disease. *Kidney Int., 64*(4), 1214–1221.

Özcelik, D., Nazıroglu, M., Tunçdemir, M., Çelik, Ö., Öztürk, M., & Flores-Arce, M. F., (2012). Zinc supplementation attenuates metallothionein and oxidative stress changes in kidney of streptozotocin-induced diabetic rats. *Biol. Trace Elem. Res., 150*(1–3), 342–349.

Palm, K., Stenberg, P., Luthman, K., & Artursson, P., (1997). Polar molecular surface properties predict the intestinal absorption of drugs in humans. *Pharmaceutical Res., 14*, 568–571. https://doi.org/10.1023/A:1012188625088.

Park, C. S., Kim, W., Woo, J. S., Ha, S. J., Kang, W. Y., Hwang, S. H., et al., (2010). Green tea consumption improves endothelial function but not circulating endothelial progenitor cells in patients with chronic renal failure. *Int. J. Cardiol., 145*(2), 261, 262.

Park, S. J., Ahmad, F., Philp, A., Baar, K., Williams, T., Luo, H., et al., (2012). Resveratrol ameliorates aging-related metabolic phenotypes by inhibiting cAMP phosphodiesterases. *Cell, 148*(3), 421–433.

Qin, T., Du, M., Du, H., Shu, Y., Wang, M., & Zhu, L., (2015). Folic acid supplements and colorectal cancer risk: Meta-analysis of randomized controlled trials. *Sci. Rep., 5*, 12044.

Qin, X., Cui, Y., Shen, L., Sun, N., Zhang, Y., Li, J., et al., (2013). Folic acid supplementation and cancer risk: A meta-analysis of randomized controlled trials. *Int. J. Cancer, 133*(5), 1033–1041.

Rajput, R., Wairkar, S., & Gaud, R., (2018). Nutraceuticals for better management of osteoporosis: An overview. *J. Functional Foods, 47*, 480–490. doi: 10.1016/j.jff.2018.06.013.

Rane, B. R., & Keservani, R., (2020). Introduction to nutraceuticals and dietary supplements. In: *Nutraceuticals and Dietary Supplements* (pp. 35–56). Apple Academic Press. ISBN: 9781771888738.

Rivera, L., Morón, R., Sánchez, M., Zarzuelo, A., & Galisteo, M., (2008). Quercetin ameliorates metabolic syndrome and improves the inflammatory status in obese Zucker rats. *Obesity (Silver Spring), 16*(9), 2081–2087.

Ruggeri, R. M., Vicchio, T. M., Cristani, M., Certo, R., Caccamo, D., Alibrandi, A., et al., (2016). Oxidative stress and advanced glycation end products in Hashimoto's thyroiditis. *Thyroid, 26*(4), 504–511.

Rysz, J., Franczyk, B., Kujawski, K., Sacewicz-Hofman, I., Ciałkowska-Rysz, A., & Gluba-Brzózka, A., (2021). Are nutraceuticals beneficial in chronic kidney disease? *Pharmaceutics, 13*(2), 1–34.

Saito, M., Ueno, M., Ogino, S., Kubo, K., Nagata, J., & Takeuchi, M., (2005). High dose of garcinia cambogia is effective in suppressing fat accumulation in developing male Zucker obese rats, but highly toxic to the testis. *Food Chem. Toxicol., 43*(3), 411–419.

Sanna, A., Firinu, D., Zavattari, P., & Valera, P., (2018). Zinc status and autoimmunity: A systematic review and meta-analysis. *Nutrients [Internet], 10*(1). Available from: http://dx.doi.org/10.3390/nu10010068.

Singh, P., Kesharwani, R. K., & Keservani, R. K., (2017a). Antioxidants and vitamins: Roles in cellular function and metabolism. In: *Sustained Energy for Enhanced Human Functions and Activity* (pp. 385–407). Academic Press. ISBN:9780128054130.

Singh, P., Kesharwani, R. K., & Keservani, R. K., (2017b). Protein, carbohydrates, and fats: Energy metabolism. In: *Sustained Energy for Enhanced Human Functions and Activity* (pp. 103–115). Academic Press. ISBN:9780128054130.

Sosnowska, B., Penson, P., & Banach, M., (2017). The role of nutraceuticals in the prevention of cardiovascular disease. *Cardiovasc. Diagn. Ther., 7*(Suppl 1), S21–S31. doi: 10.21037/cdt.2017.03.20.

Sung, J., Ho, C. T., & Wang, Y., (2018). Preventive mechanism of bioactive dietary foods on obesity-related inflammation and diseases. *Food Funct., 9*(12), 6081–6095.

Svensson, M., Schmidt, E. B., Jørgensen, K. A., & Christensen, J. H., (2008). The effect of n-3 fatty acids on lipids and lipoproteins in patients treated with chronic haemodialysis: A randomized placebo-controlled intervention study. *Nephrol. Dial. Transplant, 23*(9), 2918–2924.

Tanaka, T., & Sugie, S., (2007). Inhibition of colon carcinogenesis by dietary non-nutritive compounds. *J. Toxicol. Pathol., 20*(4), 215–235.

Taylor, P., Chen, G., Wang, H., Zhang, X., Yang, S., Chen, G. U., et al., (2014). Nutraceuticals and functional foods in the management of hyperlipidemia nutraceuticals and functional foods in the management of hyperlipidemia. *Crit. Rev. Food Sci. Nutr., 54*(9), 1180–1201. doi: 10.1080/10408398.2011.62935.

Tognola, C., Alessandro, M., Milani, M., Cartella, I., Tavecchia, G., Grasso, E., et al., (2020). Nutraceuticals in chronic coronary syndromes: Preclinical data and translational experiences. *High Blood Press Cardiovasc Prev. [Internet]*, 0123456789. Available from: https://doi.org/10.1007/s40292-020-00416-8.

Toledo, E., Salas-Salvadó, J., Donat-Vargas, C., Buil-Cosiales, P., Estruch, R., Ros, E., Corella, D., et al., (2015). Mediterranean diet and invasive breast cancer risk among women at high cardiovascular risk in the PREDIMED trial: A randomized clinical trial. *JAMA Intern. Med., 175, 178*(11, 12), 1731, 1732, 1752–1760. doi: 10.1001/jamainternmed.2015.4838. Erratum In.

Tomasetti, C., Li, L., & Vogelstein, B., (2017). Stem cell divisions, somatic mutations, cancer etiology, and cancer prevention. *Science, 355*(6331), 1330–1334.

Upadhyay, J., Kesharwani, R. K., & Misra, K., (2009). Comparative study of antioxidants as cancer preventives through inhibition of HIF-1 alpha activity. *Bioinformation, 4*(6), p.233.

Vaccaro, M., Bagnato, G., Cristani, M., Borgia, F., Spatari, G., Tigano, V., et al., (2017). Oxidation products are increased in patients affected by non-segmental generalized vitiligo. *Arch Derm. Res., 309*(6), 485–490.

Vasques, C. A. R., Schneider, R., Klein-Júnior, L. C., Falavigna, A., Piazza, I., & Rossetto, S., (2014). Hypolipemic effect of garcinia cambogia in obese women: Hypolipemic effect of garcinia cambogia. *Phytother. Res., 28*(6), 887–891.

Veber, D. F., Johnson, S. R., Cheng, H. Y., Smith, B. R., Ward, K. W., & Kopple, K. D., (2002). Molecular properties that influence the oral bioavailability of drug candidates. *J Medicinal Chemistry, 45*, 2615–2623. https://doi.org/10.1021/jm020017n.

Venkatakrishnan, K., Chiu, H. F., & Wang, C. K., (2019). Extensive review of popular functional foods and nutraceuticals against obesity and its related complications with a special focus on randomized clinical trials. *Food Funct., 10*(5), 2313–2329.

Vernaglione, L., Cristofano, C., & Chimienti, S., (2008). Omega-3 polyunsaturated fatty acids and proxies of cardiovascular disease in hemodialysis: A prospective cohort study. *J. Nephrol., 21*(1), 99–105.

Wang, S., Liang, X., Yang, Q., Fu, X., Zhu, M., Rodgers, B. D., et al., (2017). Resveratrol enhances brown adipocyte formation and function by activating AMP-activated protein kinase (AMPK) α1 in mice fed high-fat diet. *Mol. Nutr. Food Res., 61*(4), 1600746.

Wang, Y., Li, S., & Li, C., (2019). Perspectives of new advances in the pathogenesis of vitiligo: From oxidative stress to autoimmunity. *Med. Sci. Monit., 25*, 1017–1023.

Wildman, R. E. C., (2007). *Handbook of Nutraceuticals and Functional Foods* (2nd edn.). New York, CRC Press.

Willett, W. C., (1995). Diet, nutrition, and avoidable cancer. *Environ. Health Perspect., 103*, 165–170.

Willett, W. C., (2002). Balancing life-style and genomics research for disease prevention. *Science, 296*, 695–698.

World Health Organization. *Cancer*. Available online: www.who.int/news-room/fact-sheets/detail/cancer (accessed on 05 June 2023).

Yang, J. Y., Della-Fera, M. A., Rayalam, S., Ambati, S., Hartzell, D. L., Park, H. J., et al., (2008). Enhanced inhibition of adipogenesis and induction of apoptosis in 3T3-L1 adipocytes with combinations of resveratrol and quercetin. *Life Sci., 82*(19, 20), 1032–1039.

Zhang, G., Wang, Y., Zhang, Y., Wan, X., Li, J., Liu, K., et al., (2012). Anti-cancer activities of tea epigallocatechin-3-gallate in breast cancer patients under radiotherapy. *Curr. Mol. Med., 12*, 163–176.

Zhen-Yu, C., Jiao, R., & Ma, K. Y., (2008). Cholesterol-lowering nutraceuticals and functional. *J. Agricultur. Food Chem., 56*(19), 8761–8773.

Zhu, W., Qin, W., Zhang, K., Rottinghaus, G. E., Chen, Y. C., Kliethermes, B., et al., (2012). Trans-resveratrol alters mammary promoter hypermethylation in women at increased risk for breast cancer. *Nutr. Cancer, 64*, 393–400.

Zhu, Y., Bai, J., Zhang, Y., Xiao, X., & Dong, Y. (2016). Effects of bitter melon (*Momordica charantia* L.) on the gut microbiota in high fat diet and low dose streptozocin-induced rats. *International Journal of Food Sciences and Nutrition, 67*(6), 686–695.

CHAPTER 2

ANTIOXIDANTS ACTIVITY OF MICRONUTRIENTS AND MACRONUTRIENTS

MOUSUMI BHATTACHARYA

Nutritionist, Kolkata, West Bengal, India

ABSTRACT

Antioxidants prevent or may delay the cell damage caused by free radicals. Free radical synthesis is the process of the human body's reaction against environmental and other pressures. They are also called free-radical scavengers. Antioxidants help neutralize free radicals in our body, thereby stimulating overall health. Antioxidants protect against the cell damage caused by free radicals, called oxidative stress. Anti-oxidants may be natural or artificial in origin. Some of the plant-based food rich in natural antioxidants (phytonutrients) there are two types of antioxidants: endogenous, produced by the body, and exogenous, obtained mainly from food items. The Human body's antioxidant defense mechanism keeps a regular check of free radicals.

2.1 INTRODUCTION

Antioxidant means chemicals that prevent oxygen consumption. The use of antioxidants in some of the most significant industrial processes, some of this process includes the prevention of metal corrosion, the vulcanization of rubber, and the fuel polymerization in internal ignition engines,

Micronutrients and Macronutrients as Nutraceuticals. Prakash Chandra Gupta, PhD,
Sayan Bhattacharyya, MBBS, MD, Nisha Sharma, PhD, MPharm, Rajesh K. Kesharwani, PhD,
Raj K. Keservani, PhD, MPharm (Eds.)
© 2024 Apple Academic Press, Inc. Co-published with CRC Press (Taylor & Francis)

was studied in the early 19th and 20th centuries. Industrial chemicals added during manufacture to prevent oxidation of polymers and fuels, as well as preservatives in food and cosmetics, are known as antioxidants (Dabelstein et al., 2000). Antioxidants have become increasingly important for public health following the recent pandemic. Antioxidants are alloys that suppress the oxidation chemical reaction. Ascorbic acid (Vitamin C), and thiols are some of the examples of antioxidants. Oxidative stress is balanced in the plant and animal body by the complex system of overlapping antioxidants, e.g., glutathione (GSH) (Dabelstein et al., 2000).

The normal functioning of a cell depends on a proper balance between peroxident and antioxidants. Various free radicals are produced in this balancing process. Free radicals scavenging properties of antioxidants delay or inhibit cell damaging (Halliwell, 1995). Antioxidants with a smaller molecular weight interact safely with free radicals, bringing the chain reaction to a halt before critical molecules are harmed. GSH, ubiquinol, and uric acid (UA) are examples of antioxidants that are created during the normal metabolism process in the body (Shi et al., 1999). Vitamins A, C, and E are among the lightest anti-oxidants in the diet. Some of the important micronutrients (vitamins) antioxidants include vitamin E (-tocopherol), vitamin C (ascorbic acid), and B-carotene (Levine et al., 1999). As the body cannot manufacture these micronutrients, they need to be supplied in the diet. If free radicals are not quenched by antioxidants, they can cause damage to the cells, proteins, DNA, and RNA (Srilakshmi, 2019).

2.2 FACTORS AFFECTING THE PRODUCTION OF FREE RADICALS

Two types of factors that increase the production of free radicals in the body include internal factors like inflammation and external factors like pollution, exposure to UV rays, and cigarette smoke. The oxidative stress is already associated with some cancer, cardiac disease, stroke, arthritis, respiratory illness, emphysema, immunodeficiency, Parkinson's disease (Keservani et al., 2015), and some other inflammatory and ischemic disorders (Nordqvist, 2020).

External factors: Free radical produces in the body by the natural process of metabolism. Oxidative stress happens in the body when the number of free radicals exceeds the body's ability to eliminate or nullify them (med. libretexts.org, 2016).

Below are activities and processes which can induce oxidative stress (Nordqvist, 2020):

- Excessive exercise;
- Smoking;
- Environmental pollution;
- Mitochondrial activity;
- High levels of vitamin C;
- High levels of oxygen;
- Exposure to chemicals like pesticides and some medications, including chemotherapy;
- Ionizing radiation;
- Asbestos;
- Industrial solvents;
- Ozone;
- Consumption of certain foods, such as refined and processed foods, trans fats, artificial sweeteners, and certain colors and additives;
- Tissue trauma as a result of inflammation and injury;
- Ischemia and reperfusion damage.

2.2.1 TRACE MINERALS LIKE IRON, COPPER

They accomplish this by being free radicals or by giving energy that causes electrons to flow more swiftly between atoms. By damaging the free radical detoxification systems and the mechanisms involved in healing oxidative damage, excessive exposure to free radicals from the environment can cause disease (med.libretexts.org, 2016).

Classification: Antioxidants are split into two categories depending on whether they are water-soluble (hydrophilic) or lipid-soluble (lipophilic). Water-soluble antioxidants react with oxidizing agents in cell cytosol and blood plasma, while lipid-soluble antioxidants preserve the cell membranes against lipid peroxidation (LPO) (Sies, 1997). The body can produce these chemicals, or they can be received from food. Different antioxidants are found at different levels in human fluids and tissues, with some, like ubiquinone or GSH, being concentrated in cells and others, such as UA, being more widely distributed. Some antioxidants are only found in a few organisms, yet they are important in illnesses and can act as virulence factors (Vertuani et al., 2004).

2.3 ANTIOXIDANT DEFENSE SYSTEM

The antioxidant defense system in the body has exogenous and endogenous antioxidants which work together at the molecular level to shield the cellular membranes, proteins, DNA, and RNA from the damaging effects of free radicals. Exogenous antioxidants include ascorbic acid, tocopherols, beta carotene, and non-nutrients such as polyphenols and flavonols, which are received by dietary intake (Keservani et al., 2020).

The compounds referred to as the first line of defense, which are primary antioxidants or secondary antioxidants are listed below (Mattil, 1947):

Primary Antioxidant	Secondary Antioxidant
Vitamin E	Copper
Vitamin C	Zinc and Selenium
Carotenoids	Glutathione reductase
Flavonoids	N-acetylcysteine
Polyamines	Glucose-6-phosphate dehydrogenase
Melatonin	Ceruloplasmin
Glutathione	Transferrin
Ubiquinone	Metallothionein
Lipoic acid	Albumin
Uric acid	Bilirubin
Superoxide dismutase	Nutraceuticals
Glutathione peroxidase	–
Catalase	–

Source: Reprinted/Adapted from Mattil, 1947

Some of them can be synthesized endogenously, while others must be obtained through diet. Several antioxidant enzymes require trace elements from the diet as cofactors. Copper, zinc, and manganese are required by cytoplasmic and mitochondrial SODs to catalyze the removal of superoxide radicals produced by the cell. CAT, which requires iron and cytosolic glutathione peroxidase ($GSHP_X$) removes hydrogen peroxide (H_2O_2). This enzyme, which is selenium-dependent, also removes potentially toxic lipid hydroperoxides from the cell (Srilakshmi, 2019).

2.3.1 *VITAMIN E*

Vitamin E is a phenolic antioxidant that is fat-soluble. The hydrogen from the hydroxyl (HO) group on the ring structure rapidly donates itself to free radicals, rendering them unreactive. It blends with lipid molecules and protects them from oxidation since it is lipid soluble. It's also the body's first line of defense against LPO, which is a type of lipid oxidation. Peroxide lipid derivatives are produced during this process. Vitamin E protects cell membranes from free radical oxidation in this way. When free radicals attack the lipids that makeup cell membranes, they set off a chain reaction that can be extremely harmful. Vitamin E is among the most important chain-blocking antioxidants in our bodies, as it can prevent these chain reactions from occurring in the first place. Polyunsaturated lipids are the most vulnerable to oxidation.

Vitamin E's antioxidant properties play a crucial role in inhibiting the formation of lipofuscin, a pigment that typically accumulates within tissues as we age. Vitamin E is also involved in protecting vitamin A from oxidative damage by sparing the trace element selenium. Vitamin E also protects organisms from the effects of nitrosamines, a powerful tumor promoter (Srilakshmi, 2019).

2.3.1.1 FUNCTIONS OF VITAMIN E AS AN ANTIOXIDANT PROPERTY

- Vitamin E protects against the peroxidation of polyunsaturated fatty acids (PUFAs) in different tissues and membranes.
- It protects RBC from hemolysis by oxidizing agents.
- Vitamin E helps in protecting and maintaining proper reproductive function.
- In cellular respiration through the electron transport chain vitamin E is required.
- It prevents the oxidation of vitamin A and carotenes.
- It increases the synthesis of haem by enhancing the activity of enzymes alpha-amino-levulinic acid (ALA) synthase and ALA dehydratase.
- Vitamin E is also responsible for adequate synthesizing nucleic acid.
- It delays the onset of cataracts in the eyes.
- Vitamin E helps in decreasing LDL oxidation, prevents the uptake of oxygenated LDL by endothelium and monocytes/macrophages, decreases expression of inflammatory cell adhesion molecules, and inhibits platelet aggregation and thrombus formation.
- Vitamin E also helps in decreasing endothelial dysfunction.
- Vitamin E is required in preserving creatine in skeletal muscle.

- Vitamin E also helps in the absorption of amino acids in the intestine.

Food sources of vitamin E: Dark green leafy vegetables, Alfalfa, vegetable oils, nuts, whole grains, and rice bran oil are one of the best sources of vitamin E. It's also found in small amounts in lettuce, grasses, and the embryos of many seeds. In the case of plant sources, vitamin E is usually found in conjunction with chlorophyll in plant chloroplasts. Plant foods are typically higher in vitamin E content than animal foods (Srilakshmi, 2019).

2.3.2 VITAMIN C

Vitamin C is one of the most important hydrosoluble antioxidants in cells (L-ascorbic acid) (Upadhyay et al., 2009). Vitamin C is a powerful reducing agent and free radical scavenger in biological systems, scavenging oxidizing free radicals and damaging oxygen-derived species such as the hydroxyl radical, H_2O_2, and singlet oxygen (Hacisevki, 2009; Arrigoni & De Tullio, 2002). Vitamin C is susceptible to oxidation as it is a reducing agent that functions in the body as an antioxidant. Usually, mammals can synthesize vitamin C from glucose except a few including humans lack the liver enzyme gulonolactone-oxidase, which is required to catalyze one step of this process. Vitamin C is involved in a number of activities in plants, including growth, planned cell death, pathogenic responses, hormonal responses, flowering process, and senescence, as well as protection against environmental challenges (Tolbert et al., 1975; Pehlivan, 2014). Vitamin C acts as a redox buffer, reducing, and neutralizing the reactive oxygen. It also acts as a cofactor for enzymes that control photosynthesis, hormonal biosynthesis, and other antioxidants regenerating, and it also regulates cell division and growth, participates in signal transduction, and plays a role in various physiological processes, including immune stimulation, collagen synthesis, hormones, neurotransmitters, and iron absorption. Vitamin C is also responsible for the detoxification of heavy metals. Scurvy is caused by a severe deficiency of vitamin C, but low levels of vitamin C lead to symptoms such as increased susceptibility to infections, teeth loosening, mouth, and eye dryness, hair loss, dry itchy skin, exhaustion, and sleeplessness. Vitamin C, on the other hand, may act as a prooxidant in the presence of transition metals such as iron and copper, initiating a variety of hazardous radical processes. Vitamin C can operate as a powerful, effective, and inexpensive antioxidant whilst acting as a radical promoter (Pehlivan, 2017).

Vitamin C (ascorbic acid) is an important cofactor of α-ketoglutarate-dependent dioxygenases. Prolyl hydroxylases, for example, are involved in collagen manufacturing as well as the downward regulation of hypoxia-inducible factor (HFI)-1, a transcription factor that controls several genes involved in tumor growth, energy metabolism, neutrophil function, and apoptosis. Inhibition of the HIF system by vitamin C may give alternate or additional techniques for limiting tumor, infection, and inflammation progression. Vitamin C assists the endothelial nitric oxide synthase (eNOS) to function by recovering eNOS cofactor tetrahydrobiopterin, which is important for arterial elasticity and blood pressure management. Plant evidence suggests that Vitamin C plays a role in covalent adduct production with electrophilic secondary metabolites. Vitamin C can also act as the source of the signaling molecule, H_2O_2, and as Michael's donor to form covalent adducts with endogenous electrophiles, present in plants (Traber & Stevens, 2011).

2.3.2.1 FUNCTION

2.3.2.1.1 ROLE OF VITAMIN C IN LIPID PEROXIDATION (LPO)

The radical chain reaction LPO is an example of a radical chain reaction. The radical process is initiated by the removal of hydrogen atoms from bisallylic C-H bonds by reactive oxygen species (ROS) produced by a variety of sources, including the electron transport chain, xanthine oxidase, MPO, and NADPH oxidase (Traber & Stevens, 2011). Because many of the enzymes that produce ROS are entrenched in lipid bilayers, and because the bisallylic C-H link in PUFAs is relatively weak compared to other C-H bonds, lipids are frequently good targets for oxygen radicals. Carbon-centered lipid radicals mix with molecular oxygen to create peroxyl radicals, which may participate in the radical propagation reaction if not neutralized by tocopherol in membranes. Lipid hydroperoxides are chemically unstable and can produce a variety of LPOs, including 2-alkenals, epoxides, and malondialdehyde if they are not reduced to hydroxy-fatty acids by GSH-dependent reductases (Traber & Stevens, 2011).

Vitamin C protects against LPO by scavenging ROS and decreasing lipid hydroperoxyl radicals by one electron through the redox cycle of vitamin E. Furthermore, data from our laboratory suggests that vitamin C may play a part in the process of protecting cells from LPO-derived 2-alkenes. When vitamin C-deficient THP-1 human cells were exposed to LPO 4-hydroxy-2(E)-nonenal (HNE), protein carbonylation was significantly reduced relative to THP-1 cells that were not pre-incubated with vitamin C (Huvaere & Skibsted, 2015; Peterson & Dwyer, 1998). Ascorbate's protective effects were connected to an increase in GSH-HNE conjugate synthesis and phase I metabolites, as measured by LC-MS/MS, as well as increased GSH conjugate transit from the cells into the medium (Huvaere & Skibsted, 2015; Papatheodorou & Weiss, 2007).

1. **Collagen Formation:** Collagen is a significant structural protein found in connective tissue, bone, teeth, cartilage, skin, and scar tissue (which binds cells and tissues together). Vitamin C is required by the connective tissue fibroblast cells (which are responsible for collagen synthesis) and the bone-forming osteoblasts. It is estimated that collagen constitutes about one-quarter of all the protein in the body. Collagen is formed from a precursor protein known as tropocollagen by the hydroxylation of the amino acids, proline, and lysine within tropocollagen. Vitamin C is required to allow these essential hydroxylation reactions to proceed (Srilakshmi, 2019).
   The enzymes that catalyze these hydroxylation reactions require the direct participation of (ferrous iron and oxygen. The role of vitamin C is to act as a reducing agent that preserves the iron in the ferrous state, preventing its oxidation to the ferric state. Any vitamin C deficiency causes improper collagen synthesis, which leads to slowed wound healing, capillary disruption, and faulty bone and tooth formation. One of the first effects of any impairment of collagen synthesis is small pinpoint hemorrhages which result from weakness in the membranes that line the blood capillaries and in the fibers that hold cells underneath the surface of the skin. These weaknesses allow blood to escape into enlarged intercellular spaces, accounting for the capillary bleeding associated with scurvy. These subcutaneous hemorrhages appear most often in areas subjected to mechanical stress, such as the gums, which become soft, spongy, and prone to bleeding.
   When collagen production is disrupted, the matrix formation is compromised, and the body's ability to accumulate the calcium and phosphorus required for healthy bone mineralization is harmed. As a result, the bone weakens and might even become deformed. When the supporting cartilage, which is also primarily collagen, gets weaker, bones might become dislodged from their joints.

When a vitamin C shortage occurs, the dentin layer of the tooth does not form regularly. This causes teeth to be structurally weak, making them more susceptible to mechanical stress and decay. When sufficient vitamin C is present, skin grafts to restore damaged tissue recover more quickly.

2. **Carnitine Synthesis:** Carnitine cannot be synthesized without vitamin C. Carnitine is a nitrogen-containing chemical molecule that aids in the transfer of fatty acids into mitochondria, where they are oxidized to provide energy to cells (Srilakshmi, 2019).
3. **Neurotransmitter Synthesis:** The copper-containing enzyme dopamine oxygenase, which catalyzes the oxidation of dopamine to generate the neurotransmitter norepinephrine, requires vitamin C to function. Vitamin C appears to play a part in the hydroxylation process of tryptophan during serotonin production (5-hydroxy tryptamine). The occurrence of high vitamin C concentrations in brain and adrenal tissues is most likely due to vitamin C's role in neurotransmitter production.
4. **Activation of Hormones:** Many peptide hormones and hormone-releasing factors are made from precursor molecules that are then enzymatically changed to become active. Bombesin (human gastrin-releasing peptide), calcitonin, gastrin, oxytocin, thyrotropin, corticotropin, vasopressin, and growth hormone-releasing peptide are all activated by vitamin C. High quantities of ascorbic acid are seen in the adrenal glands, especially during times of stress. The hydroxylation processes in the production of corticosteroid hormones require vitamin E (Srilakshmi, 2019).
5. **Cellular Respiration:** Vitamin C is involved in the electron transport chain. Some components, e.g., cytochrome oxidase exhibit optimal activity when ascorbic acid is present.
6. **Tryptophan Metabolism:** Vitamin C is necessary for hydroxylation from tryptophan (hydroxylase enzyme) to hydroxytryptophan in serotonin synthesis.
7. **Immunologic Functions:** Vitamin C improves immunoglobulin production and promotes leukocyte phagocytic activity.
8. **Prevention of Cataracts:** Vitamin C lowers the risk of cataract formation in the eye.
9. **Cholesterol Metabolism:** The synthesis of bile acids from cholesterol (enzyme-7 alpha-hydroxylase) is enhanced by ascorbic acid.

10. **Maturation of Erythrocytes:** The conversion of methemoglobin to hemoglobin is aided by vitamin C. Ascorbic acid is required for the conversion of hemoglobin to bile pigments. In association with tetra-hydro folic acid, ascorbic acid is involved in the maturation of erythrocytes.
11. **Drug Detoxification:** Vitamin C is necessary for the proper functioning of the body's drug-detoxifying metabolic processes. These include the mixed-function oxidase system and the flavin-monooxygenase system in the liver.
12. **General Antioxidant:** Because of the normal metabolic processes and exposure to medications and environmental contaminants, the body produces a number of harmful oxidizing agents. These oxidizing chemicals can be converted to innocuous compounds by a variety of enzymes and antioxidant-reducing agents (such as vitamin E, carotene, and vitamin C). Many forms of oxidizing free radicals can react with vitamin C and be scavenged. It can also regenerate reduced forms of vitamin E, which have antioxidant properties. Vitamin C is believed to play an important role in maintaining the structure of blood vessels, and its antioxidant properties may help to avoid tissue damage that leads to cardiovascular disease (CVD) (Keservani et al., 2016a, b; Srilakshmi, 2019).
13. **Iron Metabolism:** Vitamin C is a reducing agent, meaning it helps absorption by keeping ferric ions in ferrous form. On the more alkaline duodenal pH, vitamin C creates soluble complexes with ferric ions, which retain iron solubility. Vitamin C aids in the transfer of iron from blood plasma to ferritin, which is then stored in the liver, as well as the release of iron from ferritin as needed. The involvement of ascorbate in iron metabolism is linked to both increased absorption and intracellular iron-binding protein metabolism. Calcium absorption is also aided by vitamin C, which prevents calcium from being incorporated into insoluble complexes. Vitamin C transforms folic acid's inactive form into its active forms, dihydrofolic acid, and tetra-hydrofolic acid, as well as stabilizes the active form.
14. **Good Sources of Vitamin C:** Ascorbic acid is widely present in plant foods, particularly in fresh fruits and vegetables, especially green leafy vegetables. Amla (Myirobalan) is the richest natural source of vitamin C. Guava, orange, lime, and green leafy vegetables like drumstick leaves and agathi are also excellent sources of vitamin C. Meat, milk, cereals, and pulses are not-so-good sources of ascorbic

acid. Ascorbic acid is the vitamin most vulnerable to air oxidation of all the vitamins. One of the typical properties of this vitamin is its intense reducing power, and hence it is oxidized rapidly in air. It is for this reason that when vegetables become dry and hence it is oxidized, stale, or cut and exposed to air most of the vitamin C originally present is destroyed. When dry pulses and beans are allowed to germinate, vitamin C is present in both the grain and the growing sprout, with the former containing around 85% and the latter containing the remainder. During famines or prolonged droughts where fresh vegetables or fruits cannot be obtained, sprouted grains can be used as a source of dietary vitamin C. Sprouted green gram has almost 3 times more vitamin C than sprouted Bengal gram. 17–20 trig of vitamin C is during germination per 100 g of pulses.

The majority of the vitamin C in fresh fruits and vegetables is destroyed when they are heated or dried. Amla stands out among these fruits not only because of its high acidity and vitamin C content but also because it contains compounds that help to keep the vitamin from being destroyed by heat or drying. The ascorbic acid content of citrus fruit and tomato juices possibly can be preserved by the canning process. Freezing has no negative consequences for this vitamin. Vitamin C can be easily oxidized in the presence of Copper and Iron, but not in the presence of aluminum. As a result, foods prepared with copper utensils lose ascorbic acid quickly. Alkalis quickly destroy it, although it is relatively stable in a mild acid solution. Baking soda, by contrast, has a negative effect, but cooking by steam removes very little ascorbic acid. Cooking reduces ascorbic acid content by more than 60–75%. The vitamin C content of food is maximized by the following processes:

- Harvesting at the peak of maturity;
- Storing in a cool, moist environment;
- Restricting exposure to air and sunlight;
- Avoiding soaking food in water;
- Cooking food in a minimal amount of water or even better, cooking by microwave;
- Cooking the food in pieces as large as possible;
- Consuming fruits immediately after cutting;
- Avoiding making juices or macerating;
- Eating vegetable preparations immediately after cooking;
- Wherever possible fresh raw vegetables should be used for obtaining enough vitamin C (Srilakshmi, 2019).

2.3.3 CAROTENOIDS

Carotenoids are natural pigments produced by plants that are responsible for the vibrant colors of a variety of fruits and vegetables. Beta carotene is a provitamin-A carotenoid, which means it may be transformed into vitamin A by the body (retinol) (Olsen et al., 2020). The foods we eat contain plenty of carotenoids, and the majority of them have antioxidant properties (Paiva & Russell, 1999). Plants rely on carotenoids to defend themselves from photooxidative reactions. It does not actively participate in photosynthesis but rather passes the energy it receives to chlorophyll, as well as acting as a protective factor for chlorophyll due to its high antioxidant action, which shields organic molecules from oxidation (www.worldofmolecules.com, 2022). They're also good antioxidants, scavenging singlet molecular oxygen and peroxyl radicals. Carotenoids are a component of the antioxidant defense system in humans. They work along with other antioxidants to produce a synergistic effect. Carotenoids in mixtures are more potent than single molecules. Human studies are increasingly showing that carotenoids protect the skin from photooxidative damage (Stahl & Sies, 2003). Carotenoids are most likely engaged in the scavenging of two ROS, singlet molecular oxygen ($1O_2$) and peroxyl radicals, among the different defense methods. They're also good deactivators of electronically stimulated sensitizer molecules, which are involved in the production of radicals and singlet oxygen (Stahl & Sies, 2003).

2.3.3.1 FUNCTION

2.3.3.1.1 Better Cognitive Function

Beta carotene might improve our cognitive function, agreeing to a few ponders, due to its antioxidant effects. A Cochrane Review–2018 of eight research on antioxidants, including beta carotene, discovered some cognitive and memory advantages related to beta carotene supplementation (Olsen et al., 2020).

But there is sufficient evidence to suggest that consuming fruits and vegetables in general, particularly those high in beta-carotene, can help prevent cognitive decline and dementia.

2.3.3.1.2 Good Skin Health

The antioxidant properties of beta-carotene may also aid in skin health. According to research, eating enough antioxidant micronutrients, such as beta carotene, can boost the skin's defenses against UV radiation and maintain general skin health and beauty. According to the study, dietary beta-carotene gives significantly less UV protection than topical sunscreen (Keservani et al., 2020).

2.3.3.1.3 Lung Health

Beta carotene, which is produced by the body, aids in the correct functioning of the lungs. Furthermore, persons who consume a lot of beta-carotene-rich foods may have a reduced chance of developing some type of cancer, such as lung cancer. Taking beta-carotene supplements, on the other hand, may increase the chance of developing lung cancer in smokers.

2.3.3.1.4 Eye Health

Carotenoids, such as beta carotene, are found in foods and may help to improve eye health and protect against diseases like age-related macular degeneration (ARMD), which causes vision loss. High blood levels of carotenoids, such as beta carotene, have been shown to decrease the risk of advanced ARMD by up to 35%. Furthermore, studies have reported that eating foods rich in beta-carotene, such as fruits and vegetables, can reduce the incidence of ARMD in smokers.

2.3.3.1.5 Reduction in Risk of Certain Cancers

According to research, eating a diet high in antioxidant foods like beta-carotene can help protect against the development of some cancers. This includes:

- Breast cancer before menopause;
- Pancreatic cancer;
- Lung carcinoma.

2.3.3.2 SOURCES

Red, orange, and yellow fruits and vegetables, as well as dark leafy green vegetables and other green vegetables, are rich in beta-carotene. Cooked carrots have more carotenoids than raw carrots, according to research. The bioavailability of carotenoids might be improved by adding olive oil. Because beta-carotene is a fat-soluble molecule, it is more easily absorbed when consumed with fat.

2.3.3.2.1 The Foods with the Highest Beta Carotene Content

Vegetables with dark green leaves, including kale and spinach, sweet potatoes, carrots, broccoli, butternut squash, cantaloupe, red, and yellow peppers, apricots, broccoli, peas, romaine lettuce, and strawberries.

Paprika, cayenne, chili, parsley, cilantro, marjoram, sage, and coriander are examples of herbs and spices that contain beta-carotene (www.healthline.com, 2021). Beta carotene is thought to be an antioxidant as well as an immune system enhancer. Cryptoxanthin, alpha-carotene, zeaxanthin, lutein, and lycopene are members of the antioxidant carotenoid family. These minerals, unlike beta-carotene, are not converted to vitamin A at significant levels (Stahl & Sies, 2003).

2.3.4 FLAVONOIDS

Fruits, cereals, vegetables, tea, wine, flowers, roots, bark, and stems all contain flavonoids, which are a group of natural compounds with varying phenolic structures. These natural products are well-known for their health benefits, and scientists are working to extract the flavonoids, which are active chemicals. Flavonoids are currently thought to be an important component in a wide range of nutraceutical, pharmacological, medical, and cosmetic products. This is due to their ability to control important cellular enzyme processes, as well as their anti-oxidative, anti-inflammatory, anti-mutagenic, and anti-carcinogenic capabilities. With the finding of the low cardiovascular mortality rate and the prevention of CHD by flavonoids (Keservani et al., 2014), research on flavonoids (Keservani et al., 2014) has gotten a boost. Flavonoids are plentiful in plant-based foods and beverages such as fruits, vegetables, tea, chocolate, and wine, and are thus referred to as dietary flavonoids (Panche et al., 2016). Some flavonoids have hormone-like properties, and they look a lot like steroid hormones, especially estrogen. Fruits and vegetables, tea, red wine, and cereals all contain flavonoids (Srivastava & Bezwada, 2015). Hormone-like flavonoids with steroid-like qualities, particularly estrogenic functions, are widely known for their ability to prevent a wide range of chronic diseases, as well as having numerous neuroprotective benefits on the brain. Flavonoids (Keservani et al., 2010a, b) have been employed in the food business for many years because of their ability to preserve foods, add color and flavor, and create dietary supplements, among other things (Routray & Orsat, 2012).

2.3.4.1 ANTIOXIDANT ACTIVITY

The most talked-about attribute of all flavonoids is their dynamic antioxidant power, which aids in free radical scavenging by acting as potential reducing agents and protecting the organism from oxidative processes (Routray & Orsat, 2012). Flavonoids' antioxidant properties are due to a combination of iron chelating and free radical scavenging effects, as well as inhibition of oxidase enzymes such as cyclooxygenase, MPO, lipoxygenase, NADPH oxidase, and xanthin oxidase, inhibiting the formation of reactive oxygen and organic hydroperoxide species (Jimenez et al., 2009). Furthermore, flavonoids, such as phospholipase A2, might block enzymes by indirectly engaging themselves in the oxidative process. At the same time, flavonoids can activate other antioxidant enzymes such as CAT and SOD.

Their functions can be grouped in two categories: (i) general (non-specific activity – in relation to the presence of phenolic compounds) and (ii) specific activity (depending on the specific chemical and structural features of the active substance), which are presented in table (Fraga et al., 2010):

Nonspecific Activity	Specific Activity
Activation of antioxidant enzymes.	Interaction with enzymes.
Reactive oxygen species neutralization.	Anti-inflammatory activity.
Absorption of ultraviolet radiation.	Affinity for estrogenic receptors.
Inhibition of enzymes.	Microorganism growth inhibition.
Chelating trace metals.	Influence on regulating systems and transmission of tissue signals.
Inhibition of radical reactions.	Impact on the cardiovascular system.
Antioxidant properties.	Interaction with receptors.
Interactions with membranes.	Interaction with transcription factors.

Source: Reprinted/Adapted from Fraga et al., 2010

2.3.4.2 APPLICATIONS IN FOOD

Over the past few years, many polyphenols have been used as natural antioxidants in oils and fats to prevent the oxidation of lipids, to protect food and beverages from exposure to light, to extend food shelf life, as a supplement to animal feed to improve the health of animals, to protect animal products, as an antimicrobial agent in foodstuffs, and as a health functional ingredient in foods and dietary supplements; these applications can be attributed primarily to antioxidant and antimicrobial properties (Arct & Pytkowska, 2008; Yilmaz, 2006; Huvaere & Skibsted, 2015). In recent years, the usage of synthetic chemicals in the food sector has decreased. This is primarily due to our desire to limit the risk of developing diseases linked to the intake of synthetic products. Flavonoids are phytochemicals that humans are unable to produce. These chemicals, on the other hand, could be employed as food additives to promote human health and boost their levels (Peterson & Dwyer, 1998; Rodriguez et al., 2015).

Many fruits and vegetables contain large levels of flavonoids. Natural antioxidants and flavonoids have been identified as some of the most essential micronutrients that may be used in industry to lower the usage of synthetic compounds in foods while also improving human health because of their ability to cure a variety of diseases. Because of their antibacterial and antioxidant properties, these bioactive chemicals can be employed to extend the shelf life and preserve numerous foods (Pandey & Rizvi, 2009). Furthermore, interest in natural products has grown in recent years, owing to a reduction in the overuse of synthetic antioxidants in food (such as BHA and BHT) and an increase in the use of natural compounds such as antioxidants or food components, which have lower toxicity than synthetic products (Soobrattee et al., 2005; www.who.int, 2022).

2.3.5 POLYAMINES

The polyamines putrescine, spermine, and spermidine are antioxidants that are involved in a variety of biological activities, including cell proliferation and differentiation. Dietary polyamines play a crucial role in human health, particularly in intestine maturation and immune system differentiation and development. Polyamines' antioxidant and anti-inflammatory properties can be used to prevent chronic conditions like CVD. Food is a significant source of polyamines in addition to endogenous production. However, there are no recommendations for daily polyamine intake, it is known that polyamine requirements are high during periods of rapid cell proliferation (for example, during the neonatal period). Furthermore, because de novo polyamine synthesis declines with age, dietary sources of polyamines become more important in an aging population. Polyamine daily consumption varies according to published estimates, owing to differences in dietary patterns and data-gathering approaches. Polyamines can be available in a variety of foods at varying levels. Putrescine may have a microbiological origin, but spermine and spermidine are naturally present in food. Spermidine is the most common polyamine in plant-based foods, but spermine is more common in animal-derived foods (Munoz-Esparza et al., 2019).

2.3.5.1 POLYAMINES AND HEALTH

Polyamines are important for cell growth and proliferation, as well as the stabilization of DNA's negative charges, RNA transcription, protein synthesis, immune response modulation, apoptosis, ion channel regulation, particularly by inhibiting potassium channels, and antioxidant activity (Larque et al., 2007; Ruiz-Cano et al., 2012; Gomez-Gallego et al., 2017; Moinard et al., 2005).

Polyamines' antioxidant activity is mostly observed in membrane lipids and nucleic acids. Because of its increased amount of positive charges, spermine is a polyamine with potent antioxidant properties. Metal chelation, which limits the creation of hydroperoxides and slows the generation of secondary oxidation products, is the major mechanism of polyamine antioxidant action (Moinard et al., 2005; Eliassen et al., 2002). Polyamines, particularly in lipophilic conditions, may also be able to remove free radicals.

2.3.5.1.1 Potential Effects of Polyamines

In the postnatal stage – polyamines are important in people, particularly in the early stages of life, according to several studies. It is well-documented that the need for polyamines increases with rapid cell proliferation, notably in the neonatal stage (Larque et al., 2007; Ruiz-Cano et al., 2012; Eliassen et al., 2002; Kalac, 2014). Requirements are also higher during surgery or during wound-healing and aging phases (Larque et al., 2007; Kalac, 2014; Lovaas, 1991).

Polyamines (spermine and spermidine) speed up the process of proliferation and maturation of the digestive tract and play a major role in the immune system's differentiation and development (Ruiz-Cano et al., 2012; Eliassen et al., 2002; Lovaas, 1991). Furthermore, these chemicals can aid in the modulation of the inflammatory response due to their antioxidant characteristics (Moinard et al., 2005; Toro-Funes et al., 2013).

➢ Sources: Breast milk is the initial source of polyamines in the diet. Other foods that are high in polyamines include peas, mushrooms, hazelnuts, pistachios, spinach, broccoli, cauliflower, and green beans. The fruit group has the lowest levels of polyamines. For example, reported values for spermidine in apples, oranges, cherries, pears, or tangerines are less than 21 nmol/g and 1.98 nmol/g for spermine (Munoz-Esparza et al., 2019).

2.3.6 MELATONIN

Melatonin, sometimes known as the sleep hormone, is produced mainly by the pineal gland in the presence of darkness. It controls man's circadian cycle. Its intrinsic features, such as high cell permeability, the capacity to easily cross the placental and blood-brain barriers, and its role as an endogenous reservoir for free radical scavengers with additional indirect activities, make it a good candidate for use as an adjuvant in biomedicine. Melatonin exerts its potent anti-inflammatory effects possibly by inhibiting inflammasome activation (Perez-Cano et al., 2010).

Melatonin has now been demonstrated to be a multifunctional antioxidant and free radical scavenger. The hydroxyl radical, peroxynitrite anion, and nitric oxide (NO) are among the free radicals and reactive oxygen intermediates that it detoxifies. A number of antioxidant enzymes, such as glutathione peroxidase (GPx), GSH reductase, glucose-6-phosphate dehydrogenase, GSH reductase, and SOD, are also likely to be stimulated. NO synthase, on the other hand, is a pro-oxidative enzyme. Melatonin also overcomes all morphophysiological barriers, such as the blood-brain barrier and the placenta, to reach every cell. Its antioxidant efficiency is boosted by these properties. The membrane lipids and nuclear DNA both have been found to be protected against oxidative damage by melatonin. Melatonin has been found to be resistant to macro-molecular lesions and malfunctions associated with free radicals in every experiment model in which it has been tested (Perez-Cano et al., 2010).

- Sources: Melatonin has been found in a wide range of foods, including fungus, animals, and plants, throughout the previous few decades. Melatonin-rich food groups include eggs and fish from animals, whereas the highest melatonin content is found in nuts from plants. Melatonin is also found in a variety of mushrooms, grains, and germinating legumes and seeds. It has been shown that the melatonin concentration in human serum can elevate significantly after the consumption of melatonin-containing food (Perez-Cano et al., 2010).

2.3.7 GLUTATHIONE (GSH)

Glutathione, commonly known as GSH, is a tripeptide made up of glycine, cysteine, and glutamic acid that is an endogenous component of cellular metabolism. It's usually found in the liver (Kalac, 2006). It is the most promising low molecular-weight antioxidant synthesized in cells. GSH plays a major role in the removal of many reactive species (Kalac, 2006). GSH is involved in food metabolism, antioxidant defense, and the control of cellular metabolic functions ranging from gene expression, DNA, and protein synthesis to signal transmission, cell proliferation, and apoptosis (Kalac, 2006). Electrophiles that would ordinarily attach to proteins or nucleic acids, causing cellular damage and genetic changes, are bound by GSH conjugation (facilitated by a family of GSH transferase enzymes) (Kalac, 2006).

- **Detoxification:** GSH is composed of three amino acids: cysteine, glutamic acid, and glycine. The principal binding substrates for GSH are heavy metals and fat-soluble poisons, making them water-soluble and appropriate for renal elimination. GSH is one of the body's most powerful antioxidants. An excess of phase I hepatic activity can deplete GSH and expose the body to toxin-induced damage (Gomez-Gallego et al., 2008).

GSH has been shown to protect human retinal pigment epithelium (RPE) cells from injury. After consuming the necessary amino acids and sulfur-containing meals, GSH is produced in the liver. This underappreciated water-soluble compound acts as an antioxidant and regenerator of vitamin E and carotenoids, as well as an intracellular enzyme (Singh et al., 2017). GSH is found in meat, fruits, fish, and vegetables, a major enhancer of GSH synthesis is vitamin C (Gomez-Gallego et al., 2008).

2.3.8 UBIQUINONE

Coenzyme Q10 (ubiquinone) works in a similar way as vitamin E. It promotes skin repair and regeneration in a variety of ways. It comes in powder or liquid form, and the color ranges from orange to dark red. After the age of 30, the skin's coenzyme levels drop, resulting in a slower rate of collagen and elastin production. It's also known as a bio-molecular indicator of aging. In nature, ubiquinone is a highly lipophilic compound with bioavailability difficulties. It is a powerful antioxidant that is widely used in cosmetic formulations as a protective ingredient in the form of liposomes, lipid nanoparticles (NPs), and emulsions to prevent skin aging and photoaging (Biol-N'garagba & Louisot, 2003) This molecule is our single-endogenously generated lipid soluble antioxidant, found in all membranes and outperforming other antioxidants in terms of both quantity and efficacy. Because of its proximity to oxidative events and the powerful regeneration by continual reduction at all locations, its protective action extends to lipids, proteins, and DNA (Biol-N'garagba & Louisot, 2003).

2.3.9 LIPOIC ACID

Alpha-lipoic acid, which is required for mitochondrial dehydrogenase processes, has received a lot of attention very recently as an antioxidant. Lipoate reacts with ROS such as superoxide radicals, hydroxyl radicals (OH), hypochlorous acid, peroxyl radicals, and the reduced single oxygen, dihydrolipoate. It also protects membranes through interactions with vitamin C and GSH, which may recycle vitamin E. In addition to its antioxidant properties, dihydrolipoate can also exert pro-oxidant effects by reducing iron. In a number of oxidative stress scenarios, such as ischemia-reperfusion injury and diabetes, alpha-lipoic acid treatment has been found to be helpful. Both alpha-lipoic acid and dihydrolipoic acid attach hydrophobically to proteins like albumin, preventing glycation processes. It can help in preventing cataracts, HIV activation, dementia, and radiation damage. Lipoate can also function as a redox regulator for proteins like myoglobin, thioredoxin, prolactin, and NF-kappa B transcription factor (Soda, 2010).

2.3.10 URIC ACID (UA)

Uric acid (UA) is the end product of purine metabolism in humans, which resulted from the loss of uricase activity caused by several mutations in its gene during the Miocene period, resulting in humans having greater UA levels than other mammals. In addition, 90% of the UA filtered by the kidneys is reabsorbed rather than expelled (Reiter et al., 1999). UA is involved in a complicated reaction with a few oxidants, and it may have some protective properties in some situations. UA, on the other hand, cannot scavenge all radicals, such as superoxide. UA acts as an antioxidant only in the hydrophilic environment, which appears to be a major limitation of the antioxidant functions of UA. UA reactions with oxidants can produce additional radicals, which can cause a radical chain reaction and oxidative damage to cells. Along with this, UA itself and/or downstream radicals can improve, as a biologically active pro-inflammatory factor, the production of intracellular oxidants through the ubiquitous NADPH oxidase-dependent pathway. This can result in redox-dependent intracellular signaling and, in some conditions, oxidative stress. So, these facts taken together may explain its oxidant-antioxidant paradox (Tarocco et al., 2019). As an antioxidant and neuroprotector, UA has some obvious negative effects as well as some less well-known positive ones. High plasma UA levels are a risk factor for CVD and Metabolic Syndrome, as well as a trigger for gout and renal calculi. Reduced excretion, increased synthesis, or both are the major causes of elevated plasma UA (Reiter et al., 1999). Dietary purines lead to the formation of UA within the body.

- **Sources:** Purine-rich foods include tuna, haddock, trout, anchovies, sardines, mussels, and herring, as well as excessive alcohol, yeast, bakery products, including beer and liquor, high-fat foods like bacon, red meat (including veal), and dairy products, organ meats like liver and sweetbreads, and sugary foods and beverages. Ham and beef, poultry, oyster, shrimp, crab, and lobster all contain moderate quantities of purine.

2.3.11 SUPEROXIDE DISMUTASE (SOD)

SOD is the first detoxification enzyme and the most powerful antioxidant inside the cell. It's an important endogenous antioxidant enzyme that works as part of the body's initial line of defense against ROS. It catalyzes the dismutation of two molecules of superoxide anion (O_2) into H_2O_2 and molecular oxygen (O_{22}), reducing the danger of the superoxide anion. The enzyme is essential for cellular health since it protects body cells against damaging oxygen radicals, free radicals, and other free radicals that cause aging and cell death. With age, SOD levels reduce, while free radical production increases. It has been reported that having SOD supplements on a daily basis can help to protect the immune system, reduce the risk of disease, and slow down the aging process (Meng et al., 2017).

2.3.12 GLUTATHIONE PEROXIDASE (GPX)

Glutathione peroxidase (GPx) is an antioxidant enzyme containing selenium that transforms GSH to GSH disulfide and degrades H_2O_2 and lipid peroxides to lipid alcohols and water, respectively. By providing reducing equivalents for various metabolic pathways, reduced GSH plays a key role in the control of the intracellular redox state of vascular cells (Tabet & Touyz, 2007; Lubos et al., 2011). H_2O_2 and lipid peroxides are not detoxified in the absence of sufficient GPx activity or GSH levels, and transition metals (Fe^{2+}) may convert them to lipid peroxyl radicals and OH radicals, respectively. In low-level oxidative stress, the GPx/GSH system is regarded to represent a key defense (Forman et al., 2009). Importantly, GPx-1 has many cellular functions: to protect cells from oxidative damage; to regulate metabolism and mitochondrial function; and to control cellular processes, such as apoptosis, growth, and signaling by modulating intracellular levels of H_2O_2 and the overall intracellular redox balance. The complex effects of GPx-1 in biological systems appear to stem from the delicate balance between intracellular oxidants and antioxidants and the deleterious effects of shifting the balance too far in either direction, which can result in oxidative or reductive stress (Gad, 2014). GSH is produced by the human body, but it can also be obtained through diet. Avocados, spinach okra, and asparagus are some of the greatest dietary sources.

2.3.13 CATALASE (CAT)

Catalase (CAT) is an antioxidant enzyme that catalyzes the breakdown of H_2O_2 to water and oxygen in practically all living tissues exposed to oxygen (like plants, animals, and bacteria) (Bentinger et al., 2007). The enzyme catalyzes the breakdown or reduction of H_2O_2 to water and molecular oxygen, completing the detoxifying process started by SOD. It employs either iron or manganese as a cofactor (Salguero, 2007; Abel, 2018). It's abundant in cells, where it's always looking for H_2O_2 molecules to eat. The enzyme CAT is extremely efficient. In less than a second, it can decompose millions of H_2O_2 atoms. The enzyme is primarily found in peroxisomes, although it is not found in mammalian cells' mitochondria (Meng et al., 2017). CAT is one of the most important antioxidant enzymes which mitigate oxidative stress to a considerable degree by converting cellular H_2O_2 into water and oxygen. Many age-related degenerative disorders, such as diabetes mellitus, hypertension, anemia, Alzheimer's disease (AD), Parkinson's disease, vitiligo, bipolar disorder, cancer, and schizophrenia, are thought to be linked to CAT deficiency or malfunction (Shegokar, 2016). CAT deficiency may play a role in the graying of human hair. The body produces H_2O_2, which is then destroyed by CAT. If CAT levels decrease, H_2O_2 cannot be broken down properly. H_2O_2 interferes with the production of melanin pigment that gives color to hair (Packer et al., 1995; Sautin & Johnson, 2008).

2.3.14 COPPER

Copper (Cu) is a trace element that is essential for humans and animals. Copper in the body alternates between cuprous (Cu^{2+}) and cupric (Cu^{3+}) forms, with the Cu^{2+} form accounting for the majority of the copper. Copper's unique capacity to receive and transfer electrons explains its critical role in oxidation-reduction (redox) reactions and free radical scavenging (Batta, 2016). Even though Hippocrates prescribed copper compounds to treat diseases as early as 400 B.C. (Ighodaro & Akinloye, 2018), scientists are still learning more about copper's functions in the human body (Tabet & Touyz, 2007).

Many oxidation-reduction processes involving copper-containing oxidases require copper as a cofactor. Copper enzymes play a role in connective tissue maturation, iron metabolism, energy production, and neurotransmission, among other physiologic pathways (Batta, 2016). Copper is a zinc-balancing mineral that is required for the formation of hemoglobin, the oxygen-transporting protein. Copper helps to prevent acne by assisting in the proper functioning of the prostate and the activity of the other oily glands. Copper is required for the proper functioning of nerves and joints. Copper is abundant in organ meat, seafood, nuts, seeds, wheat bran cereals, and whole-grain goods.

2.3.14.1 Antioxidant Functions of Copper

2.3.14.1.1 Superoxide Dismutase (SOD)

Superoxide dismutase (SOD) is an antioxidant that catalyzes the conversion of superoxide radicals (ROS – also known as free radicals) to H_2O_2, which can then be reduced to H_2O by other antioxidant enzymes (Traber & Stevens, 2011). Copper is found in two types of SOD: (i) copper/zinc SOD, which is found in most body cells, including red blood cells (RBC); and (ii) extracellular SOD, which is a copper-containing enzyme found in high concentrations in the lungs and low concentrations in plasma (Sies, 1997).

2.3.14.1.2 Ceruloplasmin

Ceruloplasmin has two different ways of acting as an antioxidant. Free iron and copper ions are powerful free-radical damage catalysts. Ceruloplasmin binds to copper and inhibits free copper ions from catalyzing oxidative damage. Ceruloplasmin's ferroxidase activity (oxidation of ferrous iron) aids iron loading onto transferrin, its transport protein, and may prevent free ferrous ions (Fe^{2+}) from engaging in potentially detrimental free-radical processes (Lubos et al., 2011).

2.3.14.1.3 Regulation of Gene Expression

Copper concentrations in the cell may affect protein synthesis by stimulating or inhibiting the transcription process of particular genes. Copper may influence gene expression by increasing intracellular oxidative stress levels. In response to oxidative stress, a good number of signal transduction pathways are triggered, which can lead to an increase in the expression of genes included in the detoxification of ROS (Batta, 2016).

2.3.14.1.4 Safety

1. **Toxicity:** Copper poisoning is uncommon among the general public. Acute copper poisoning has been reported as a result of beverage contamination due to storage in copper-containing containers, along with contaminated water. Acute copper toxicity symptoms include nausea, vomiting, abdominal pain, and diarrhea; these symptoms assist in preventing further copper ingestion and absorption. Severe liver damage, renal failure, coma, and death are more serious indications of acute copper intoxication. The likelihood of liver damage from long-term exposure to lesser amounts of copper is of greater nutritional concern.

2.3.15 ZINC AND SELENIUM

Zinc (Zn) and selenium (Se), both important minerals, provide antioxidant activities to animal cells through separate processes. Selenium is necessary for Selenium-dependent GPx, a category of water-soluble enzymes, to catalyze the destruction of water-soluble and membrane-bound hydroperoxides. Selenium-dependent GPx activities are reduced in diet deficient in Se; nevertheless, Se intakes over the minimum amount for normal growth result in a minor to moderate increase in Selenium-dependent GPx activities. Selenium can be classified as a general antioxidant because of its enzymatic nature as an antioxidant, limiting peroxide levels in cells by decomposing hydroperoxides. Zn, on the other hand, is a non-essential cofactor for any antioxidant enzyme and only acts as an antioxidant at particular places. Although Zn is required for the structure of the enzyme Cu, Zn SOD and the activity of this enzyme are not diminished in Zn insufficiency, although it is frequently suppressed with high Zn intakes. In two ways, Zn can act as a site-specific antioxidant. Firstly, Se competes with Fe and Cu to bond with cellular membranes and some proteins, displacing these redox-active metals and increasing their availability for binding to metallothionein (MT) and ferritin, respectively. Second, Zn protects proteins from oxidation by binding to their sulfhydryl groups. Zn status does not directly affect tissue peroxide levels but may shield specific molecules from peroxidative and oxidative damage (Vasak, 2005).

2.3.16 GLUTATHIONE (GSH) REDUCTASE

It's part of a chain of enzymes that keeps GSH in its reduced state. GSH reductase is a flavoprotein in nature and is used to catalyze the reduction of glutathione disulfide (GSSG), where NADPH participates as an electron donor. Because GR is involved in the regeneration of a reduced form of GSH during the detoxification of peroxides and free radicals generated in mitochondria, it is thought that this enzyme is vital for antioxidant protection and the cell's redox potential (Nandi et al., 2019). It aids in the maintenance of normal function and the prevention of oxidative stress in human cells. GSH reductase can scavenge OH, singlet oxygen, and a wide range of electrophiles.

2.3.17 N-ACETYLCYSTEINE (NAC)

NAC is largely appreciated for its ability to produce antioxidants. NAC is required for the production and replenishment of GSH, in addition to two other amino acids, glycine, and glutamine. Its antioxidant capabilities are also beneficial in the therapy of a variety of other oxidative stress-related diseases, including heart disease, infertility, and several psychiatric disorders. N-acetyl cysteine (NAC) is a pharmacological antioxidant and cytoprotectant that is frequently used. It has been shown to reduce endogenous oxidants and protect cells from a variety of pro-oxidative stimuli. Because NAC is a poor oxidant scavenger, the chemical mechanisms underlying its antioxidative actions have remained unknown (Chelikani et al., 2004). NAC is a crucial component of the detoxification process in our bodies. It can help prevent adverse environmental and pharmacological toxic effects. In fact, clinicians routinely administer intravenous (IV) NAC to persons who have taken too much acetaminophen to prevent or mitigate renal and liver damage. Because of its antioxidant and anti-inflammatory properties, NAC can be used in the treatment of a variety of liver illnesses. Likewise, an animal survey suggested that NAC could help to lessen the negative consequences of schizophrenia, such as social disengagement, lethargy, and short attention spans. NAC pills can also help cocaine addicts avoid relapse by reducing withdrawal symptoms. Furthermore, preliminary research suggests that NAC may reduce marijuana and nicotine usage and cravings. Many of these illnesses have little or no therapeutic options that are currently effective. Individuals suffering from these illnesses may benefit from NAC expectorant, which loosens mucus in the airways. NAC replenishes GSH levels in the lungs and decreases inflammation in the bronchial tubes and lung tissue as an antioxidant. Long-term inflammation and oxidative damage of lung tissue occur in chronic obstructive pulmonary disease patients (COPD), causing airways to tighten, resulting in shortness of breath and cough. COPD symptoms, exacerbations, and lung deterioration have all been improved with NAC supplementation. The ability of NAC to replace GSH and control glutamate levels in the brain can help to improve brain health. Glutamate, a brain neurotransmitter, is involved in a wide range of learning, behavior, and memory activities, whereas GSH, an antioxidant, helps protect brain cells from oxidative damage. Because NAC helps in regulating glutamate levels and replenish GSH, it might help people with brain and memory ailments. NAC may potentially reduce the risk of heart disease by lowering oxidative damage to heart tissues. It's also been found to boost NO synthesis, which aids in vasodilation, and this can lessen the risk of heart attacks (Wikipedia, 2022).

2.3.18 GLUCOSE-6-PHOSPHATE DEHYDROGENASE

It is the primary enzyme present in the pentose phosphate pathway, which is a depiction of an alternate route for glucose metabolism and produces NADPH for fatty acid synthesis and ribose residue for nucleic acid biosynthesis, nucleotide, and so on, is glucose 6 phosphate dehydrogenase (G-6PDH). G-6PDH is a NADP-dependent enzyme that catalyzes the conversion of glucose 6 phosphate to 6 phosphogluconate via the formation of 6 phosphogluconolactone and the reaction occurs in the cytosol (Tappia et al., 1998). The peroxisomal fractions produce only 3% to 4% of the cellular activity of NADP+ G-6-P DH (Wood et al., 2009). G6PD (glucose-6-phosphate dehydrogenase) is a cytosolic enzyme that produces NADPH, an electron donor molecule that helps neutralize harmful oxidizing chemicals (Linder & Hazegh-Azam, 1996).

2.3.19 CERULOPLASMIN

Ceruloplasmin is a ferroxidase enzyme found in the human gene. Ceruloplasmin is the most common copper-carrying protein in the blood, and it also helps with iron metabolism. Ceruloplasmin is a protein with particular domains that facilitate cellular energy production while also inhibiting the development of oxygen radicals. The protein's complicated shape and structure, which includes strategically placed copper ions that may both supply and take electrons from substrates including oxygen, iron, and iron-binding proteins, allows it to fulfill dual functions. Copper is required for ceruloplasmin's wide range of actions, which enhance iron metabolism. Iron metabolism is disrupted by a deficiency or mutation process in the ceruloplasmin gene that prevents copper incorporation in ceruloplasmin (Tumlund et al., 2006).

2.3.20 TRANSFERRIN

Transferrins are glycoproteins reported in vertebrates that bind to iron (Fe) and therefore facilitate its transfer via the blood plasma (Erdman et al., 2012). They are made in the liver and contain two Fe^{3+} ion binding sites (Johnson et al., 1992). Transferrins transports iron from the spleen, liver, and bone marrow to other organs through the blood. It is a vital biochemical indicator of the body's iron level. Transferrin regulates the amount of iron released by reticuloendothelial cells and the amount of iron taken up by the bone marrow. Once iron is bonded to transferrin, it is transferred to the bone marrow to be used in the creation of hemoglobin and erythrocyte parts. Perspiration, epithelial cell desquamation, and menstruation are all ways in which the human body loses iron. Iron loss is unavoidable, and there are no ways to control it. As a result, iron homeostasis is highly reliant on absorption regulation, which occurs mostly in the proximal intestine (Mattill, 1947). The iron-bound transferrin is essential for iron distribution to the body's various cells (Mattie et al., 2008).

2.3.21 METALLOTHIONEINS (MTS)

It's part of a group of metal-binding proteins found in bacteria, plants, invertebrates, and vertebrates (Bremmer, 1998; Colak & Zoric, 2019). The detoxification of heavy metals, the homeostasis of essential metals, and cellular antioxidant defense are only some of its biological roles. The fundamental physiological role of MT is to keep vital trace metal elements like Zinc and Cu in the body in equilibrium, storage, and transport. Heavy metals in the body, for example, can bind to MT as well. Zinc, on the other hand, has a higher binding capacity to MT than other divalent metals, owing to its important role in the body. Therefore, MT is also known as a zinc-binding protein. Because of its numerous thiol groups, MT has been classified as an antioxidant molecule in the body's cellular antioxidant defense mechanism (Keservani & Kesharwani, 2015). Early research demonstrates that MT is a major zinc-binding protein that is involved in various biochemical and cytochemical aspects of zinc-associated homeostasis in humans, including immunology. A number of experimental experiments have demonstrated that MT may serve as an antioxidant molecule against reactive oxygen and nitrogen species (ROS/RNS). Data from a cell-free system model, for example, has shown that MT can operate as a free radical scavenger (Ezerina et al., 2018).

2.3.22 ALBUMIN

Albumin, the most abundant present protein in the plasma, has important anti-oxidant properties. The molecule acts by trapping free radicals and having a number of binding sites. Function related to changes in redox status, albumin structure, and its favorable antioxidant capabilities can be altered in healthy or pathological situations. Albumin is the most common plasma protein that is affected by oxidative stress (He et al., 2018). Albumin binds a wide range of molecules and was sometimes also referred to as a "sponge" or "steamer" of the circulatory system. Albumin's ligand-binding abilities are responsible for many of its antioxidant properties. Albumin's tendency to bind substances such as metal ions, fatty acids, medicines, and hormones is widely documented. The albumin structure's flexibility allows it to easily adapt to ligands, and its three-domain configuration enables a multitude of binding sites (www.healthline.com, 2021). Albumin's ability to bind homocysteine, a sulfur-containing amino acid produced by the degradation of methionine residue, could be another source of antioxidant activity. Plasma homocysteine levels are a well-known risk factor for atherosclerosis, and it may act via oxidizing LDL (Bhardwaj, 2012; Peters, 1996; Roche et al., 2008).

2.3.23 BILIRUBIN (BR)

Bilirubin (BR) was once thought to be nothing more than the "end product" of heme catabolism. Because fetal erythrocytes split and it has a temporary inability to combine with glucuronic acid, it has significant amounts in healthy newborns. It is a powerful antioxidant and anti-inflammatory in nature and more effective than vitamin E in preventing the oxidation process of lipids and other kinds of oxidation (Harris, 2019; Crichton & Charloteaux-Wauters, 1987). It's also been suggested that it's one of the serum's primary defense systems against oxidative damage (Hall et al., 2002; Anderson & Shah, 2013). Furthermore, it has been shown to protect the cardiovascular system, neurological system, hepatobiliary system, pulmonary system, and immunological system (Stocker et al., 1987; Belanger et al., 1997; Gopinathan et al., 1994).

2.3.24 NUTRACEUTICALS

Nutraceuticals, also known as phytochemicals, are typical bioactive, chemical components that have been extracted from their specific nutrition lattice and have beneficial effects to promote health and preventing illnesses. Nutraceuticals are foods or parts of foods that have an important role in maintaining health, increasing health, balancing susceptibility, and, as a result, anticipating, and treating chronic illnesses. It has both therapeutic and physiological effects.

Nutraceuticals include prebiotics, probiotics, dietary fiber, polyunsaturated fats, antioxidants, and other natural or herbal foods (Rane et al., 2020a–c). Home-grown goods, various supplements, and dietary supplements are major components of nutraceuticals, which aid in the prevention of various infections as well as the pathophysiology of the disease (Das et al., 2012).

2.3.24.1 ROLE OF NUTRACEUTICALS IN OUR HEALTH

The idea that plant bio-active chemicals have antioxidant and other health-promoting characteristics has resulted from extensive scientific research that has linked foods of plant roots and well-being. Nutraceuticals with diverse biological activities towards the human body are well known for their ability to deliver health benefits (Das et al., 2012).

2.3.24.2 ANTIOXIDANT ACTIVITY

Nutraceuticals have the ability to scavenge free radicals. Antioxidant-rich nutraceuticals can help to prevent neurodegenerative illnesses including Parkinson's disease and AD. Grapes, parsley, turmeric, spinach, onion, garlic, rosemary, and rose are all rich in nutraceuticals and have strong antioxidant properties.

2.3.24.2.1 Gastrointestinal Health

Probiotics, for instance, are herbal nutraceuticals that play a special role in intestinal health. Oxidative and antigenic stress in the gastrointestinal tract may be decreased with nutraceuticals. Flavonoids and polyphenols have antioxidant properties and can help with gastroprotection and cytoprotection.

2.3.24.3 RENAL AND EXCRETORY HEALTH

They increase urinary bladder health, sphincter tone, and balance, as well as support healthy urinary oxalate excretion and have a protective effect on the kidneys. Retaining calcium, developing calcium crystals and oxalate, and maintaining healthy microbial flora in the urinary tract are all important. Important nutraceuticals include potassium citrate, IP6, lutein, lycopene, magnesium citrate, pycnogenol pine bark, pygeum, and astaxanthin.

Reproductive Health

Nutraceutical food supplements such as Flax seed oil, Vitamin B6, Vitamin B12, ubiquinone Q10, Pycnogenol, and Fish oil are useful for improving both male and female reproductive potential. It reduces male infertility, boosts sperm count, improves sperm motility, and treats sperm dysfunction in men. They also protect sperm from oxidative damage. It minimizes menstrual issues, lowers the risk of preterm labor, protects eggs in the fallopian tubes, and supports embryonic growth and development in females.

2.3.24.4 STEM CELL GROWTH

Blueberries, green tea, carnosine, catechins, vitamin D3, PUFA, and essential amino acids are all nutraceuticals that have an important impact on the proliferation and multiplication of stem cells. They also aid in tissue regeneration and healing by activating and attracting endogenous stem cells at the damage site.

2.3.24.5 PROLONGING LIFE SPAN

Blueberry, green tea, catechins, carnosine, vitamin D3, PUFA, and essential amino acids are all nutraceuticals that have a major impact on the proliferation and multiplication of stem cells. They aid in the extension of a healthy life span. Citrus fruits and soybeans include nutraceuticals that affect epigenetic changes, autophagy, and necrosis (Das et al., 2012).

2.4 FOOD GROUPS HAVING ANTIOXIDANT PROPERTIES

2.4.1 WHOLE GRAINS

Whole grains are high in vitamin E and selenium. Selenium is kept in a reduced state by vitamin E. Selenium is found in cereals and millets as selenomethionine. When compared to inorganic selenium, it has a bioavailability of 85–100% through dietary sources. Grain bran contains a lot of phenolic acids. Whole grains are rich in antioxidants such as tocopherols and selenium, along with polyphenols such as phytic acid, inositols, lignins, and phytoestrogens. They also contain copper, zinc, and manganese, all of which are important antioxidants (Singh et al., 2017a, b; Srilakshmi, 2019).

2.4.2 PULSES

Pulses like Bengal gram and red gram have the highest levels of selenium, while green gram has the lowest. The highest phenolic content is found in black gram dal, followed by rajma and green gram dal. Soybean contains isoflavones and phytoestrogens, and black soybean seed coat contains anthocyanin. As a result, pulses are high in antioxidants.

2.4.3 NUTS AND OIL SEEDS

Nuts contain a lot of phytosterols and other phytochemicals that can help reduce cholesterol levels in the bloodstream. Phytochemicals found in nuts include flavonoids, phenolic compounds, luteolin (an important antioxidant), ellagic acid, and tocopherols. Walnuts have the highest antioxidant activity. Ground nuts and sesame have a high amount of phenolic content.

Fruits and vegetables: Plant sources produce vitamins, especially vitamin C, carotene, and tocopherols. Plants produce phytochemicals that have antioxidant properties (Sharma et al., 2020).

Nutrient	Sources
B-carotene	Green leafy vegetables, ripe yellow fruits, and vegetables like papaya, musk melon, mango, pumpkin, and carrots.
Vitamin C	Oranges, lemons, sweet lime, guava, gooseberry, sprouted pulses.
Vitamin E	Cereals, cereal products, oil seeds, and nuts.
Selenium and Zinc	Meats, seafood, and cereals.
Copper	Oysters, liver, mushrooms, nuts, and chocolate.
Iron	Cereals, millets, pulses, meat, liver, green leafy vegetables.

Phytochemicals	Source
Carotenoids-α-carotene, β-carotene, lutein, lycopene, cryptoxanthin, zeaxanthin.	Mango, papaya, passion fruit, raspberry, spinach, carrots, tomatoes, curry leaves, and lettuce.
Carotenoid-lycopene	Tomatoes, watermelon, red peppers, pink grapes. Tomatoes that have been cooked provide more health benefits. Cooking the tomato breaks down the cell walls, releasing more lycopene. Cooking oils allows the lycopene to move easily into the body.
Carotenoid-lutein	Green leafy vegetables
Anthocyanins	Red grapes, pomegranate
Allium compounds	Onion, garlic
Glucosinolates	Broccoli, cabbage, Brussels sprouts
Polyphenols	Found in all plant foods, especially strawberries, raspberries, blackberries, and walnuts.
Flavonoids	Apples, pears, cherries, grapes, citrus fruits, onions, kale, broccoli, lettuce, and garlic.
Phytoestrogens	Soyabean, yam, cabbage, green beans, peas, potatoes, carrots, barley, rice, wheat, chickpeas, apples, cherries, and pomegranate.
Terpenes	Citrus fruits, carrots, parsley, broccoli, cabbage, and cucumber.

2.4.4 WHEY PROTEIN

Whey proteins act as an antioxidant by boosting tissue GSH levels and preventing tumor growth in a variety of ways. A low molecular weight ultrafiltration permeate derived from milk or whey contains water-soluble antioxidant components.

2.4.5 OILS

Rice bran oils have a high concentration of unsaponifiable components like tocotrienols and oryzanol. These are antioxidants. Palm oil is high in tocopherols and tocotrienols. Olive oil contains a higher amount of vitamin E.

2.4.6 SPICES

Spices also have lots of antioxidant properties as mentioned below:

Spices	Antioxidant
Saffron, Annoto	Carotenoids
Ginger	It can stimulate protective enzymes involved in xenobiotic metabolism.
Turmeric	Curcumin diferuloylmethane is present in turmeric which reduces the mutagenicity in smokers, promotes mechanisms of detoxification and production of GSH, inhibits lipid peroxidation, scavenge superoxide anion and hydroxyl radicals. It also possesses anti-inflammatory properties.
Clove	Eugenol, -allyl-methoxyphenol, the active principle of clove oil has antioxidant properties.
Grapes	Grape seeds contain large amounts of flavonoid substances called polyphenols.
Coriander seeds and mustard	Contains selenium

2.4.7 TEA

According to NIN studies, flavonoids in green and black tea have stronger antioxidant properties than standard antioxidant vitamins. Other powerful polyphenols found in black tea include thearubigins and theaflavins. Green tea is high in polyphenols (particularly catechins and gallic acid) and phytochemical compounds. These compounds have the potential to boost the antioxidant capacity of green tea polyphenol (GTP). Metal ions like iron and copper can be chelated through GTP.

By inhibiting redox-sensitive transcription factors, they can also act as an antioxidant. Pro-oxidant enzymes such as inducible nitric oxide synthase (iNOS), lipoxygenase, cyclooxygenases (COXs), and xanthine oxidase are inhibited.

Antioxidant enzymes such as GSH transferases and SOD are activated. As a result, increasing your flavonoid intake lowers the risk of death from coronary heart disease (CHD).

2.4.8 COFFEE

Caffeine is a hydroxyl radical scavenger that outcompetes oxygen in electron reactions and prevents the formation of ROS by neutralizing or eliminating them. Caffeine is expected to lower the risk of ROS-mediated diseases such as CVD, type 2 diabetes, arthritis, or cancer, such as liver, pancreatic, or colorectal cancer. Robusta coffee has two times as much caffeine as arabica coffee, which has a more pleasant aroma. Filtered coffee contains a lot of antioxidants.

2.4.9 RED WINE

Resveratrol, a naturally occurring chemical present in red wine, can help in slowing down the aging process.

2.4.10 HERBS

Herbs have antioxidant properties as well, thanks to the presence of several phytochemicals. Phytochemicals such as monoterpenes, sesquiterpenes, and flavonoids can be found in basil, dill, fennel, and marjoram. Diterpenoids and ursolic acid are found in rosemary and sage. Coumarins, phthalides, polyacetylenes, and terpenoids are found in anise caraway, celery seeds, and cilantro (Sharma et al., 2020; Rane et al., 2020c; Srilakshmi, 2019).

2.5 CONCLUSIONS

Foods high in antioxidants may help reduce the risk of various conditions (including some cancers and heart disease). Antioxidants may scavenge free radicals from the body's cells, reducing or avoiding oxidative damage. Antioxidants taken from whole meals, rather than extracted from food and provided in pill form, are increasingly being shown to be more effective (Vasak, 2005; Coyle et al., 2002). Inflammation is the body's natural response to repair injuries and ward off infection. Chronic inflammation, on the other hand, is less desirable and can lead to unfavorable health outcomes. Dietary antioxidants have been demonstrated to be useful in delaying or avoiding the onset of most chronic inflammatory disorders. Natural antioxidants are easily absorbed into the human system from the gut. Antioxidant-rich foods should be included in our diet to meet the antioxidant requirement and prevent degenerative disorders (Srilakshmi, 2019).

KEYWORDS

- **antioxidant defense system**
- **antioxidants**
- **free radicals**
- **lipid peroxidation**
- **nutraceuticals**
- **nutrients**
- **vitamin**

REFERENCES

Abel, R. Jr., (2018). Age-related macular degeneration. In: *Integrative Medicine* (pp. 838–846. e1). Elsevier.

Anderson, E. R., & Shah, Y. M., (2013). Iron homeostasis in the liver. *Comprehensive Physiology, 3*(1), 315.

Arct, J., & Pytkowska, K., (2008). Flavonoids as components of biologically active cosmeceuticals. *Clinics in Dermatology, 26*(4), 347–357.

Arrigoni, O., & De Tullio, M. C., (2002). Ascorbic acid: Much more than just an antioxidant. *Biochimicaet Biophysica Acta (BBA)-General Subjects, 1569*(1–3), 1–9.

Atiya, A. M., Strandvik, B., Sabel, K. G., Palme, K. C., Strömberg, R., & Yngve, A., (2014). Polyamine levels in breast milk are associated with mothers' dietary intake and are higher in preterm than full-term human milk and formulas. *Journal of Human Nutrition and Dietetic, 27*(5), 459–467.

Batta, A., (2016). Uric acid is a significant antioxidant. *EJPMR, 3*(8), 242–248.

Bélanger, S., Lavoie, J. C., & Chessex, P., (1997). Influence of bilirubin on the antioxidant capacity of plasma in newborn infants. *Neonatology, 71*(4), 233–238.

Bentinger, M., Brismar, K., & Dallner, G., (2007). The antioxidant role of coenzyme Q. *Mitochondrion, 7*, S41–50.

Beta Carotene Molecule [Internet]. Worldofmolecules.com. Available from: https://www.worldofmolecules.com/antioxidants/bcarotene.htm (accessed on 05 June 2023).

Bhardwaj, S. B., (2012). Alcohol and gastrointestinal tract function. *Bioactive Food as Dietary Interventions for Liver and Gastrointestinal Disease: Bioactive Foods in Chronic Disease States,* 81.

Biol-N'garagba, M. C., & Louisot, P., (2003). Regulation of the intestinal glycoprotein glycosylation during postnatal development: Role of hormonal and nutritional factors. *Biochimie., 85*(3, 4), 331–352. PubMed Abstract | CrossRef Full Text | Google Scholar.

Bremner, I., (1998). Manifestations of copper excess. *The American Journal of Clinical Nutrition, 67*(5), 1069S–1073S.

Cardiovascular Diseases (CVDs) [Internet]. Who.int. Available from: https://www.who.int/news-room/fact-sheets/detail/cardiovascular-diseases-(cvds) (accessed on 05 June 2023).

Chelikani, P., Fita, I., & Loewen, P. C., (2004). Diversity of structures and properties among catalases. *Cellular and Molecular Life Sciences CMLS, 61*(2), 192–208.

Čolak, E., & Žorić, L., (2019). Antioxidants and age-related macular degeneration. In: *Handbook of Nutrition, Diet, and the Eye* (pp. 85–106). Academic Press.

Coyle, P., Philcox, J. C., Carey, L. C., & Rofe, A. M., (2002). Metallothionein: The multipurpose protein. *Cellular and Molecular Life Sciences CMLS, 59*(4), 627–647.

Crichton, R. R., & Charloteaux-Wauters, M., (1987). Iron storage and transport. *Eur. J. Biochem., 164,* 485–506.

Dabelstein, W., Reglitzky, A., Schütze, A., & Reders, K., (2000). Automotive fuels. *Ullmann's Encyclopedia of Industrial Chemistry.*

Das, L., Bhaumik, E., Raychaudhuri, U., & Chakraborty, R., (2012). Role of nutraceuticals in human health. *J. Food Sci. Technol. [Internet], 49*(2), 173–183. Available from: http://drsanjayagrawal.in/role-of-nutraceuticals-on-human-health/ (accessed on 05 June 2023).

Eliassen, K. A., Reistad, R., Risøen, U., & Rønning, H. F., (2002). Dietary polyamines. *Food Chemistry, 78*(3), 273–280. CrossRef Full Text | Google Scholar.

Erdman, Jr. J. W., Macdonald, I. A., & Zeisel, S. H., (2012). *Present Knowledge in Nutrition.* John Wiley & Sons.

Ezeriņa, D., Takano, Y., Hanaoka, K., Urano, Y., & Dick, T. P., (2018). N-acetyl cysteine functions as a fast-acting antioxidant by triggering intracellular H_2S and sulfane sulfur production. *Cell Chemical Biology, 25*(4), 447–459.

Forman, H. J., Zhang, H., & Rinna, A., (2009). Glutathione: Overview of its protective roles, measurement, and biosynthesis. *Molecular Aspects of Medicine, 30*(1, 2), 1, 2.

Fraga, C. G., Galleano, M., Verstraeten, S. V., & Oteiza, P. I., (2010). Basic biochemical mechanisms behind the health benefits of polyphenols. *Molecular Aspects of Medicine, 31*(6), 435–445.

Gad, S. C., (2014). Glutathione. In: Wexler, P., (ed.), *Encyclopedia of Toxicology* (p. 751). San Diego, CA: Elsevier.

Gómez, G. C., RosBerruezo, G., Bernal, C. M. J., Pérez, C. D., & PeriagoCastón, M. J., (2008). Papel de laspoliaminasen la alimentación: Importancia de laspoliaminasen la alimentacióninfantil. *Archivos Latinoamericanos de Nutrición, 58*(2), 117–125.

Gómez-Gallego, C., Kumar, H., García-Mantrana, I., Du Toit, E., Suomela, J. P., Linderborg, K. M., Zhang, Y., et al., (2017). Breast milk polyamines and microbiota interactions: Impact of mode of delivery and geographical location. *Annals of Nutrition and Metabolism, 70*(3), 184–190. PubMed Abstract | CrossRef Full Text | Google Scholar.

Gopinathan, V., Miller, N. J., Milner, A. D., & Rice-Evans, C. A., (1994). Bilirubin and ascorbate antioxidant activity in neonatal plasma. *FEBS Letters, 349*(2), 197–200.

Hacisevki, A., (2009). An overview of ascorbic acid biochemistry. *Journal of Faculty of Pharmacy of Ankara University, 38*(3), 233–255.

Hall, D. R., Hadden, J. M., Leonard, G. A., Bailey, S., Neu, M., Winn, M., & Lindley, P. F., (2002). The crystal and molecular structures of diferric porcine and rabbit serum transferrins at resolutions of 2.15 and 2.60 Å, respectively. *Acta Crystallographica Section D: Biological Crystallography, 58*(1), 70–80.

Halliwell, B., (1995). How to characterize an antioxidant: An update. In: *Biochemical Society Symposium* (Vol. 61, pp. 73–101).

Harris, Z. L., (2019). Ceruloplasmin. In: *Clinical and Translational Perspectives on Wilson Disease* (pp. 77–84). Academic Press.

He, R., Oliveira, J. L., Hoyer, J. D., & Viswanatha, D. S., (2018). Molecular hematopathology. In: *Hematopathology* (p. 712–760.e18). Elsevier.

Huvaere, K., & Skibsted, L. H., (2015). Flavonoids protecting food and beverages against light. *Journal of the Science of Food and Agriculture, 95*(1), 20–35.

Ighodaro, O. M., & Akinloye, O. A., (2018). First line defense antioxidants-superoxide dismutase (SOD), catalase (CAT) and glutathione peroxidase (GPX): Their fundamental role in the entire antioxidant defense grid. *Alexandria Journal of Medicine, 54*(4), 287–293.

Jiménez, C. I., Martínez, E. Y., & Fonseca, J. G., (2009). Flavonoides y sus acciones antioxidantes. *Revista de la Facultad de Medicina UNAM, 52*(2), 73–75.

Johnson, M. A., Fischer, J. G., & Kays, S. E., (1992). Is copper an antioxidant nutrient? *Critical Reviews in Food Science & Nutrition, 32*(1), 1–31.

Kalač, P., (2006). Biologically active polyamines in beef, pork and meat products: a review. *Meat Science, 73*(1), 1–11.

Kalač, P., (2014). Health effects and occurrence of dietary polyamines: A review for the period 2005–mid 2013. *Food Chemistry, 161*, 27–39. PubMed Abstract | CrossRef Full Text | Google Scholar.

Keservani, R. K., & Sharma, A. K., (2014). Flavonoids: Emerging trends and potential health benefits. *Journal of Chinese Pharmaceutical Sciences, 23*(12), 815.

Keservani, R. K., Kesharwani, R. K., & Sharma, A. K., (2015). Pulmonary and respiratory health: Antioxidants and nutraceuticals. In: Debasis, B., Harry, G. P., & Anand, S., (eds.), *Nutraceutical and Functional Foods in Human Life and Disease Prevention* (pp. 279–295). Chapter 17, Boca Raton, Fl. CRC Press, Taylor and Francis. ISBN: 9781482237214.

Keservani, R. K., Kesharwani, R. K., Sharma, A. K., & Ahmad, M. F., (2015). Stress, Parkinson's, Alzheimer's disease: Role of dietary supplements. In: Debasis, B., Harry, G. P., & Anand, S., (eds.), *Nutraceutical and Functional Foods in Human Life and Disease Prevention* (pp. 241–253). Chapter 14, Boca Raton, Fl. CRC, Press, Taylor and Francis. ISBN: 9781482237214.

Keservani, R. K., Kesharwani, R. K., Sharma, A. K., Vyas, N., & Chadoker, A., (2010b). Nutritional supplements: An overview. *International Journal of Current Pharmaceutical Review and Research, 1*(1), 59–75.

Keservani, R. K., Kesharwani, R. K., Vyas, N., Jain, S., Raghuvanshi, R., & Sharma, A. K., (2010a). Nutraceutical and functional food as future food: A review. *Der Pharmacia Lettre, 2*(1), 106–116.

Keservani, R. K., Sharma, A. K., & Kesharwani, R. K., (2016a). Nutraceutical and functional foods for cardiovascular health. In: *Food Process Engineering* (pp. 257–278, 291–312). Apple Academic Press, CRC Press. ISBN: 9781771884020.

Keservani, R. K., Sharma, A. K., & Kesharwani, R. K., (2016b). Medicinal effect of nutraceutical fruits for the cognition and brain health. *Scientifica,* 10. Article ID 3109254. doi: 10.1155/2016/3109254.

Keservani, R. K., Sharma, A. K., & Kesharwani, R. K., (2020). *Nutraceuticals and Dietary Supplements: Applications in Health Improvement and Disease Management*. CRC Press. ISBN: 9781771888738.

Keservani, R. K., Singh, S., Singh, V., Kesharwani, R. K., & Sharma, A. K., (2015). Nutraceuticals and functional foods in the prevention of mental disorder. In: Debasis, B., Harry, G. P., & Anand, S., (eds.), *Nutraceutical and Functional Foods in Human Life and Disease Prevention* (pp. 255–268). Chapter 15, Boca Raton, Fl. CRC Press, Taylor and Francis. ISBN: 9781482237214.

Larque, E., Sabater M. M., & Zamora, S., (2007). *Biological Significance of Dietary* Polyamines, *23*(1), 87–95.

Levine, M., Rumsey, S. C., Daruwala, R., Park, J. B., & Wang, Y., (1999). Criteria and recommendations for vitamin C intake. *JAMA, 281*(15), 1415–1423.

Libretexts, (2016). *8.2: Generation of Free Radicals in the Body [Internet].* Medicine LibreTexts. Libretexts. Available from: https://med.libretexts.org/Bookshelves/Nutrition/Book%3A_An_Introduction_to_Nutrition_(Zimmerman)/08%3A_Nutrients_Important_as_Antioxidants/8.02%3A_Generation_of_Free_Radicals_in_the_Body (accessed on 05 June 2023).

Linder, M. C., & Hazegh-Azam, M., (1996). Copper biochemistry and molecular biology. *The American Journal of Clinical Nutrition, 63*(5), 797S–811S.

Løvaas, E., (1991). Antioxidative effects of polyamines. *Journal of the American Oil Chemists' Society, 68*(6), 353–358. Cross Ref Full Text | Google Scholar.

Lubos, E., Loscalzo, J., & Handy, D. E., (2011). *Glutathione Peroxidase-1 in Health and Disease: From Molecular Mechanisms to Therapeutic Opportunities, 15*(7), 1957–1997. doi: 10.1089/ars.2010.3586.

Marklund, S. L., (1984). Extracellular superoxide dismutase and other superoxide dismutase isoenzymes in tissues from nine mammalian species. *Biochemical Journal, 222*(3), 649–655.

Mattie, M. D., McElwee, M. K., & Freedman, J. H., (2008). Mechanism of copper-activated transcription: Activation of AP-1, and the JNK/SAPK and p38 signal transduction pathways. *Journal of Molecular Biology, 383*(5), 1008–1018.

Mattill, H. A., (1947). *Antioxidants. Annual Review of Biochemistry, 16*(1), 177–192.

Meng, X., Li, Y., Li, S., Zhou, Y., Gan, R. Y., Xu, D. P., & Li, H. B., (2017). Dietary sources and bioactivities of melatonin. *Nutrients, 9*(4), 367.

Miller, R. A., & Britigan, B. E., (1997). Role of oxidants in microbial pathophysiology. *Clin. Microbiol. Rev., 10*(1), 1–8.

Moinard, C., Cynober, L., & De Bandt, J. P., (2005). Polyamines: Metabolism and implications in human diseases. *Clinical Nutrition, 24*(2), 184–197.

Muñoz-Esparza, N. C., Latorre-Moratalla, M. L., Comas-Basté, O., Toro-Funes, N., Veciana-Nogués, M. T., & Vidal-Carou, M. C., (2019). Polyamines in food. *Frontiers in Nutrition, 6*, 108.

Nandi, A., Yan, L. J., Jana, C. K., & Das, N., (2019). Role of catalase in oxidative stress-and age-associated degenerative diseases. *Oxidative Medicine and Cellular Longevity*.

Nordqvist, J., (2020). *Vitamin C: Why We Need it, Sources, and How Much is Too Much [Internet].* Medicalnewstoday.com (accessed on 05 June 2023).

Olsen, N., (2020). *Beta Carotene: Benefits, Foods to Eat, and More [Internet*]. Healthline. Available from: https://www.healthline.com/health/beta-carotene-benefits (accessed on 05 June 2023).

Packer, L., Witt, E. H., & Tritschler, H. J., (1995). Alpha-lipoic acid as a biological antioxidant. *Free Radical Biology and Medicine, 19*(2), 227–250.

Paiva, S. A., & Russell, R. M., (1999). β-carotene and other carotenoids as antioxidants. *Journal of the American College of Nutrition, 18*(5), 426–433.

Panche, A. N., Diwan, A. D., & Chandra, S. R., (2016). Flavonoids: An overview. *Journal of Nutritional Science, 5*.

Pandey, K. B., & Rizvi, S. I., (2009). Plant polyphenols as dietary antioxidants in human health and disease. *Oxidative Med. Cell Longev., 2*, 270–278.

Papatheodorou, L., & Weiss, N., (2007). Vascular oxidant stress and inflammation in hyperhomocysteinemia. *Antioxidants & Redox Signaling, 9*(11), 1941–1958.

Pehlİvan, F. E., (2014). Effects of exogenously applied ascorbic acid on red cabbage cotyledons subjected to copper excess. *Fresenius Environmental Bulletin., 23*(8), 1812–1815.

Pehlivan, F. E., (2017). Vitamin C: An antioxidant agent. *Vitamin C, 2*, 23–35.

Perez-Cano, F. J., González-Castro, A., Castellote, C., Franch, A., & Castell, M., (2010). Influence of breast milk polyamines on suckling rat immune system maturation. *Developmental & Comparative Immunology, 34*(2), 210–218. PubMed Abstract | CrossRef Full Text | Google Scholar.

Peterson, J., & Dwyer, J., (1998). Flavonoids: Dietary occurrence and biochemical activity. *Nutrition Research, 18*(12), 1995–2018.

Rane, B. R., Bharath, M. S., Patil, R. R., Keservani, R. K., & Jain, A. S., (2020b). Novel approaches in nutraceuticals. In: Rajesh, K. K., Raj, K. K., & Anil, K. S., (eds.), *Enhancing the Therapeutic Efficacy of Herbal Formulations Through Novel Drug Delivery Systems* (pp. 241–266). Chapter 11, IGI Global International Publisher, Pennsylvania, USA. ISBN13: 9781799844532.

Rane, B. R., Patil, A. S., Keservani, R. K., & Jain, A. S., (2020a). Novel approaches in herbal formulation. In: Rajesh, K. K., Raj, K. K., & Anil, K. S., (eds.), *Enhancing the Therapeutic Efficacy of Herbal Formulations Through Novel Drug Delivery Systems* (pp. 43–68). Chapter 2, IGI Global International Publisher, Pennsylvania, USA. ISBN13: 9781799844532.

Rane, B. R., Tadavi, S. A., & Keservani, R. K., (2020c). Naturopathy. In: Anil, K. S., Raj, K. K., & Surya, P. G., (eds.), *Herbal Product Development* (pp. 321–347). Chapter 12, Apple Academic Press, CRC Press, Taylor & Francis Group. ISBN: 9781771888776.

Reiter, R., Tan, D. X., Cabrera, J., D'Arpa, D., Sainz, R., Mayo, J., & Ramos, S., (1999). The oxidant/antioxidant network: Role of melatonin. *Neurosignals, 8*(1, 2), 56–63.

Roche, M., Rondeau, P., Singh, N. R., Tarnus, E., & Bourdon, E., (2008). The antioxidant properties of serum albumin. *FEBS Letters, 582*(13), 1783–1787.

Rodríguez, E. E., Reyes, M. A., Rivera, G., Espinosa, L. G., Palos, I., & Bocanegra, V., (2015). *Alimentosfuncionales y Compuestosbioactivos, 63.* Editores P Y V México.

Routray, W., & Orsat, V., (2012). Microwave-assisted extraction of flavonoids: A review. *Food and Bioprocess Technology, 5*(2), 409–424.

Ruiz-Cano, D., Pérez-Llamas, F., & Zamora, S., (2012). Polyamines, implications for infant health. *Archivosargentinos de Pediatria, 110*(3), 244–250.

Rutjes, A. W., Denton, D. A., Di Nisio, M., Chong, L. Y., Abraham, R. P., Al-Assaf, A. S., Anderson, J. L., et al., (2018). Vitamin and mineral supplementation for maintaining cognitive function in cognitively healthy people in mid and late life. *Cochrane Database of Systematic Reviews*, (12).

Salguero, M. L., (2007). Detoxification. In: Rakel, D., (ed). *Integrative Medicine* (pp. 1123–1135). Toronto, ON, Canada: Elsevier.

Sautin, Y. Y., & Johnson, R. J., (2008). Uric acid: The oxidant-antioxidant paradox. *Nucleosides, Nucleotides, and Nucleic Acids, 27*(6, 7), 608–619.

Sharma, A. K., Keservani, R. K., & Gautam, S. P., (2020). *Herbal Product Development.* Apple Academic Press, CRC Press, Taylor & Francis Group. ISBN: 9781771888776.

Shegokar, R., (2016). What nanocrystals can offer to cosmetic and dermal formulations. *Nanobiomaterials in Galenic Formulations and Cosmetics,* 69–91.

Shi, H., Noguchi, N., & Niki, E., (1999). Comparative study on dynamics of antioxidative action of α-tocopheryl hydroquinone, ubiquinol, and α-tocopherol against lipid peroxidation. *Free Radical Biology and Medicine, 27*(3, 4), 334–346.

Sies, H., (1997). Oxidative stress: Oxidants and antioxidants. *Experimental Physiology: Translation and Integration, 82*(2), 291–295.

Singh, P., Kesharwani, R. K., & Keservani, R. K., (2017a). Antioxidants and vitamins: Roles in cellular function and metabolism. In: *Sustained Energy for Enhanced Human Functions and Activity* (pp. 385–407). Academic Press. ISBN:9780128054130.

Singh, P., Kesharwani, R. K., & Keservani, R. K., (2017b). Protein, carbohydrates, and fats: Energy metabolism. In: *Sustained Energy for Enhanced Human Functions and Activity* (pp. 103–115). Academic Press. ISBN:9780128054130.

Soda, K., (2010). Polyamine intake, dietary pattern, and cardiovascular disease. *Medical Hypotheses, 75*(3), 299–301.

Soobrattee, M. A., Neergheen, V. S., Luximon-Ramma, A., Aruoma, O. I., & Bahorun, T., (2005). Phenolics as potential antioxidant therapeutic agents: Mechanism and actions. *Mutation Research/Fundamental and Molecular Mechanisms of Mutagenesis, 579*(1, 2), 200–213.

Srilakshmi, B., (2019). *Nutrition Science* (4th edn., pp. 1–50). ISBN: 978-81-224-3223-7.

Srivastava, N., & Bezwada, R., (2015). *Flavonoids*: *The Health Boosters.* White Paper. Hillsborough NJ: Indofine Chemical Company.

Stahl, W., & Sies, H., (2003). Antioxidant activity of carotenoids. *Mol. Aspects Med. [Internet], 24*(6), 345–351. Available from: http://dx.doi.org/10.1016/s0098-2997(03)00030-x (accessed on 05 June 2023).

Stahl, W., & Sies, H., (2003). Antioxidant activity of carotenoids. *Molecular Aspects of Medicine, 1, 24*(6), 345–351.

Stocker, R., Glazer, A. N., & Ames, B. N., (1987). Antioxidant activity of albumin-bound bilirubin. *Proceedings of the National Academy of Sciences, 84*(16), 5918–5922.

Stocker, R., Yamamoto, Y., McDonagh, A. F., Glazer, A. N., & Ames, B. N., (1987). Bilirubin is an antioxidant of possible physiological importance. *Science, 235*(4792), 1043–1046.

Tabet, F., & Touyz, R. M., (2007). Reactive oxygen species, oxidative stress, and vascular biology in hypertension. In: *Comprehensive Hypertension* (pp. 337–347). Mosby.

Tarocco, A., Caroccia, N., Morciano, G., Wieckowski, M. R., Ancora, G., Garani, G., & Pinton, P., (2019). Melatonin as a master regulator of cell death and inflammation: Molecular mechanisms and clinical implications for newborn care. *Cell Death & Disease, 10*(4), 1, 2.

Tolbert, B. M., Downing, M., Carlson, R. W., Knight, M. K., & Baker, E. M., (1975). Chemistry and metabolism of ascorbic acid and ascorbate sulfate. *Annals of the New York Academy of Sciences, 258*(1), 48–69.

Toro-Funes, N., Bosch-Fusté, J., Veciana-Nogués, M. T., Izquierdo-Pulido, M., & Vidal-Carou, M. C., (2013). In vitro antioxidant activity of dietary polyamines. *Food Research International, 51*(1), 141–147. Cross Ref Full Text | Google Scholar.

Traber, M. G., & Stevens, J. F., (2011). Vitamins C and E: Beneficial effects from a mechanistic perspective. *Free Radical Biology and Medicine, 51*(5), 1000–1013.

Turnlund, J. R., Shils, M. E., Shike, M., Ross, A. C., Caballero, B., & Cousins, R. J., (2006). *Modern Nutrition in Health and Disease.* Philadelphia: Lippincott Williams & Wilkins.

Upadhyay, J., Kesharwani, R. K., & Misra, K., (2009). Comparative study of antioxidants as cancer preventives through inhibition of HIF-1 alpha activity. *Bioinformation, 4*(6), 233.

Vašák, M., (2005). Advances in metallothionein structure and functions. *Journal of Trace Elements in Medicine and Biology, 19*(1), 13–17.

Vertuani, S., Angusti, A., & Manfredini, S., (2004). The antioxidants and pro-antioxidants network: An overview. *Current Pharmaceutical Design, 10*(14), 1677–1694.

Wikipedia contributors, (2022). *Catalase [Internet].* Wikipedia, The Free Encyclopedia. Available from: https://en.wikipedia.org/w/index.php?title=Catalase&oldid=1068284788 (accessed on 05 June 2023).

Wood, J. M., Decker, H., Hartmann, H., Chavan, B., Rokos, H., Spencer, J. D., Hasse, S., et al., (2009). Senile hair graying: H_2O_2-mediated oxidative stress affects human hair color by blunting methionine sulfoxide repair. *The FASEB Journal, 23*(7), 2065–2075.

Yilmaz, Y., (2006). Novel uses of catechins in foods. *Trends in Food Science & Technology, 17*(2), 64–71.

CHAPTER 3

VITAMINS AS NUTRACEUTICALS

SUDDHASATWYA CHATTERJEE[1] and MEGHNA DUTTA[2]

[1]*Sr. Consultant, Department of Internal Medicine and Rheumatology, Apollo Multispecialty Hospitals, Kolkata, West Bengal, India*

[2]*Registrar, Critical Care Unit, Apollo Multispecialty Hospitals, Kolkata, West Bengal, India*

ABSTRACT

Vitamins are important nutraceuticals, and they are hence interesting to read and do research on. They have a number of health-promoting effects and can be obtained naturally or via supplements. In this chapter, the authors have tried to delve into the realm of vitamins and tried to explore their nutraceutical potential.

3.1 INTRODUCTION

"Let food be thy medicine and medicine be thy food" was advised by the great scholar Hippocrates around 2,500 years ago and is certainly still the tenet of today. A nutraceutical can be defined to be any entity that is a food or part of a food and also provides various medical or health benefits, including the prevention and management of many diseases. It is a broad umbrella term that is used to describe any product derived from food sources with added benefits towards well-being in addition to the basic nutritional value found in foods (Keservani et al., 2010a, b, 2014). Nutraceutical products can be considered non-specific biological therapeutic options used to promote

Micronutrients and Macronutrients as Nutraceuticals. Prakash Chandra Gupta, PhD, Sayan Bhattacharyya, MBBS, MD, Nisha Sharma, PhD, MPharm, Rajesh K. Kesharwani, PhD, Raj K. Keservani, PhD, MPharm (Eds.)
© 2024 Apple Academic Press, Inc. Co-published with CRC Press (Taylor & Francis)

the general welfare, control symptoms, and prevent malignant processes. In short, they are more than food but less than pharmaceuticals. Sadly, there is no internationally approved definition of these products therefore their judgment varies from country to country. In most countries, nutraceuticals are taken as part of dietary supplements. The prevalence of nutraceutical use is 50%–70% in developed countries and this count keeps increasing by the age. Also, women use more nutraceuticals than men.

Nutraceuticals include a mosaic of products that range from natural diets, herbal products, bio-fortified crops, and genetically modified and processed food products (Alamgir, 2004). These products typically claim to prevent chronic diseases, enrich health, hinder the aging process, and/or augment life expectancy (Mandala et al., 2010). Numerous studies have concluded that in addition to other lifestyle modifications, appropriate nutrition, nutraceutical supplements, vitamins, antioxidants, minerals, weight loss, exercise, cessation of smoking, and moderate restriction of alcohol and caffeine go a long way in the prevention and control of hypertension and other lifestyle diseases in many patients and model systems (Houston, 2010, 2013; Degu et al., 2021).

Food sources used as nutraceuticals can be broadly classified as (i) dietary fiber; (ii) probiotics; (iii) prebiotics; (iv) polyunsaturated fatty acids (PUFAs); (v) antioxidant vitamins; (vi) polyphenols; and (vii) spices (Kalia, 2005; Kokate, 2002; Keservani et al., 2020). Compared to artificially synthesized products, natural sources of vitamins have greater appeal and bioavailability. Another possible alternative for the production of vitamins is the use of natural and genetically modified microorganisms. Within the heterogeneous group of nutraceuticals, vitamins are predominantly considered dietary supplements, and a few may actually play a crucial role in the prevention and therapy of lifestyle diseases.

Vitamins were originally discovered in 1912 by Casimir Funk who devised the term "Vitamines" for them. They were soon found to prevent and cure ailments like Scurvy, Beriberi, angular stomatitis, Pellagra, Xerophthalmia, rickets, and others. Slowly many chemists isolated many other vitamins and derived the chemical stricture of those vitamins.

Vitamins are low-molecular-weight organic compounds, which are needed by the human body in minute quantities and are obligatory for various important bodily functions, and whose deficiency causes specific disease symptoms, without including other essential nutrients such as dietary minerals, essential fatty acids, or essential amino acids. Vitamins also do not comprise other essential macronutrients and micronutrients that promote health and do not provide cellular structural material and energy. There are 2 major groups of vitamins namely, fat-soluble vitamins (A, D, E, K) and water-soluble vitamins (e.g., Vitamins B complex, Vitamin C).

A vitamin is related to its corresponding deficiency state if these conditions are met, as follows:

- The alleged vitamin is a natural dietary constituent of healthy subjects.
- Individuals suffering from concerned deficiency disease consume a diet that is lacking in vitamins.
- The symptoms experienced by such persons can be relieved by the inclusion of the vitamin in the diet.
- Though not always demonstrable, the observed deficiency symptoms should have a direct relation with the biochemical role of the vitamin.

An isolated vitamin deficiency is usually hard to come across but is seen if the dietary supply is adequate but there is faulty utilization of the vitamin. Scurvy, a classic example of single vitamin deficiency, was common among seamen who did not eat fresh fruits or vegetables during their long sea voyages. This deficiency cropped up because vitamin C was destroyed by the maneuvers used to conserve food while at sea (Bhagavan, 2002).

A major role is played by nutraceuticals when people (particularly those on special diets, strict vegetarians, and vegans) eat foods enriched with vitamins, for example in the form of functional foods. This ensures that an adequate intake of all vitamins is met in the diet. One such example is vitamins extracted from natural sources such as seaweed. Seaweed is a good source of some water-soluble (B_1, B_2, B_{12}, and C) and fat-soluble (beta-carotene with vitamin A activity, vitamin E) vitamins. Besides being beneficial as anti-oxidants and having good biochemical functions, the vitamin C in seaweeds decreases blood pressure, beta-carotene reduces the chances of cardiovascular disease (CVD), and vitamin C, E, and carotenoids reduce the risk of cancer (Škrovánková, 2011). Seaweed is also an abundant source of minerals predominantly iron and iodine. The mineral composition of seaweed is on one hand dynamic based on various intrinsic and extrinsic factors but there may also be intra-species variation (Mišurcová, 2011).

3.1.1 VITAMERS

Vitamins are found in a variety of related forms known as *vitamers*. These are several related compounds exhibiting biological activities against some specific vitamin deficiency states. Thus, vitamers are substances with similar chemical compositions that exhibit a qualitatively parallel vitamin activity.

α-, β-, γ-, and δ-tocopherols and α-, β-, γ-, and δ-tocotrienols are naturally occurring vitamers of vitamin E. All these eight naturally occurring vitamers of vitamin E can be sourced from a wide variety of plant-based foods that are derived in varying amounts from different sources.

Pyridoxine, pyridoxal, and pyridoxamine as well as a 5′-phosphate derivative of each are the 6 naturally occurring vitamers of vitamin B6, all of which are found in various food items. Plant-based foods are very good sources of pyridoxine and its phosphorylated variant, pyridoxine-5′-phosphate. The most stable vitamer of vitamin B6 is pyridoxine. Pyridoxine hydrochloride is used in fortification of foods and dietary supplements.

The vitamers of vitamin B9 are characterized as pteroyl polyglutamate and contain one to six additional glutamate molecules compared to folic acid – hence, sometimes referred to as "food folates." Pteroylmonoglutamic acid, commonly known as folic acid, is one of the vitamers of vitamin B9.

3.2 VITAMINOIDS

Vitaminoids are molecules with vitamin-like activity. L-carnitine and combined carnitine (acetyl-carnitine), inositol, choline, taurine, and ubiquinone (coenzyme-Q) are considered vitaminoids. Vitaminoids have a very important role in nutrigenomics and nutraceuticals.

3.2.1 L-CARNITINE AND ACETYL L-CARNITINE (ALCAR)

Obesity is a burning public health problem all over the world and is considered a disease resulting from chronic low-grade inflammation. As L-carnitine transports fatty acids into mitochondria for oxidation, it is being promoted as a weight-loss supplement (Villani et al., 2000). A double-blinded randomized controlled clinical trial measured the Body mass index, daily dietary intake, and serum levels of free L-carnitine, interleukin-6 (IL-6), high-sensitivity C-reactive protein (Hs-CRP), and Interleukin-10 (IL-10) levels of individuals before and after certain interventions, namely, L-carnitine supplementation, aerobic training + placebo, L-carnitine supplementation + aerobic training and placebo. This study concluded that though L-carnitine supplementation alone did not affect serum IL-6, Hs-CRP, and IL-10 levels in women suffering from obesity, a favorable effect was observed in those who received L-carnitine in combination with aerobic training (Rafarf et al., 2015).

The fatigue seen in chronic fatigue syndrome (CFS) is caused by a disturbance in mitochondrial function. Decreased carnitine levels have been observed in CFS. It's well known that treatment with acetyl L-carnitine (ALCAR) has demonstrated considerable improvement in fatigue in patients with CFS (Plioplys et al., 1997). A pilot, randomized, double-blind, cross-over trial concluded that ALCAR is better tolerated and more efficacious than amantadine for the treatment of MS-related fatigue (Tomassini, 2004).

3.2.2 INOSITOL

Often referred to as vitamin B8, inositol is not a vitamin. More specifically, myoinositol is a carbocyclic sugar that is plentiful in the brain and other tissues of mammals. Inositol supplementation may aid in the treatment of certain medical conditions like anxiety disorders and fertility issues, among others.

Inositol has an effect on the pathway of neurotransmitter generation. Serotonin is among the neurotransmitters affected by inositol. There have been studies to determine the effect of inositol supplementation on the symptoms related to conditions affecting serotonin levels and the brain. These conditions include anxiety disorders like obsessive-compulsive disorder (OCD), panic disorder, and post-traumatic stress disorder (PTSD). Many studies have documented a reduction in the number of panic attacks in those suffering from panic disorders. Another study reported a considerable improvement in symptoms of OCD when this vitaminoid was supplemented.

One of the key pathophysiologies of metabolic syndrome is insulin resistance. Inositol can help manufacture the molecules that are involved in the action of insulin in our cells. Thus, inositol may be the answer to augmenting the body's sensitivity to insulin and hence reducing insulin resistance. This was greatly highlighted by the results of a study that found that inositol supplementation in post-menopausal women with metabolic syndrome improved cholesterol levels, blood pressure, and insulin sensitivity when compared to a placebo. Similar benefits of this vitaminoid on insulin sensitivity and glycemic control have been observed through studies in women with gestational diabetes. Though the results are mixed, there is evidence to state that inositol may improve the action of insulin in women suffering from polycystic ovarian syndrome (PCOS).

Women with PCOS are at increased risk of suffering from infertility and several other diseases. Reduced insulin sensitivity is a major cause of this syndrome, inositol may be a potential treatment by virtue of its ability to increase insulin sensitivity. Various studies have been witnessed on the beneficial effects of inositol in improving the function of ovaries and fertility in women with PCOS.

This vitaminoid has also been extensively explored as a treatment option for certain neuropsychiatric disorders due to its ability to affect certain neurotransmitters in the brain (Keservani et al., 2016b). Some research and small studies have reported that inositol is efficacious in reducing the symptoms of depression. However, this chemical has not displayed any effect on those who have not responded to the standard medications for depression.

Inositol naturally occurs in the human body and a variety of dietary products. Depending on the composition of our diet, the amount obtained from food varies from 1 gram to several grams. This molecule has a very good safety profile for the diverse quantities obtained from food items. Inositol supplementation of up to 4 grams is very well tolerated even in pregnancy with no adverse effects (Tinsley, 2018).

3.2.3 CHOLINE

Choline is a vitaminoid but not a true vitamin because it doesn't function as a coenzyme. It is a vitaminoid as it supplies the ingredients required to build severely important substances with a wide range of functions and interacts with folic acid and several other nutrients. This compound naturally occurs in certain foods and is available as a supplement. A limited amount is also produced in the liver but is certainly insufficient to meet our daily requirements. Two special groups are especially deficient in choline in their diet – pregnant women due to lack of choline in pre-natal supplements in addition to the low levels of choline in an average diet, and in patients on total parenteral nutrition (TPN) as choline is not typically included in the TPN formulas.

After conversion into acetylcholine, a neurotransmitter, this molecule helps in muscle contraction, and activation of response to pain, and plays a vital role in neurological functions such as memory and thinking. Metabolism of choline occurs mainly in the liver where it is converted to phosphatidylcholine, a substance that helps in the generation of those proteins that carry fat and also in the breakdown of cholesterol. This vitamin-like compound is also food for the good bacteria in our gut.

There is data to indicate that choline protects against the risk of CVD. Partnering with Vitamin B folate, choline lowers the blood levels of homocysteine by converting it to methionine. Hyperhomocysteinemia is a risk factor for cardiovascular ailments. Choline can also contribute to reducing blood pressure and stroke.

Studies show a causal link between low blood levels of choline and liver pathology. Fats are carried away from the liver by phosphatidylcholine, and so a deficiency of choline inevitably results in increased storage of fat in liver tissue, thus increasing the risk of Non-Alcoholic Fatty Liver Disease (NAFLD) which may finally progress to cirrhosis, liver cancer or liver failure. However, there is a lack of evidence on whether dietary or non-dietary supplementation of choline is the treatment of NAFLD.

Acetylcholine plays an important role in memory and thinking. Studies have shown that people suffering from Alzheimer's disease are deficient in the enzyme that converts choline to acetylcholine. Theoretically, cognitive decline can be prevented with a higher dietary intake of choline. Observational studies have supported the fact that higher levels of cognition like memory are associated with higher intakes of choline.

3.2.4 TAURINE

Taurine is a β amino acid found in very high amounts in most cells with noticeably high levels in excitable tissues. Because taurine controls basic events in the cell, thereby having a significant say on mortality, there is understandably considerable interest in its physiology. The addition of taurine to infant formula, nutritional supplements, and energy drinks has gained importance in lieu of these findings. Increased dietary consumption of taurine has reduced the risk of hypertension and hypercholesterolemia. Taurine has been classified as a conditionally essential nutrient or functional nutrient in man. Humans cannot synthesize this compound in the body and hence accumulate taurine by retention. This vitaminoid has a relationship with diminished body mass index and reduced inflammatory markers in women with obesity (Schaffer et al., 2018).

3.2.5 UBIQUINONE (COENZYME Q)

Ubiquinone also known as Coenzyme Q-10, is a naturally occurring quinone that occurs in most aerobic organisms ranging from bacteria to mammals.

Coenzyme Q10 is also known by various names like Coenzyme Q, CoQ, CoQ10, Ubiquinone, Ubiquinone-Q10, Ubidecarenone, or Vitamin Q10. The potential use of coenzyme Q10 supplements alone or in combination with other drug regimens and nutritional supplements can help in the prevention or treatment of some of the following conditions: CVDs, high blood pressure, cancer, periodontal diseases, mitochondrial disorders, radiation injury, obesity, diabetes, Parkinson's disease, acquired immune deficiency syndrome (AIDS), gastric ulcers, allergy, migraine headaches, kidney failure, muscular dystrophy, and aging (Saini, 2011).

3.3 VITAMINS E AND C: CARDIOVASCULAR EFFECTS

Despite prospective studies that suggest that decreased levels of vitamin E and C increases the risk of CVD (Stampfer et al., 1993), large-scale clinical trials did not report such significant risk reduction (Bjelakovic et al., 2007). High levels of vitamin E in plasma have actually been associated with a rise in mortality (Miller et al., 2005). Similarly, in research on risk factors of CVD, Vitamin C has been shown to moderately decrease total serum cholesterol levels, raise HDL levels, and exert a hypotensive effect, yet some epidemiological studies have been inconclusive on this aspect (Lynch et al., 1996).

A logical explanation for the limited effects of vitamin E and C may be due to the time and mode of administration, and the presence of additional non-endothelial adverse effects of therapeutic interventions such as probable blunting of ischemic myocardial "preconditioning" (Tritto et al., 2001).

3.3.1 VITAMIN E

Vitamin E is the collected term assigned to a group of fat-soluble compounds that have significant antioxidant properties and are crucial for survival. These vitamins (Singh et al., 2017; Keservani et al., 2020) were first discovered by Evans and Bishop in 1922. The fat-soluble nature of this vitamin accounts for its storage in the adipose tissue of animals and humans and is hence predominantly found in fat-rich foods. The entire group of vitamin E compounds known as tocochromanols, divided into tocopherols and tocotrienols, functionally demonstrate the biological activity of d-alpha-tocopherol. Edible vegetable oils are the richest dietary source of vitamin E, they contain various vitamers in varying proportions. Alpha- and gamma-tocopherols are found in serum and erythrocytes. Minute levels of beta- and delta-tocopherols are present in plasma. The predominance of alpha-tocopherol in humans can be explained by the more rapid metabolism of the other forms and from the alpha-tocopherol transfer protein (alpha-TTP). The binding affinity of α-tocopherol with alpha-TTP leads to its excretion in urine, whereas the other β-, δ-, and γ-tocopherols are secreted into bile and excreted in feces. The alpha-tocopherol form also collects in various non-hepatic tissues.

The functioning of vitamin E greatly relies on vitamin B3, vitamin C, selenium, and glutathione (GSH). Numerous conditions or diseases like senility, cancer, arthritis, and cataracts, are linked to oxidation. While α-tocopherol inhibits the generation of new free radicals, γ-tocopherol traps and neutralizes the existing free radicals. Hence vitamin E may play a role in preventing or delaying the onset or progression of chronic diseases associated with reactive oxygen species (ROS).

By averting the generation of oxidized phospholipids, vitamin E may be playing a major role in membrane repair. Inhibiting clumping of platelets and releasing prostacyclin from the endothelium is achieved by raising the concentration of α-tocopherol in the innermost lining of blood vessels.

Vitamin E increases the organization of membrane lipid packaging, hence allowing for more compact packaging of the membrane. It is well known that membrane phospholipids are prime targets of oxidants. This class of vitamins is also effective in preventing membrane lipid peroxidation (LPO). Raised concentrations of the alpha-vitamer in the endothelial cells inhibit the accumulation of platelets and release prostacyclin from endothelium.

3.4 VITAMIN D

Low levels of vitamin D in plasma are associated with a higher incidence of cardiovascular events, this is more noticeable in hypertensives (Wang et al., 2008). Vitamin D deficiency has a greater association with endothelial dysfunction and LPO (Sugden et al., 2008). Vitamin D deficiency has been associated with the risk of many diseases, but the exact molecular mechanism is unknown. A study on the free-living middle-aged Norwegian women population was directed at identifying various pathways of gene expression in the blood and its association with vitamin D status. A targeted pathway-level analysis revealed that vitamin D status is often associated with the immune system and immune cell function, also major signaling cascades. In this way, vitamin D has an indirect effect on cancer post-genome groups (Standahl et al., 2013).

After various studies, a positive statistical significance has been discovered between vitamin D and sarcopenia, the possibility of occurrence of colorectal polyps, modulating immune function, and detection of oral cancer. The effect of the interaction between vitamin D and physical activity on the structural and functional properties of muscles was studied through animal experiments and population surveys. Hypovitaminosis D, with limited physical activity, was found to quicken the decrease in muscle weight, and grip strength, and increased the protein expression of all three protein expressions evaluated.

People living in higher altitudes have a greater predisposition to vitamin D deficiency. Vitamin D supplementation appeared to have a protective effect against high-risk adenomatous polyps (HARPS) in individuals aged between 50 and 74 years. People living in high-latitude areas have shown statistically significant findings on the protective effect of vitamin D against the development of colorectal polyps.

Five meta-analyzes of certain high-quality studies gave the idea that a wide range of doses of vitamin D could decrease the risk of respiratory tract infections in all age groups, and this risk reduction was most noticeable in people who were vitamin D deficient at the commencement of their respective trials. Vitamin D has demonstrated improvement in immune function in a variety of diseases such as tuberculosis, influenza, and upper respiratory tract infections. Besides, it is quite rational to bridge a gap between nutritional insufficiency and immune function, thereby, enhancing the resistance to infection.

An increased risk of oral squamous cell carcinoma has been noted in subjects with vitamin D deficiency, especially when serum levels are less than 25 ng/ml. This hints towards the role of the micronutrient in preventing this form of cancer (Nutraceuticals World, 2020).

The COVID-19 pandemic has witnessed a surge of human clinical trials trying to evaluate the potential role of vitamin D concentrations on the incidence of infection and overall outcome in the affected patients. Over 100 clinical trials have been published since the onset of the ongoing coronavirus pandemic, the majority of which have evaluated the vitamin D levels in the blood, while a couple of studies exclusively focused on the consumption of vitamin D in relation to the occurrence of this disease. In this context, the Council for Responsible Nutrition (CRN) has presented 13 meta-analyzes, through its "Vitamin D and Me!" consumer awareness campaign. Though not all, in a majority of these research reviews, vitamin D assays appear to have a direct correlation with decreased incidence and severity of the SARS-CoV-2 infection (Kazemi et al., 2021). These studies nowhere indicate that vitamin D is an alternative to vaccines, masks, social distancing, or other measures that mitigate the spread of the coronavirus. Most of the researchers concluded that a positive correlation exists between vitamin D and lowered incidence of COVID-19, and many meta-analyzes found evidence of increased severity of the disease and mortality with low vitamin D levels (Petrelli et al., 2021). One meta-analysis found a reduction in mortality following vitamin D supplementation after diagnosis of the viral infection. Hence there is sufficient data to suggest that vitamin D levels may have a say, in combination with other therapies, in strengthening the immune system against the virus.

3.5 HOMOCYSTEINE

Homocysteine is a non-proteinogenic, sulfur-containing amino acid whose metabolism involves several vitamins, including vitamin B12, folate, and B2. Deficiency of one or more of these vitamins can result in hyperhomocysteinemia which apparently plays a role in CVD. Increased risk of ischemic heart disease and stroke has been reported in individuals with moderately elevated plasma homocysteine levels (The Homocysteine Studies Collaboration, 2002). Elevated levels of plasma homocysteine and vitamin B6 deficiency are independently associated with the possibility of the occurrence of coronary and cerebrovascular events (Vanuzzo et al., 2007).

Also, increased homocysteine levels are associated with reduced physical performance and muscle strength in older women. Studies are required to view whether reducing homocysteine levels can actually delay physical decline (Swart, 2013).

3.6 VITAMIN B3

Niacin improves endothelium-dependent dilatation of blood vessels. It also decreases plasma cholesterol levels and increases plasma high-density lipoprotein (HDL) levels (Altschul et al., 1955; Sorrentino et al., 2010).

3.7 VITAMIN A

Underneath the husk of a rice grain lies the outer coat, known as the aleurone layer. This, along with the underlying sub-aleurone layer, contains the majority of the macro- and micro-nutrients such as iron, zinc, and vitamin A. Whole grain consumption has been positively correlated with good health due to the presence of large amounts of fiber, minerals, vitamins, plant phenols, phytoestrogens, and various other bioactive phytochemicals (Liji, 2019). Vitamin A, a fat-soluble vitamin, is closely linked with the occurrence of xerophthalmia, night blindness, corneal ulceration, and eventually complete blindness.

3.8 VITAMIN DEFICIENCY

Some wrong food processing methods or imbalances in diet can lead to vitamin deficiency. Lack of a balanced diet may be due to psychological issues, environmental disturbances, or economic insufficiency.

Polishing of rice refers to the milling of rice to remove the husk, bran, germ, and the various nutrients present in the grain, leaving behind only the starch. Polished rice lacks moisture content, mineral, biotin, niacin, protein, and fat content compared to brown or lightly milled rice. Unpolished rice or brown rice is a form of rice in which the husk has been removed but the bran is left partially or completely intact. Whole grain rice is rich in γ-oryzanol and tocotrienols (vitamin-E precursors). Rice contains some amount of β-carotene (less than 0.2 ppm) which is restricted to the bran and hence lost during milling. High chances of vitamin A deficiency have hence been noticed in individuals who exclusively consume polished rice (Liji, 2019).

Heart failure and other CVDs have been found to be associated with low vitamin levels. Levels of water-soluble vitamins A, D, E, and K are closed related to infectious diseases.

3.9 ANTIOXIDANT VITAMINS

Vitamin C, Vitamin E, and carotenoids are collectively known as anti-oxidant vitamins. These vitamins act individually as well as synergistically in unison to prevent oxidative reactions in the body. Oxidation at the biochemical level can lead to several degenerative diseases including cancer, diseases of the heart and circulation, cataracts, etc. These vitamins exert their anti-oxidant effects via free-radical scavenging mechanisms.

Tocopherols and tocotrienols present in vitamin E act by transferring hydrogen atoms and thus they can scavenge singlet oxygen and other reactive species. This in turn prevents the peroxidation of poly-unsaturated fatty acids with the biological membrane and low-density lipoproteins (Meydani, 2000). Tocotrienols are more dynamic compared to tocopherols with regard to mobility within the biological membrane due to the presence of unsaturated side chains and therefore invade tissues with saturated fatty layers, i.e., the brain and liver. The former also have more recycling capacity and are more efficient inhibitors of liver oxidation (Watkins et al., 1999). Vitamin E and selenium act synergistically against LPO.

Ascorbic acid donates hydrogen ions to lipid radicals, satiates singlet oxygen radicals, and removes molecular oxygen. It has been established that ascorbic acid along with tocopherol synergistically scavenges aqueous radicals.

Carotenoids such as lycopene, β-carotene, lutein, and zeaxanthin are the most efficient singlet oxygen quencher in biological systems without the production of any oxidizing products. At lower oxygen concentrations, β-carotene traps peroxy free radicals in tissues. Thus, the anti-oxidant properties of vitamin E are complemented by β-carotene (Das et al., 2012).

3.10 ROLE OF NUTRACEUTICALS IN HUMAN HEALTH

Food products used as nutraceuticals can be divided into categories like dietary fiber, prebiotics, probiotics, PUFAs, antioxidants, and other different types of herbal and natural foods. These products help in fighting against some of the major health ailments of the century such as obesity (Jarouliya & Keservani, 2019), CVDs (Keservani et al., 2016a), cancer, osteoporosis, joint-related disorders, diabetes mellitus, and dyslipidemia (Rane et al., 2020a–c, Das et al., 2012; Rane & Keservani, 2020).

Nutraceuticals are the emerging category of natural products that has faded the fine line between food and drugs (Adelaja et al., 1999). Though nutraceuticals have been used by mankind for ages, only recently scientifically supported and medical evidence has allowed nutraceuticals to emerge as potentially effective. The countries with the largest nutraceutical markets include USA, UK, and Japan. They have led to a new era of medicine and health where the food industry has become a field of research.

3.11 CONCLUSION

In the past 5 years, the world has witnessed explosive growth of the multi-billion-dollar food and food product industry. Research and development are at the peak in this emerging field of vitamins as nutraceuticals. Functional foods and nutraceuticals are hence ostensibly efficacious in the prevention and treatment of a number of diseases. They can lead to the overall promotion of health due to the presence of measured concentrations of certain physiologically active compounds.

KEYWORDS

- **choline**
- **food**
- **homocysteine**
- **nutraceuticals**
- **ubiquinone**
- **vitaminoids**
- **vitamins**

REFERENCES

Adelaja, A. O., & Schilling, B. J., (1999). Nutraceutical: Blurring the line between food and drugs in the 21st century. *Mag. Food Farm. Resour. Issues, 14,* 35–40.

Alamgir, A. N. M., (2004). Vitamins, nutraceuticals, food additives, enzymes, anesthetic aids, and cosmetics. In: *Therapeutic Use of Medicinal Plants and Their Extracts: 2; Progress in Drug Research* (p. 74). Springer, Cham. https://doi.org/10.1007/978-3-319-92387-1_5.

Altschul, R., Hoffer, A., & Stephen, J. D., (1955). Influence of nicotinic acid on serum cholesterol in man. *Arch Biochem., 54,* 558, 559.

Bhagavan, N. V., (2002). "r-38 vitamin metabolism. *Medical Biochemistry* (4th edn., pp. 901–928). Academic Press. https://www.sciencedirect.com/science/article/pii/B9780120954407500408 (accessed on 05 June 2023).

Bjelakovic, G., Nikolova, D., Gluud, L. L., et al., (2007). Mortality in randomized trials of antioxidant supplements for primary and secondary prevention: Systematic review and meta-analysis. *JAMA, 297,* 842–857.

Das, L., Bhaumik, E., et al., (2012). Role of nutraceuticals in human health. *J. Food Sci. Technol., 49*(2), 173–183. Published online.

Degu, A., Abebe, A., & Engidawork, E., (2021). Methanol (80%) leaf extract of *Otostegia integrifolia* Benth (*Lamiaceae*) lowers blood pressure in rats through interference with calcium conductance. *BMC Complement Med Ther., 21*(1), 49. doi: 10.1186/s12906-021-03222-4. PMID: 33541332; PMCID: PMC7863373.

Houston, M. C., (2010). The role of cellular micronutrient analysis, nutraceuticals, vitamins, antioxidants and minerals in the prevention and treatment of hypertension and cardiovascular disease. *Ther. Adv. Cardiovasc Dis., 4*(3), 165–183. doi: 10.1177/1753944710368205. Epub 2010 Apr 16. PMID: 20400494.

Houston, M. C., (2013). The role of nutrition, nutraceuticals, vitamins, antioxidants, and minerals in the prevention and treatment of hypertension. *Altern., Ther. Health Med., 19*(1), 32–49. PMID: 23981465.

Jarouliya, U., & Keservani, R. K., (2019). Pathways leading to child obesity. In: *Global Perspectives on Childhood Obesity: Current Status, Consequences and Prevention* (2nd edn., pp. 137–146). Chapter 12, Academic Press, Elsevier. ISBN: 9780128128404.

Kalia, A. N., (2005). *Textbook of Industrial Pharmacognosy* (pp. 204–208). CBS publisher and distributor, New Delhi.

Kazemi, A., et al., (2021). Association of vitamin D status with SARS-CoV-2 infection or COVID-19 severity: A systematic review and meta-analysis. *Advances in Nutrition.*

Keservani, R. K., & Sharma, A. K., (2014). Flavonoids: Emerging trends and potential health benefits. *Journal of Chinese Pharmaceutical Sciences, 23*(12), 815.

Keservani, R. K., Kesharwani, R. K., Sharma, A. K., Vyas, N., & Chadoker, A., (2010b). Nutritional supplements: An overview. *International Journal of Current Pharmaceutical Review and Research, 1*(1), 59–75.

Keservani, R. K., Kesharwani, R. K., Vyas, N., Jain, S., Raghuvanshi, R., & Sharma, A. K., (2010a). Nutraceutical and functional food as future food: A review. *Der Pharmacia Lettre, 2*(1), 106–116.

Keservani, R. K., Sharma, A. K., & Kesharwani, R. K., (2016a). Nutraceutical and functional foods for cardiovascular health. In: *Food Process Engineering* (pp. 257–278, 291–312). Apple Academic Press. ISBN: 9781771884020.

Keservani, R. K., Sharma, A. K., & Kesharwani, R. K., (2016b). Medicinal effect of nutraceutical fruits for the cognition and brain health. *Scientifica, 2016*, 10. Article ID 3109254. doi: 10.1155/2016/3109254.

Keservani, R. K., Sharma, A. K., & Kesharwani, R. K., (2020). *Nutraceuticals and Dietary Supplements: Applications in Health Improvement and Disease Management*. CRC Press. ISBN: 9781771888738.

Kokate, C. K., Purohit, A. P., & Gokhale, S. B., (2002). Nutraceutical and cosmaceutical. *Pharmacognosy* (21st edn., pp. 542–549). Pune, India. Nirali Prakashan.

Liji, T. M., (2019). *Should We Eat Polished Rice?* News-Medical.net. Available from: https://www.news-medical.net/health/Should-We-Eat-Polished-Rice.aspx.Last (accessed on 05 June 2023).

Lynch, S. M., Gaziano, J. M., & Frei, B., (1996). Ascorbic acid and atherosclerotic cardiovascular disease. *Subcell Biochem., 25*, 331–367.

Mandala, F., Guida, R., & Guida, G. F., (2010). Nutraceuticals: What they are and how they work. *E-Journal of Cardiology Practice, 9*(12).

Meydani, M., (2000). Effect of functional food ingredients: Vitamin E modulation of cardiovascular diseases and immune status in the elderly. *Am. J. Clin. Nutr., 71*, 1665S–1668S.

Miller, E. R., et al., (2005). Meta-analysis: High dosage vitamin E supplementation may increase all-cause mortality. *Ann. Intern. Med., 142*, 37–46.

Mišurcová, L., Machů, L., & Orsavová, J., (2011). Seaweed minerals as nutraceuticals. *Adv Food Nutr Res., 64*, 371–390. doi: 10.1016/B978-0-12-387669-0.00029-6. PMID: 22054962.

Nutraceuticals World, (2020). *The Latest in Vitamin D Research.* (online). Available at: https://www.nutraceuticalsworld.com/contents/view breaking news/ 2020-04-23/the-latest-in-vitamin-d-research/ (accessed on 05 June 2023).

Petrelli, F., et al., (2021). Therapeutic and prognostic role of vitamin D for COVID-19 infection: A systematic review and meta-analysis of 43 observational studies. *J. Steroid Biochem. Mol. Biol.*

Plioplys, A. V., & Plioplys, S., (1997). Amantadine and L-carnitine treatment of chronic fatigue syndrome. *Neuropsychobiology, 35*(1), 16–23. doi: 10.1159/000119325. PMID: 9018019.

Rafarf, M., Karimi, M., & Jafari, A., (2015). Effect of L-carnitine supplementation in comparison with moderate aerobic training on serum inflammatory parameters in healthy obese women. *J. Sports Med. Phys. Fitness, 55*(11), 1363–1370. Epub 2015 Jan 20. PMID: 25600905.

Rane, B. R., & Keservani, R., (2020). Introduction to nutraceuticals and dietary supplements. In: *Nutraceuticals and Dietary Supplements* (pp. 35–56). Apple Academic Press. ISBN: 9781771888738.

Rane, B. R., Bharath, M. S., Patil, R. R., Keservani, R. K., & Jain, A. S., (2020b). Novel approaches in nutraceuticals. In: Rajesh, K. K., Raj, K. K., & Anil, K. S., (eds.), *Enhancing the Therapeutic Efficacy of Herbal Formulations Through Novel Drug Delivery Systems* (pp. 241–266). Chapter 11, IGI Global International Publisher, Pennsylvania, USA. ISBN13: 9781799844532.

Rane, B. R., Patil, A. S., Keservani, R. K., & Jain, A. S., (2020a). Novel approaches in herbal formulation. In: Rajesh, K. K., Raj, K. K., & Anil, K. S., (eds.), *Enhancing the Therapeutic Efficacy of Herbal Formulations Through Novel Drug Delivery Systems* (pp. 43–68). Chapter 2, IGI Global International Publisher, Pennsylvania, USA. ISBN13: 9781799844532.

Rane, B. R., Tadavi, S. A., & Keservani, R. K., (2020c). Naturopathy. In: Anil, K. S., Raj, K. K., & Surya, P. G., (eds.), *Herbal Product Development* (pp. 321–347). Chapter 12, Apple Academic Press, Taylor & Francis Group. ISBN: 9781771888776.

Saini, R., (2011). Coenzyme Q10: The essential nutrient. *J Pharm Bioallied Sci., 3*(3), 466, 467. doi: 10.4103/0975-7406.84471. PMID: 21966175; PMCID: PMC3178961.

Schaffer, S., & Kim, H. W., (2018). Effects and mechanisms of taurine as a therapeutic agent. *Biomol. Ther. (Seoul), 26*(3), 225–241. doi: 10.4062/biomolther.2017.251. PMID: 29631391; PMCID: PMC5933890.

Singh, P., Kesharwani, R. K., & Keservani, R. K., (2017). Antioxidants and vitamins: Roles in cellular function and metabolism. In: *Sustained Energy for Enhanced Human Functions and Activity* (pp. 385–407). Academic Press. ISBN:9780128054130.

Škrovánková, S., (2011). Seaweed vitamins as nutraceuticals. *Adv. Food Nutr. Res., 64*, 357–369. doi: 10.1016/B978-0-12-387669-0.00028-4. PMID: 22054961.

Sorrentino, S. A., Besler, C., Rohrer, L., et al., (2010). Endothelial-Vaso protective effects of high-density lipoprotein are impaired in patients with type 2 diabetes mellitus but are improved after extended-release niacin therapy. *Circulation, 121*, 110–122.

Stampfer, M. J., Hennekens, C. H., Manson, J. E., et al., (1993). Vitamin E consumption and the risk of coronary artery disease in women. *N. Engl. J. Med., 328*, 1444–1449.

Standahl, O. K., Rylander, C., Brustad, M., Aksnes, L., & Lund, E., (2013). Plasma 25 hydroxyvitamin D level and blood gene expression profiles: A cross-sectional study of the Norwegian women and cancer post-genome cohort. *Eur. J. Clin. Nutr., 67*(7), 773–778.

Sugden, J. A., Davies, J. I., Witham, M. D., et al., (2008). Vitamin D improves endothelial function in patients with type 2 diabetes mellitus and low vitamin D levels. *Diabet. Med., 25*, 320–325.

Swart, K. M., Enneman, A. W., Van, W. J. P., Van, D. S. C., et al., (2013). Homocysteine and the methylenetetrahydrofolate reductase677C à T polymorphism in relation to muscle mass and strength, physical performance and postural sway. *Eur. J. Clin. Nutr., 67*(7), 743–748.

The Homocysteine Studies Collaboration, (2002). Homocysteine and risk of ischemic heart disease and stroke: A meta-analysis. *JAMA, 288*, 2015–2022.

Tinsley, G., (2018). *5 Evidence-Based Health Benefits of Inositol.* [online] Healthline. Available at: https://www.healthline.com/nutrition/inositol-benefits#TOC_TITLE_HDR_2 (accessed on 05 June 2023).

Tomassini, V., Pozzilli, C., Onesti, E., et al., (2004). Comparison of the effects of acetyl L-carnitine and amantadine for the treatment of fatigue in multiple sclerosis: Results of a pilot, randomized, double-blind, crossover trial. *J. Neurol. Sci.*

Tritto, I., & Ambrosio, G., (2001). Role of oxidants in the signaling pathway of preconditioning. *Antioxid. Redox Signal, 3*, 3–10.

Vanuzzo, D., Pilotto, L., Lombardi, R., et al., (2007). Both vitamin B6 and total homocysteine plasma levels predict long-term atherothrombotic events in healthy subjects. *Eur. Heart J., 28*(4), 484–491.

Villani, R. G., Gannon, J., Self, M., & Rich, P. A., (2000). L-carnitne supplementation combined with aerobic training does promote weight loss in moderately obese women. *Int. J. Sport Nutr. Exerc. Metab., 10*(2), 199–207. doi: 10.1123/ijsnem.10.2.199. PMID: 10861338.

Wang, T. J., Pencina, M. J., Booth, S. L., et al., (2008). Vitamin D deficiency and risk of cardiovascular disease. *Circulation, 117*, 503–511.

Watkins, T. R., Bierenbaurn, M. I., & Giampalala, A., (1999). Tocotrienols: Biological and health effects. In: Papas, A. M., (ed.), *Antioxidant Status, Diet, Nutrition and Health* (pp. 479–496). Boca Raton: CRC Press.

CHAPTER 4

MICROBES AND MICROBIAL PRODUCTS AS NUTRACEUTICALS

SAYAN BHATTACHARYYA

Associate Professor, Microbiology, All India Institute of Hygiene and Public Health (AIIH&PH), Kolkata, West Bengal, India

ABSTRACT

Nutraceuticals are nutrient molecules having some pharmacological properties and applications. Many microbes and microbial products or metabolites can be used as nutraceuticals. They impart many health benefits to the host and kill pathogenic bacteria in the gut. This is a very new and interesting area of public health and Nutrition research. In this chapter detailed information is given about this.

4.1 INTRODUCTION

Nutraceuticals are those compounds that possess drug-like activity. They are generally derived from foods or edible objects. They can also be obtained from microorganisms and their metabolic products. Of late, there is a lot of interest that has developed in these natural molecules. This is for their enormous potential in the prevention and treatment of infectious diseases.

Globally, there has been an ever-increasing demand for these nutraceutical compounds for both the prevention and treatment of many human ailments. This has led to the emergence of nutraceuticals as a multi-billion-dollar industry. However, limitations in supply and difficulties in extracting them

Micronutrients and Macronutrients as Nutraceuticals. Prakash Chandra Gupta, PhD, Sayan Bhattacharyya, MBBS, MD, Nisha Sharma, PhD, MPharm, Rajesh K. Kesharwani, PhD, Raj K. Keservani, PhD, MPharm (Eds.)
© 2024 Apple Academic Press, Inc. Co-published with CRC Press (Taylor & Francis)

from natural resources like plants, animals, or fungi, restrict their large-scale use (Keservani et al., 2010a, b). Metabolic engineering by production from microbes has hence come up as an eco-friendly alternative for producing these nutraceutical moieties from simple sources of Carbon (Wang et al., 2016). Microbial platforms have been used in the form of the most widely used bacteria like *Escherichia coli* and *Saccharomyces cerevisiae*. They have been engineered as good factories in the cell for the manufacturing of many diverse and complex value-added chemicals like plant-derived chemicals, prebiotics, polysaccharides, and poly-amino acids.

Microorganisms like Escherichia coli, Corynebacterium glutamicum, Yarrowia lipolytica, and *Saccharomyces cerevisiae* have been modified and engineered to produce nutraceuticals like polyunsaturated fatty acids (PUFAs). Other nutraceuticals are also produced, like carotenoids, polyphenols, alkaloids, non-proteinogenic forming amino acids, and poly-amino acids. Thus, they are very helpful as a source of Nutraceuticals (Dhar et al., 2020).

Recently, many novel food ingredients called nutraceuticals have been introduced in the food and pharmaceutical market. These include the components that have proven beneficial effects on human health, like low-calorie sugars and B-group vitamins (Singh et al., 2017). Lactic acid bacteria (LAB), especially *Lactococcus lactis*, have also been documented to be ideal sources for manufacturing these vital nutraceuticals (Hugenholtz et al., 2002).

The seawater or marine environment can be an excellent source of nutraceuticals (Keservani et al., 2020). Approximately 70% of Earth is formed of water. Naturally, marine microorganisms and other microbes in the sea form 50% of Earth's microbial diversity (Sharma et al., 2020). However, they are being explored only now (Gupta et al., 2019). Marine microorganisms, coral, and algae are important sources of bioactive compounds like antioxidants, Vitamins, enzymes, and antibiotics that can be used to treat many infectious and non-infectious diseases and conditions. Bioceuticals or Bioactive metabolites from marine microorganisms can be used as active pharmacological ingredients or APIs in many medicinal products and nutraceuticals.

Microalgae contain Vitamins A, C, H, B1, B2, and B6 which have many significant biological functions. They can be utilized as food additives and for food fortification also. *Stenotrophomonas* spp., bacteria isolated from marine sponges and sea urchins, are excellent sources of nutraceutical molecules that can inhibit many Gram-positive bacteria. However, this bacterium can also have some nutraceutical-like properties in itself, and more research in this context is awaited. *Pseudoalteromonas phenolica*, a bacterium found in the seawater has been found to inhibit MRSA or Methicillin-resistant *Staphylococcus aureus* by dint of its brominated biphenyl compounds.

Fungi like *Ampelomyces* spp. obtained from the sea has been shown to exhibit antimicrobial properties (Gupta et al., 2019). *Pseudallescheria* spp., which is a marine fungus obtained from seawater also has many antibacterial properties. Dioxopiperazine compounds obtained from this fungus can inhibit MRSA. Aspyrone derivatives obtained from marine *Exophiala* spp. can inhibit the growth of *Staphylococcus aureus*. ***Lachnum papyraceum*** from the sea has been shown to have antinemat helminthic effects.

4.2 USE OF *LACTOBACILLI* AND *BIFIDOBACTERIUM* AS NUTRACEUTICALS

Many beneficial microorganisms like molds, yeast, and lactobacilli are commonly used to convert raw food matters into a number of fermented products. Bioprocessing technology has progressed a lot in this manner. Now one can see the exclusive production of food or feed ingredients or such types of related processing techniques. Bacteria belonging to the genus *Lactobacillus* are microorganisms of special and renewed interest in this regard. This is because of their long history of use and many beneficial effects. Lactobacilli were among the first microbes employed by humans for processing foodstuffs and also for the preservation of food. They do so by inhibiting the invasion by other microorganisms which cause foodborne illness or food spoilage. The genus *Lactobacillus* is thus really considered essential now for present-day food and feed technologies. This is because of a renewed interest in their beneficial effects or functional aspects (Bernardeau et al., 2006).

The first such historical evidence of the fermentation of milk and the use of lactobacilli for this very purpose can be traced way back to the period of Ptolemy in Ancient Egypt. There it was mentioned on stelae and in hieroglyphics and many other sculptures. The ancient Egyptians used to keep goats, cattle, and sheep in their homes. The milk was preserved in egg-shaped jars made of earth, which were plugged with grass in order to protect it from insects. It was consumed shortly after milking. Owing to the hot weather, in which the milk was likely to get spoiled in a few hours, it can be guessed that milk not meant for immediate use was processed into something of similar nature, like quark or yogurt-like labaneh. In spite of its long history of use, however, the full beneficial effects of fermented milk have been documented and studied only very recently. The renowned Russian-born biologist, Elie Metchnikoff, is renowned for his seminal studies on these lactic acid-producing bacteria found in milk and milk products. He postulated that the long lifespan of many Bulgarians and that of village-dwellers in the Caucasus Mountains, in present-day Azerbaijan, can be attributed to their regular consumption of fermented milk and milk products, like yogurt and kefir (Bernardeau et al., 2006).

Lactobacillus acidophilus has been shown to have antioxidant effects. Both the intact bacterial cells and metabolites produced by it are strong antioxidants, which can further its use as Nutraceutical (Patil et al., 2018). Lactobacilli produce a number of bioactive compounds that can be used as nutraceuticals, like vitamins, gamma-aminobutyric acid (GABA), enzymes, bacteriocins, bioactive peptides, conjugated linoleic acid (CLA), and many exopolysaccharides (Linares et al., 2017). Some bacterial strains which belong to the genera *Lactobacillus* and *Bifidobacterium* can also function as an extra source of B complex vitamins, like Thiamine, riboflavin, cobalamin, folate, and biotin during dairy-product fermentation. Deficiencies in riboflavin (vitamin B_2) or thiamine (vitamin B_1) can cause both liver and skin problems. They can also cause various changes in brain glucose metabolism. In this regard, *L. casei* strain KNE-1 has been documented to produce thiamine and riboflavin in several fermented milk-based drinks. *B. infantis* CCRC14633 and *B. longum* B6 strains have also been shown to produce vitamins like riboflavin and thiamine during soymilk fermentation. Lactobacilli can also have and produce many nutraceuticals that may help prevent the recurrence of peptic ulcers. They aid in killing *Helicobacter pylori* and also inhibit the relapse events of Inflammatory bowel diseases.

For weight control – obesity (Jarouliya & Keservani, 2019) is a highly significant risk factor for acquiring many diseases like type II DM, hypertension, coronary heart disease (CHD), and certain categories of cancer. Many microorganisms like probiotic bacteria can be used as a natural cure for weight control, and help by preventing the development of obesity, and improving energy metabolism (Patil et al., 2018). Some specific strains of the LAB have the capacity to lower serum cholesterol by the breakdown of bile in the gut, thereby inhibiting its re-absorption. *Lactobacillus* (*L. sporogenes* and *L. acidophilus* NCFB 1748) and *Bifidobacterium* genus members thus play a very crucial role in weight management by exerting an anti-obesity effect in experimental models and humans also (Patil et al., 2018; Linares et al., 2017). *Lactobacillus sporogenes* have the unique ability to lower cholesterol levels. They produce a significant lowering of serum low-density lipoprotein (LDL) levels and a small but significant elevation of high-density lipoprotein (HDL) cholesterol (Gupta et al., 2015). Probiotic milk containing bacteria like*Lactobacillusgasseri* SBT2055 strain can significantly reduce various parameters of obesity like body weight, BMI, waist circumference, and hip diameter in men.

Composition of the nutraceuticals extracted from bacteria – generally the nutraceuticals from bacteria are composed mostly of antioxidants like polyphenols. The most important of these polyphenols are tocopherols and tocotrienols.

4.3 ROLE OF MICROBIAL NUTRACEUTICALS IN THE PREVENTION OF FOOD SPOILAGE

Nutraceuticals derived from bacteria can prevent spoilage of food by producing different organic acids like propionic acid, acetic acid, and formic acid, and by inhibition of various pathogens. They also remove bacterial toxins (Sharma et al., 2012). Natural wines are also sometimes fortified with different strains of *Lactobacillus* spp., which can then convert malic acid into lactic acid. This can add to the stability of expensive wines (Sharma et al., 2012).

Bacteriocins are low molecular weight compounds, which are obtained from bacteria like Enterobacteriaceae and *Lactobacillus* spp.

Lack of adequate nutrients in the environment can incite the production of a plethora of bacteriocins by microbes in a situation of competition for space as well as resources. Bacteriocins are really abundant in various bacteria and have a lot of diversity in them. These bacterial genes can encode various ribosomally synthesized antimicrobial peptides (AMPs) or proteins. These proteins can then kill other related microbes (narrow spectrum) or non-related ones (broad spectrum) as a component of the inherent weaponry of bacteria. They can disintegrate the nucleus or cell wall or cytoplasmic membrane of these other types of bacteria. Sometimes they are purposefully and intentionally added to various types of foods to prevent their spoilage and enhance shelf life. Most commercial preservatives are synthesized by way of chemical synthesis. Long-term consumption of such synthetic preservatives can have many harmful effects on the human body. Also, it is not really legal to employ or encourage the use of antibiotics in food products. Bacteriocins produced by Gram-positive or Gram-negative bacteria are gene-encoded peptides or proteins, that are fit enough to be used as natural preservatives in different types of food products. Due to the susceptibility of these bacteriocins to lysis by some protease enzymes, harmless bacteriocins are perhaps digested. Thus, non-functional small peptides and amino acids are the types of foods abundant in bacteriocins, which are digested well in the gastrointestinal tract (GIT). In food technology, nisin, produced by *Lactococcus lactis*, was the first antibacterial peptide derived from the LAB. It is also a commercial bacteriocin that can be employed as a preservative in foods against contamination by microorganisms. It is sold in the market as Nisaplin. It is, by far the only bacteriocin that is approved for use as a preservative in food products by the Food and Drug Administration (FDA) of the USA or USFDA. It has been licensed as a food additive in more than 45 different countries across the world. Nisin is used in different fermented foods like kimchi, mashed potatoes, and freshly-cut food products. Another commercially used bacteriocin compound is Pediocin PA-1, which is sold in the market as Alta® 2341. It can quell the growth of the bacterial pathogen *Listeria monocytogenes* in meat. Enterocin AS-48 is used commonly in foods like cider, fruit, and vegetable juices, and also in different canned vegetables for inhibiting contamination by microbes. Enterocin CCM4231 and EJ97 are used, respectively in soy milk and zucchini purée for suppressing bacterial contamination (Yang et al., 2014; Settiani et al., 2008).

4.3.1 USE OF BACTERIOCINS FOR KILLING OTHER PATHOGENIC BACTERIA

A new class II low molecule mass bacteriocin which is called OR-7 has also been well reported in scientific literature. It has been extracted and purified from the *Lactobacillus salivarius* strain which I termed NRRL B-30514. These molecules show strong inhibitory activity against the Gram-negative gut pathogen *Campylobacter jejuni*. OR-7 is stable even when treated with lysozyme and lipase, and is also stable to heat at 90°C. It is also stable over a pH ranging from 3.0 to 9.1 (Yang et al., 2014).

4.3.2 USE OF BACTERIOCINS AS ANTICANCER AGENTS

Cancer is a top-killer disease now. Bacteriocins may indeed be very helpful in this regard. In the case of cancer chemotherapy, some researchers are of the opinion that bacteriocins show good activity against tumor cells. Because these bacteriocins can be naturally and also lawfully added to foods, they can suitably also be used as promising anti-tumor drug candidates. Some bacteriocins, like the pore-forming colicins A and E1, can also inhibit the growth of the human standard fibroblast cell line MRC5 and 11 different types of human tumor cell lines (Yang et al., 2014). Colicin D, E2, E3, and pore-forming colicin A can exert an effect on the viability of the murine leukemia cells P388. Thepore-forming toxins colicin E1 and colicin E3 can suppress v-myb-transformed chicken monoblasts (Yang et al., 2014). Nisin can also help allay the growth of cancer cells. It can also cure squamous cell carcinoma of the head and neck in mice models, according to scientific literature.

4.4 MORE ABOUT INDIVIDUAL BACTERIOCINS

1. **Nisin:** It is a class I bacteriocin which is produced by some strains of the bacterium *Lactococcuslactis* (Jozala et al., 2015). It is also called a lantibiotic molecule, having a molecular weight of 3.5 kiloDalton (kDa). It is heat-stable and is also non-immunogenic. It was legally used in England since 1953 and okayed for use by the Food and Agricultural Organization (FAO) in the year 1969 (Jozala et al., 2015). It was approved by the US FDA way back in the year 1988 and has been conferred a generally regarded as safe (GRAS) certification by the USFDA (Kamarajan et al., 2015). It has shown good activity against Gram-positive bacteria and their spores and is usually less effective on Gram-negative bacteria and fungi. It forms a pore on the cytoplasmic membrane of the bacteria and fungi. It also inhibits spores of bacteria by means of its dehydroalanine residue, which interacts with the sulfhydryl moieties in freshly formed spores. It is also relatively non-toxic to human cells, with LD50 akin to table salt. Nisin assimilation in the body is dependent on pH. It can be applied to foods like butter, cheese, and sausages. It is also used for treating conditions like dermatitis, gastric ulcers, and colonic infections. In dentistry, it is sometimes used as toothpaste and endodontic treatment (Jozala et al., 2015).

   Nisin produces its antitumor effects on HNSCC, by means of the cation transport regulator homolog 1 (CHAC1). The latter is a proapoptotic cation transport regulator. Nisin may also function through a simultaneous CHAC1-independent influx of extracellular Calcium (Kamarajan et al., 2015).

Nisin can also be used to potentiate the efficacy of existing antibiotics against multi-drug-resistant bacterial pathogens. It exhibits potent synergistic anti-biofilm activity against *E. faecalis* in combination with other agents like Penicillin, ciprofloxacin, and chloramphenicol (Meade et al., 2020).

2. **Salmocins:** These are bacteriocins that are derived from *Salmonella* which can also be expressed in plants. They have pore-forming and nuclease activity. Salmocins have been used in chicken meat in order to prevent colonization by other *Salmonella* serotypes (Hahn-Lobmann et al., 2019). They are class III bacteriocins.
3. **Helveticin:** Helveticin J and M are also class III bacteriocins (Hahn-Lobmann et al., 2019). This molecule was first purified from *Lactobacillus helveticus* in 1986. Helveticin J is an antimicrobial compound that is functional at neutral pH under aerobic or anaerobic conditions. It is susceptible to damage by proteolytic enzymes and heat (30 min at 100°C) and has a bactericidal mechanism of action (Joerger et al., 1986). Helveticin-M is a novel Class III bacteriocin synthesized by *Lactobacillus crispatus* which has shown appreciable antimicrobial activity against bacteria like *Staphylococcus aureus*, *S. saprophyticus*, and *Enterobacter cloacae* (Sun et al., 2018). Helveticin-M can damage the cell wall of Gram-positive bacteria and disorganize the outer membrane of Gram-negative bacteria. Thus, it can alter the cell surface structure. Helveticin-M also destroys the inner membrane, as shown by the leaking of the intracellular ATP from the bacterial cells and the depolarization of the membrane potential of the target bacteria (Sun et al., 2018). However, Helveticin M shows a bacteriostatic mode of action.

   Recent studies have revealed that Helveticin M can also damage the cell wall of Gram-positive bacteria and also the outer membrane of Gram-negative bacteria, and thus it has efficacy against both bacterial categories.
4. **Lactacin:** This is any of a family of pore-forming peptidic toxins having approximately 75 + 65 amino acid residues, that can permit the passage of small solutes, including water (Oxford Dictionary of Biochemistry and Molecular Biology, 2008). Lactacin is a bacteriocin molecule that is secreted by *Lactobacillus* spp. It can also inhibit other Lactobacilli and *Enterococcus* spp. In this respect, Lactacin F is more thermostable as compared to Lactacin B (Muriana et al., 1991). Lactacin F is also hydrophobic and can also suppress

Staphylococcus aureus and *Aeromonas hydrophila*. Lactacin B is also rather susceptible to proteolytic enzymes.

The Class II acidocin called CH_5 is synthesized by the Gram-positive bacillus *Lactobacillus acidophilus*. It has a limited profile of activity against Gram-positive bacteria. Chumchalova et al. in 2004 demonstrated appreciable activity of Lactacin against eight different strains of the genus *Lactobacillus*. Another acidocin, produced by *L. acidophilus* strain DSM 20079, has been described by Deraz et al. in 2005. It is a one-component peptide of low molecular weight (6.6 kDa), which is extremely heat-stable. It is active over a wide pH range (Parada et al., 2007).

5. **Leucocin:** Leucocin A, which is produced by the Gram-positive coccus *Leuconostoc gelidum*, has a molecular weight of about 4 kDa (37 amino acids). It belongs to a group of small antibacterial peptide molecules that are produced by the LAB. *Leuconostocpseudomes-enteroides* 607 produces leucocin C-607 along with some other yet unknown bacteriocin molecules. This Leucocin C-607 bacteriocin has a potentantagonistic effect on *Listeriamonocytogenes* (Chen et al., 2018).

   Mesentericin Y105 is a bacteriocin having 37 residues. It is secreted by *Leuconostoc mesenteroides* strain Y105 which shows antagonistic activity against a number of pathogenic Gram-positive bacteria like *Enterococcus faecalis* and *Listeria monocytogenes* (Parada et al., 2007).

   Bacteriocins from *Pediococcus* species are called pediocins. They were described to be protein in nature. They are inhibitory to many other bacteria. Pediocin F, which is one of the bacteriocins produced by *P. acidilactici* that can be retrieved from sausage that is fermented, is a small peptide molecule. It has a molecular weight of about 4.5 kDa. Pediocin F has been shown to be effective against a number of bacteria that are linked with food spoilage and other food-associated health hazards. Pediocin F is found to be susceptible to proteolytic enzymes and is also resistant to damage by heat and organic solvents. It is also active over a wide range of pH (Parada et al., 2007).

   Enterococcus faecium CRL35, a strain that is isolated from regional Argentinean cheese, also secretes a bacteriocin molecule called enterocin CRL35. It has potent inhibitory activity against the food-borne bacterial pathogen *Listeria monocytogenes*, and by its efficacy, this peptide molecule holds a lot of promise as an antimicrobial agent in foods.

4.5 NEW DEVELOPMENTS

Liposomal encapsulation can put an end to the degradation of bacteriocins that are used in food, by the protease enzymes of the gut. This technique has been employed in recent times in food science (Naskar et al., 2021). Researchers have also reported that nisin-loaded liposomal carriers can inhibit two major food-borne bacterial pathogens, viz. *L. monocytogenes* and *S. aureus* (Naskar et al., 2021). Chitosan is another very important nanoparticle. It is widely used along with various other bacteriocin molecules. It has now found many biomedical applications. Chitosan is really an ideal candidate for the delivery of bacteriocins and related applications owing to its non-toxic, biocompatible, and biodegradable nature. Moreover, its antibacterial and anticancer actions, along with its unique ability to deliver drugs directly to their target sites is well documented (Naskar et al., 2021). The delivery of bacteriocins in a concentrated manner or form, like those encapsulated in nanoparticles (NPs), can undoubtedly also enhance their overall effectiveness. This also mitigates the potentially toxic side effects of these bacteriocins (Zimina et al., 2020).

4.6 NEWER BACTERIOCINS

Phage tail-like bacteriocins (tailocins) are really large protein structures (20–100 kDa). They are composed of 8 to 14 different polypeptide subunits that show a lot of resemblance with the bacteriophage tail modules. Tailocins are encoded in the bacterial genome by a cluster of many genes which are about 40 kbp in length (Zimina et al., 2020).

Microcins are low-molecular-weight (<10 kDa) peptide molecules that are highly stable. They are involved in competitive interactions between bacteria that are the members of the family *Enterobacteriaceae.* Generally, these bacteriocins are very well known to be resistant to protease enzymes, extremes of pH, and also extremes of temperatures (Zimina et al., 2020).

Lantibiotics are low-molecular-weight peptides that are lower than 5 kDa in size. They consist of lanthionine and/or methyllanthionine residues. Lantibiotics are synthesized by Gram-positive bacteria. Among the newly discovered groups of lantipeptides which also have appreciable antibacterial activity are lipolantins. This new class of compounds is typified by the presence of an avionin moiety and the N-terminal guanidino fatty acid. A key member of this group, microvionin, is a bacteriocin compound that is derived from the culture extracts of the bacterium *Microbacterium arborescens* strain 5913. It is active against several disease-producing bacteria, namelyMethicillin-resistant *Staphylococcus aureus* (MRSA) and *Streptococcus pneumoniae* (Zimina et al., 2020). Thiopeptide molecules have a plethora of biological activities. They have potent antibacterial, antiparasitic, antiviral, and immunosuppressive effects. Antibacterial thiopeptides can interfere with bacterial protein synthesis by way of binding to the 50S ribosome subunit or maybe elongation factors (Zimina et al., 2020). Thiopeptide molecules are generally said to be very active in the nanomolar range. Still, their poor water solubility and added low bioavailability makes them very difficult to be utilized in clinical settings, in spite of their high bioactivity. The thiopeptide derivative GE2270A is, at this time, the only bacteriocin of this type that is undergoing clinical trials for the treatment of gastrointestinal infections caused by *Clostridium difficile* (Zimina et al., 2020).

A group of bacteriocins that are similar in structure to the thiopeptide molecules are the modified thiazole/oxazole-microcins-boromycins. Their characteristic properties include the presence of a macrocyclic amidine moiety and also a decarboxylated C-terminal thiazole structure. They also have many rare or uncommon β-methylated amino acid residues. These botromycins discovered newly, are produced by the Gram-positive bacteria belonging to the genus *Streptomyces* spp. They are very potent molecules, known to be active against multidrug-resistant bacteria, like MRSA and vancomycin-resistant enterococci or VRE (Zimina et al., 2020).

Sactibiotics are produced by bacteria belonging to the genus *Bacillus*. They are unique sulfur-containing peptide molecules. In addition to having many antibacterial properties, they may also possess spermicidal and hemolytic properties. The first such compound of this class, called subtilosin A, had been discovered about 30 years back (Zimina et al., 2020).

4.7 NUTRACEUTICALS PRESENT IN UPCOMING POPULAR BEVERAGES

Kombucha tea is a popular drink made by fermentation of black tea by a mixture of different fungi and bacteria. It is also fondly known as "Mushroom Infusion" and "The Champagne of Life." Just as sourdough bread is prepared out of a mother or starter culture, the kombucha zoogloea is prepared from a special live culture called a SCOBY (symbiotic colony of acidic acid bacteria and yeast). The SCOBY culture has historically been passed from one family of bacteria to another. This SCOBY is added to various brewed teas like black tea. However, occasionally green tea has also been employed for this purpose. It is widely used for curing a range of different disorders like memory loss, premenstrual syndrome (PMS), joint pain (called rheumatism), anorexia, aging, cancers, constipation, hypertension, arthritis, and hair regrowth. It is also now quite popular as an immune-boosting agent. Chemical assays with kombucha beverages have shown the presence in it, of a variety of different compounds, like organic acids, like acetic, gluconic, and glucuronic acids (GlcUA). However, other molecules like citric, L-lactic, malic, malonic, tartaric, oxalic, pyruvic, succinic, and usnic acids. They may also be found as active components of Kombucha tea. It may also contain various sugars (like sucrose, glucose, and fructose), B complex group of vitamins (like Vitamin B_1, B_2, B_6, B_{12}, C), amino acids, biogenic amines, purine nucleotides, proteins, some pigments, lipids, hydrolytic enzymes, ethanol, acetic acid bacteria and LAB, carbon dioxide (CO_2), minerals (like manganese, nickel, copper, Iron, zinc, cobalt, Chromium, and Cadmium), anions (fluoride, bromide, chloride, iodide, nitrate, phosphate, and sulfate), D-saccharic acid-1,4-lactone (DSL), polyphenols, and many different metabolic products of yeasts and bacteria are also abundantly found in Kombucha tea (Martínez et al., 2018). It is also very rich in Vitamins B1 and B6, minerals like Copper and Manganese, and several polyphenol compounds. Also, it contains a compound called DSL which is derived from D-glutaric acid (that is a product of the GlcUA pathway). It has many documented powerful detoxifying and antioxidant properties as well (Martínez et al., 2018).

Kombucha tea also has a substantial quantity of Ethanol. Moderate alcohol consumption has a very well-known protective effect against CHD. Actually, it increases the high-density lipoprotein (HDL) cholesterol concentrations. Because the daily consumption of ethanol in a lower quantity can protect the human body from CVDs, the regular use of kombucha, having a low concentration of ethanol, may also have an associated role in preventing CVDs. Kombucha tea also has many organic acids. These can regulate the sleep pattern well. It is also a rich source of glucuronic acid. The latter has a well-documented detoxifying action against many drugs, chemicals, bilirubin (BR), as well as several pollutants and an excess of steroid hormones (Martínez et al., 2018). Glucuronic acid also increases the bioavailability of the polyphenols found in Kombucha tea. Glucuronic acid is required for carrying out many important bodily functions. This is because it is a constituent of various essential polysaccharides found in the body, like glycosaminoglycans (GAGs). GAGs are a series of compounds formed by dimers consisting of an amino sugar (D-Glucosamine or D-Galactosamine) and a uronic acid moiety (D-glucuronic acid or L-iduronic acid), with the exception of keratan sulfate. The latter has a galactose molecule instead of an acid. They are bound to sulfate groups in variable ratios and also bind covalently to the proteins, forming proteoglycans. Nevertheless, hyaluronic acid is not sulfated. Different GAGs are formed, depending on the constituents of the dimer. GAGs molecules are part of the extracellular matrices of all the body organs and they have myriad functions. Glucuronic acid is part of the following GAGs, like hyaluronic acid, chondroitin sulfate, heparin, and dermatan sulfate. They all have well-defined structural functions, except for heparin, which is a non-structural GAG. Hyaluronic acid can function as a lubricant and also a shock absorber. It is present in high amounts in the vitreous fluid of the eye, conjunctive tissue, and synovial fluid of joints and cartilage.

Kombucha tea also has cellulose and there is the formation of a head cap resembling mushrooms. Organic acids synthesized during the fermentation process while making Kombucha can decrease the pH of tea. This leads to a lack of Oxygen induced by the acidity. Due to this, the number of possible pathogenic microorganisms, if any, is reduced, thus rendering it very safe for consumption, in spite of being of microbial origin (Martínez et al., 2018). Chondroitin sulfate is present mainly in bones and cartilage. It binds strongly to the collagen and keeps the fibers in a strong network. It also helps to prevent many joint problems and is helpful for the relief of osteoarthritis. Heparin is an intracellular compound and a very strong anticoagulant. It is produced by the mast cells. Dermatan sulfate is found in higher concentrations in the blood vessel endothelium, connective tissues, skin, cartilage, bones, and corneal tissue. Moreover, Glucuronic acid is a precursor of L-ascorbic acid or vitamin C, which is found in Kombucha beverages. This is because this tea is made from L-gulonic acid, which is involved in the metabolic pathway of Glucuronic acid. Thus, the Glucuronic acid concentration in kombucha tea increases its antioxidant levels as well.

However, all said and done, there are some potential and serious adverse health effects of Kombucha tea. These should hence always be kept in mind before prescribing or consuming it. There have been reports of many adverse effects, like gastric upset, infections, and many allergic reactions that can also be seen in kombucha tea drinkers. Kombucha tea is often brewed in homes under considerably non-sterile conditions. Thus, contamination is quite likely to take place. On the other hand, when ceramic pots that are not properly manufactured have been used for brewing, many cases of lead poisoning have occurred. This is because the acids in the tea can cause the leaching of lead from the ceramic glaze (Bauer et al). The caffeine which is present in Kombucha tea can actually worsen the symptoms of diarrhea and irritable bowel syndrome (IBS). It can also potentially lower blood sugar levels. Patients taking antidiabetic medications should hence be very cautious about having Kombucha. Kombucha tea, especially the batches made at home, where it is not easy to maintain a germ-free environment, can easily get contaminated with fungus or mold (*Aspergillus* spp.) and bacteria like *B. anthracis*. In Iran, many people incurred anthrax infections after consuming Kombucha tea (rxlist.com, 2021). Kombucha tea is also reported to be not so safe during pregnancy and also breastfeeding. There may be some other drug interactions also. Kombucha tea is known to contain alcohol. Disulfiram (Antabuse) diminishes the metabolism of alcohol. Consuming kombucha tea along with disulfiram (or Antabuse) can thus lead to profuse headaches, flushing, vomiting, and other unpleasant reactions in the persons consuming it. Also, Kombucha tea may exert many cytotoxic actions. Kombucha prepared with green tea can produce a greater degree of cytotoxic effect on cell lines like A549 and Hep-2. On the contrary, the kombucha beverage which is manufactured from black tea showed a moderate cytotoxic activity.

Kombucha and related foods are hence now termed functional foods or FOSHU (This term stands for Food for Specified Health Uses). This has also been approved by Japan's Consumer Affairs Agency, which is a leader in regulating functional foods (Martinez et al., 2021).

Kefir is also a very popular beverage and food having a creamy consistency and acidic taste. Kefir and other related products are famed nutraceutical dairy products. They are produced through fermentation by yeasts and bacteria of nutrients that are present naturally in grains of kefir. Kefir can also be prepared from a wide variety of milk, like cow, buffalo, and camel milk. The nutritional benefits conferred by this self-carbonated beverage arise due to the presence of many vital nutrients, in the form of carbohydrates, proteins, vitamins, minerals, and several other nutraceutical components. Kefir has many putative health-promoting properties like antimicrobial activity, anticarcinogenic properties, and good gut health. It also leads to better control of serum glucose and cholesterol, control of lactose intolerance, and a better immune system that can be achieved through regular consumption (Ahmed et al., 2013). Kefir has got higher levels of vitamin B_1, B_2, and folic acid in it, as compared to milk. Right across the world, kefir has been associated with some mythical healing powers and also a number of unsubstantiated, nutritional, and therapeutical claims (Stepaniak et al., 2002). It can also inhibit pathogens like EHEC 0157:H7 and *Listeria monocytogenes*. Production of Nisin by *Lactococcus* is stimulated by *Kluyveromyces* spp., both of which have been found in kefir. Kefiran and lipids obtained from these kefir grains can also be well regarded as powerful anti-stress food components. It can also be of immense help in lowering Lactose intolerance.

Plant products and essential oils can also be used to enrich the beneficial bacteria that are already present in Kefir. Flaxseed extract has also been found to improve the survival and multiplication of *Lactobacillus plantarum* in Kefir (Aiello et al., 2020).

A detailed discussion of Kefir and Kombucha tea has been given in another chapter by the same author in this book. Readers may refer to that chapter also.

Fermented foods other than kefir are also reputed for their powerful nutraceutical-like properties. Most of the patented microorganisms used for such commercial food fermentation belong to the group LAB. They are Gram-positive, acid-tolerant bacteria which can synthesize lactic acid as the major end-product of carbohydrate fermentation. This lactic acid can contribute to the organoleptic properties of this food. It is also said to create a gut environment that reduces the establishment of pathogenic bacteria in the gut. Fermented plant products are also thus gaining widespread popularity these days. Fermented soy products include, but are not limited to, products like miso, tempeh, soy paste, and soy sauce. The health benefits related to soy include the possible reduction of age-related and also hormone-related diseases. The phytochemicals called isoflavones have now received most of the attention for such actions. When soy is fermented, there is a chemical conversion of the isoflavone glycoside precursors genistin and daidzin into more active isoflavone compounds like genistein and daidzein, respectively. Another potential healthy product is a fermented soy extract which is a combination of a *Lactobacillus* strain and *Saccharomyces* species. A dose-response study with this type of fermented soy extract is needed for understanding its full utility in the prevention and/or treatment of inflammatory diseases or various health conditions like cancer, asthma, infection, and autoimmunity (Borrensen et al., 2012). Fermented rice and rice bran are other plant-based products that have also been documented to prevent and treat many diseases. Red yeast rice (RYR) is prepared after fermenting rice with the yeast *Monascus purpureus*. Its nutritional profile includes unsaturated fatty acids, sterols, B-complex vitamins, and monacolins having robust antioxidant properties. Research has shown its health benefits, involving the treatment of cholesterol, type II DM and cardiovascular diseases (CVDs) (Keservani et al., 2016). It can also possibly help in the prevention of cancer. 'Red mold rice' is another *Monascus*-fermented rice preparation that has been used in Chinese cuisine since antiquity for enhancing flavor and other medicinal applications. Modern health claims of another beverage, Red mold tea include prevention of CVDs, cancer, and Alzheimer's disease (AD) (Lee et al., 2011).

BioBran® is a rice bran functional food product that had been developed by using a water-soluble rice bran fiber (Keservani et al., 2015; Rane & Keservani, 2020) component. It is involved in immunomodulation and is believed to enhance overall health. It has also been found to improve quality of life. Also, novel metabolite profiles with bioactivity have been documented in rice bran that has been fermented with the yeast *Saccharomyces bouldardii*. There is now clinching evidence for the role of rice bran phytochemical agents in the presence and absence of fermentation for conferring disease-fighting activities. LAB and other microbes having fermentive abilities have also been purported to improve food safety by the reduction of toxic environmental substances like cyanide, which are found on the surface of foods that come from pesticide application.

Curd or yogurt (also called Dahi in Hindi and Dadhi in Sanskrit) has a lot of nutritional benefits. Yogurt which is produced from camel milk by fermentation with *Lactobacillus rhamnosus* strain PTCC 1637 has a much higher antioxidant activity as compared to cow milk. This is because of the higher proline content seen in camel milk caseins. Lactobacilli act on casein, and when added to milk, this can lead to the formation of high amounts of CLA. CLA are the major antioxidants that are found in milk fat along with the fat-soluble vitamins A and E, β-carotene, and Coenzyme Q10. Folates or folic acid have strong antioxidant effects which can protect the human body against free radical-induced damage. It has been observed that Lactobacilli are able to accumulate late in milk. Hence, they have tremendous potential to be used as functional cultures in fermented dairy products. This is done to mitigate artificial supplementation with synthetic folic acid. *Streptococcus thermophilus*, one of the most important industrial dairy starter bacteria that can synthesize folates, is considered a moderate Folate producer. When milk is converted into yogurt, the amount of folate (vitamin B_9) may be elevated to values anywhere above 200 μg/L. This again is dependent upon the starter cultures used and the storage conditions. These can also help meet the recommended dietary Folate allowance of man (400 μg DFE).

Products that contain mesophilic LAB species and *Lactococcus* spp. in particular as starter cultures, also have a markedly high vitamin K content in them. This is of great importance because observational studies have demonstrated a positive association between menaquinone (MK) intake or Vitamin K intake and bone and cardiovascular health. Dairy products are also considered an important source of Vitamin B_{12}. The latter is required for the maintenance of nervous system functions and the formation of hematopoietic cells. Its content can be elevated up to 10 folds by fermentation of milk and milk products (Melini et al., 2019).

It has been seen that the antioxidant activity and quality of fermented milk can be increased by ultrasound treatment of milk before fermentation. Ultrasound pre-treatment of milk is found to increase its antioxidant activity. This is because ultrasonication leads to an increase in hydrolysis of the disaccharide lactose. This then implies a higher content of sugars that are now available for the growth of LAB. It also increases the multiplication of viable cells and also helps in releasing antioxidant components.

Fermented milk may show a blood pressure-lowering effect as well. This activity is seen because anti-hypertensive peptides are produced from milk proteins by means of enzymatic breakdown by various digestive enzymes.

It can also be due to the proteinases produced by *lactobacilli* during the process of fermentation. Bread and other grain-based fermented products have high levels of phenolic compounds, GABA, and other bioactive compounds. Fermentation of grain-based food leads to an increase in the total phenolic compound content of bread. This is due to the biological conversion of phenolic compounds from their conjugated forms into free forms. The hydrolytic activity of enzymes produced by the fermenting bacteria or fungi used thus facilitates the structural breakdown of the grain cell walls. This also results in greater bio-accessibility and bio-availability of both the bound and conjugated phenolic compounds (Melini et al., 2012). An increase in total phenolic content has also been seen in quinoa and buckwheat which has been fermented with *P. pentosaceus* and *Lactobacillus paracasei.* For some raw materials, the choice of the fermenting microbe can alter or influence the final content of the antioxidant compounds present in the product. When quinoa is fermented with *L. paracasei*, for example, its net antioxidant activity 8s is found to be higher than when fermented by *P. Pentosaceus*. This is in contrast to buckwheat whose fermentation with the aforesaid strains shows no major difference. The bread which is prepared with wheat sourdough produced by *L. brevis* CECT 8183 and a commercial protease enzyme has got a higher total antioxidant activity. This is due to the presence of GABA and other small peptide molecules (<3 kDa). Fermentation of rye malt sourdoughs with *L. reuteri* can, in turn, lead to the formation of peptides having strong antioxidant activity. It can also lead to the accumulation of ACE-inhibitory peptides in the body (Gobbetti et al., 2014).

Anti-diabetic effects – antidiabetic properties of the various types of fermented foods have been proven by a number of tests, both *in vitro* and *in vivo*. However, the complete mechanisms leading to this effect is not yet known precisely. It has been suggested that phenolic compounds, antioxidant molecules, and GABA, which are present in various types of breads may be responsible for their putative anti-diabetic activities. Also, the application of sourdough fermentation for bread-making allows the production of bread with a low glycemic index (GI) value and reduced starch digestibility, due to the synthesis of organic acids (Melini et al., 2019). Folate levels can be elevated in oat and barley matrices by means of fermentation with the help of some microbial strains, like *S. cerevisiae* ALKO743, *Candida milleristrain* ABM4949, and *Pseudomonas* sp. ON8 and *Janthinobacterium* sp. RB4.

Fermentation of food grains has also been shown to play a stellar role in lowering the FODMAPs. The latter consists of oligosaccharides like fructans and galactans, disaccharides (here lactose), monosaccharides (in this case fructose), and polyol compounds (like sorbitol and mannitol). Therefore, fermented grain-based products can make up a major part of the staple diet of people suffering from IBS.

Fruits (Keservani et al., 2016b) and vegetables are known to be highly perishable food substances. Fermentation has been employed in all corners of the world to bestow food products or beverages with a much-extended shelf-life. However, this tradition of consumption of fermented fruit and vegetable products is more commonplace in Asian than in Western cuisines. This is highlighted by the huge number of traditional fermented food products of Asia which are available in the market like *sauerkraut*, *khalpi*, *kimchi*, *tempeh*, *gundruk*, and *sinki*. As regards fruits, many scientists report some alterations in the antioxidant potential and also phenolic contents in fruit juices after being subjected to fermentation by bacteria like *Lactobacillus* spp. (Melini et al., 2019). People have even noted an increase in the antioxidant levels of apricot juice after fermentation by singular as well as mixed cultures of the probiotic Gram-positive bacteria *Lactobacillus* and *Bifidobacterium* spp.

As regards fermented vegetables, fermentation of tomato by LAB shows that the total phenolic and flavonoid (Keservani & Sharma, 2014) concentrations are diminished after 4-week fermentation, but lycopene and antioxidant activities increase.

Lactococcus lactis isolated from *dahi* or Indian curd, produces Nisin Z that can kill *L. monocytogenes* and *S. aureus*. Several LAB species isolated from traditional Romanian fermented fruits and vegetables have good antimicrobial activity against *L. monocytogenes*, *E. coli, Salmonella*, and *Bacillus* spp. (Tamang et al., 2016).

Fish and meat – dry-fermented sausages are a very popular part of the diet of many people in Western nations. However, dietary guidelines worldwide have recommended people lower their consumption of meat. Demand for healthier meat products having less fat and cholesterol levels, better fatty acid profiles, and added health-promoting ingredients has gone up with time (Keservani et al., 2010a, b).

4.7.1 EFFECT OF FERMENTED FOODS ON MENTAL HEALTH

Bacteria in general, like *Bifidobacterium* (e.g., *B. breve*, *B. longum*, and *B. infantis*) and *Lactobacillus* (like *L. helveticus*, *L. rhamnosus*, and *L. plantarum*), that are found in many popular fermented foods, have been demonstrated to improve psychiatric illnesses and behavioral problems, like depression, anxiety, obsessive-compulsive disorder (OCD) and memory problems. They can also help attenuate stress responses in the body. They can help maintain optimum body levels of brain-derived neurotrophic factors (*BDNFs*) which are neuropeptides in nature. These are known to be lowered in cases of depression (Keservani et al., 2015).

4.7.2 COUNTRY GUIDELINES

As regards the use of fermented foods, a few countries of the world, like Australia, Kenya, South Africa, Oman, India, Sri Lanka, Qatar, and Bulgaria, have so far declared their recommendations for the consumption of various such food items in respective national dietary guidelines. In these places, many local and traditional fermented food items are widely available in the market and are quite popular also. They represent an element rooted in their national culture. However, there is still a need for fundamental research by means of studies like randomized, controlled, clinical trials. These are required in order to quantify the reproducible effects of various fermented foods in the different clusters of the populations. This is also needed to finally justify their entry into the national food guidelines. Identifying a strong and positive relationship between fermented foods and good health is thus a major pre-requisite for such step-by-step inclusion of this food category in dietary habits of the masses, and thereafter in the national dietary guidelines. Moreover, the chemical and microbiological compositions of these fermented foods should be precisely and accurately known in order to appreciate fully and also guarantee their claimed effects on health and well-being.

4.8 DISCUSSION

Microbes and their different metabolic products can be the source of vital nutraceuticals in various types of foods like fermented food and others. These nutraceuticals may be either macronutrients or micronutrients and have many important antimicrobial and preservative properties. They also help in maintaining the normal functioning of the different tissues and organs of the body. They can also help inhibit many dangerous bacterial pathogens. Fermented food matters can be a major source of dietary nutraceuticals. The intake of various fermented foods and probiotics and their subsequent effects in the enhancement of the gut microbiome can also produce a modulatory effect on the functioning of the brain as well as the CNS (central nervous system). All these properties are also being utilized for preserving food and preventing food contamination. To date, not much research has been carried out in this aspect. These can thus well be said to be the topics for future research and require more exploration.

4.9 CONCLUSION

Many microbial products can be quite effectively used as powerful nutraceutical compounds. They can confer a number of health benefits but can also be risky at times. This subject definitely needs more dedicated research.

ACKNOWLEDGMENTS

I heartily acknowledge the help I got from my beloved wife, Dr. Satarupa for being my support, and also Dr. Rajesh Kumar Kesharwani in assisting me in bringing out this treatise on micronutrients, macronutrients, and nutraceuticals.

KEYWORDS

- **bacteriocins**
- ***Bifidobacterium***
- **diversity**
- ***Lactobacilli***
- **microbes**
- **microbial products**
- **nutraceuticals**
- **pathogenic bacteria**

REFERENCES

Ahmed, Z., Wang, Y., Ahmad, A., Khan, S. T., Nisa, M., Ahmad, H., & Afreen, A., (2013). Kefir and health: A contemporary perspective. *Crit. Rev. Food Sci. Nutr., 53*(5), 422–434.

Aiello, F., Restuccia, D., Spizzirri, U. G., Carulo, G., Leporini, M., & Loizzo, M. R., (2020). Improving kefir bioactive properties by functional enrichment with plant and agro food extracts. *Fermentatio, 6*, 83–98.

Bauer, B. A. What is Kombucha Tea? Does it Have Any Health Benefits? https://www.mayoclinic.org/healthy-lifestyle/consumer-health/expert-answers/kombucha-tea/faq-20058126 (accessed on 05 June 2023).

Bernardeau, M., Guguen, M., & Jean, P. V., (2006). Beneficial lactobacilli in food and feed: Long-term use, biodiversity and proposals for specific and realistic safety assessments. *FEMS Microbiol. Rev., 30*(4), 487–513.

Borrensen, E. C., Henderson, A. J., Kumar, A., Weir, T. L., & Ryan, E. P., (2012). Fermented foods: Patented approaches and formulations for nutritional supplementation and health promotion. *Recent Pat. Food Nutr. Agric., 4*(2), 134–140.

Chen, Y. S., Wu, H. C., Kuo, C. Y., Chen, Y. W., Ho, S., & Yanagida, F., (2018). Leucocin C-607, a novel bacteriocin from the multiple-bacteriocin-producing *Leuconostoc pseudomesenteroides* 607 isolated from persimmon. *Probiotics Antimicrob Proteins, 10*(2), 148–156. doi: 10.1007/s12602-017-9359-6.

Dhar, H., Chaudhary, A., & Rana, V., (2020). Designer microbes for nutraceutical application. In: Sharma, T. R., Deshmukh, R., & Sonah, H., (eds.), *Adv. Agri.-Food Biotechnol.* Springer, Singapore. https://doi.org/10.1007/978-981-15-2874-3_11.

Gobbetti, M., Rizzello, C. G., Di Cagno, R., & De Angelis, M., (2014). How the sourdough may affect the functional features of leavened baked goods. *Food Microbiol., 37*, 30–40. doi: 10.1016/j.fm.2013.04.012.

Gupta, C., & Prakash, D., (2019). *Nutraceuticals from Microbes of Marine Sources (Book Chapter).* https://www.intechopen.com/chapters/69810 (accessed on 05 June 2023).

Gupta, C., Prakash, D., & Gupta, S., (2015). Microbes as nutraceuticals for obesity and weight control. *World Nutraceutical Conference and Expo.* https://www.longdom.org/proceedings/microbes-as-nutraceuticals-for-obesity-and-weight-control-30989.html (accessed on 05 June 2023).

Hahn-Lobmann, S., Stephan, A., Schulz, S., Schneider, S., Shaverskyi, A., Tuse, D., et al., (2019). Colicins and salmocins – new classes of plant-made non-antibiotic food antibacterials. *Front Plant Sci.* https://doi.org/10.3389/fpls.2019.00437.

Hugenholtz, J., & Smid, E. J., (2002). Nutraceutical production with food-grade microorganisms. *Curr. Opin. Biotechnol., 13*(5), 497–507. doi: 10.1016/s0958-1669(02)00367-1.

Jarouliya, U., & Keservani, R. K., (2019). Pathways leading to child obesity. In: Debasis, B., (ed.), *Global Perspectives on Childhood Obesity: Current Status, Consequences and Prevention* (2nd edn., pp. 137–146). Chapter 12, Academic Press, Elsevier. ISBN: 9780128128404.

Jian, W., Sanjay, G., Mattheos, A. G. K., & Yajun, Y., (2016). Microbial production of value-added nutraceuticals. *Current Opinion Biotechnol., 37*, 97–104. ISSN 0958-1669, https://doi.org/10.1016/j.copbio.2015.11.003.

Joerger, M. C., & Klaenhammer, T. R., (1986). Characterization and purification of helveticin J and evidence for a chromosomally determined bacteriocin produced by *Lactobacillus helveticus* 481. *J. Bacteriol., 167*(2), 439–446. doi: 10.1128/jb.167.2.439-446.1986.

Jozala, A. F., Novaes, L. C. L., & Pessoa, A. Jr., (2015). *Nisin.* https://www.intechopen.com/chapters/48931 (accessed on 05 June 2023).

Kamarajan, P., Nisin, Z. P., Hayami, T., Matte, B., Liu, Y., Danciu, T., et al., (2015). Nisin ZP, a bacteriocin and food preservative, inhibits head and neck cancer tumorigenesis and prolongs survival. *PLoS One, 10*(7), e0131008. doi: 10.1371/journal.pone.0131008.

Keservani, R. K., & Sharma, A. K., (2014). Flavonoids: Emerging trends and potential health benefits. *Journal of Chinese Pharmaceutical Sciences, 23*(12), 815.

Keservani, R. K., Kesharwani, R. K., Sharma, A. K., & Jarouliya, U., (2015). Dietary supplements, nutraceutical and functional foods in immune response (immunomodulators). In: Debasis, B., Harry, G. P., & Anand, S., (eds.), *Nutraceutical and Functional Foods in Human Life and Disease Prevention* (pp. 343–358). Chapter 20, Boca Raton, Fl. CRC Press, Taylor and Francis. ISBN: 9781482237214.

Keservani, R. K., Kesharwani, R. K., Sharma, A. K., Vyas, N., & Chadoker, A., (2010b). Nutritional supplements: An overview. *International Journal of Current Pharmaceutical Review and Research, 1*(1), 59–75.

Keservani, R. K., Kesharwani, R. K., Vyas, N., Jain, S., Raghuvanshi, R., & Sharma, A. K., (2010a). Nutraceutical and functional food as future food: A review. *Der Pharmacia Lettre, 2*(1), 106–116.

Keservani, R. K., Sharma, A. K., & Kesharwani, R. K., (2016a). Nutraceutical and functional foods for cardiovascular health. In: *Food Process Engineering* (pp. 257–278, 291–312). Apple Academic Press, CRC Press. ISBN: 9781771884020.

Keservani, R. K., Sharma, A. K., & Kesharwani, R. K., (2020). *Nutraceuticals and Dietary Supplements: Applications in Health Improvement and Disease Management.* CRC Press. ISBN: 9781771888738.

Keservani, R. K., Singh, S., Singh, V., Kesharwani, R. K., & Sharma, A. K., (2015). Nutraceuticals and functional foods in the prevention of mental disorder. In: Debasis, B.,

Harry, G. P., & Anand, S., (eds.), *Nutraceutical and Functional Foods in Human Life and Disease Prevention* (pp. 255–268). Chapter 15, Boca Raton, Fl. CRC Press, Taylor and Francis. ISBN: 9781482237214.

Kombucha Tea. https://www.rxlist.com/kombucha_tea/supplements.htm (accessed on 05 June 2023).

Lactacin (overview), (2008). *Oxford Dictionary of Biochemistry and Molecular Biology* (2nd edn.). https://www.oxfordreference.com/view/10.1093/oi/authority.20110803100046927 (accessed on 05 June 2023).

Lee, C. L., & Pan, T. M., (2011). Red mold fermented products and Alzheimer's disease: A review. *Appl. Microbiol. Biotechnol., 91*(3), 461–464.

Linares, D. M., Gomez, C., Renes, E., Fresno, J. M., Tornadijo, M. E., Ross, R. P., et al., (2017). Lactic acid bacteria and bifidobacteria with potential to design natural bio-functional health-promoting dairy foods. *Front. Microbiol.* https://doi.org/10.3389/fmicb.2017.00846.

Martínez, J. L., Suárez, L. V., Jayabalan, R., Oros, J. H., & Escalante-Aburto, A., (2018). A review on health benefits of kombucha nutritional compounds and metabolites. *CyTA – J Food, 16*(1), 390–399. doi: 10.1080/19476337.2017.1410499.

Meade, E., Slattery, M. A., & Garvey, M., (2020). Bacteriocins, potent antimicrobial peptides and the fight against multi drug resistant species: Resistance is futile? *Antibiotics, 9*(1), 32. https://doi.org/10.3390/antibiotics9010032.

Melini, F., Melini, V., Luziatelli, F., Ficca, A. G., & Ruzzi, M., (2019). Health-promoting components in fermented foods: An up-to-date systematic review. *Nutrients, 11*(5), 1189. doi: 10.3390/nu11051189.

Muriana, P. M., & Klaenhammer, T. R., (1991). Purification and partial characterization of lactacin F, a bacteriocin produced by *Lactobacillus acidophilus* 11088. *Applied Environ. Microbiol., 57*(1), 114–121.

Naskar, A., & Kim, K. S., (2021). Potential novel food-related and biomedical applications of nanomaterials combined with bacteriocins. *Pharmaceutics, 13*(1), 86. https://doi.org/10.3390/pharmaceutics13010086.

Parada, J. L., Caron, C. R., Medeiros, A. B. P., & Soccol, C. R., (2007). Bacteriocins from lactic acid bacteria: Purification, properties and use as biopreservatives. *Braz. Arch Biol. Technol., 50*(3). https://doi.org/10.1590/S1516-89132007000300018.

Patil, A., Pawar, S., & Disouza, J., (2018). Granules of unistrain *Lactobacillus* as nutraceutical antioxidant agent. *Int. J. Pharm. Sci. Res., 9*(4).1594–1599. doi: 10.13040/IJPSR.0975-8232. https://ijpsr.com/bft-article/granules-of-unistrain-lactobacillus-as-nutraceutical-antioxidant-agent/ (accessed on 05 June 2023).

Rane, B. R., & Keservani, R., (2020). Introduction to nutraceuticals and dietary supplements. In: *Nutraceuticals and Dietary Supplements* (pp. 35–56). Apple Academic Press. ISBN: 9781771888738.

Settanni, L., & Corsetti, A., (2008). Application of bacteriocins in vegetable food biopreservation. *Int. J. Food Microbiol., 121*, 123–138. 10.1016/j.ijfoodmicro.2007.09.001.

Sharma, S., & Manisha, K., (2012). Bacterial nutraceuticals: A beneficial health supplement. *CIBTech J. Biotechnol., 1*(2, 3), 13–17.

Singh, P., Kesharwani, R. K., & Keservani, R. K., (2017). Antioxidants and vitamins: Roles in cellular function and metabolism. In: *Sustained Energy for Enhanced Human Functions and Activity* (pp. 385–407). Academic Press. ISBN:9780128054130.

Stepaniak, L., & Fetlinski, A., (2002). Fermented milk: Kefir. In: *Encyclopedia Daily Sci.* Elsevier, https://www.sciencedirect.com/topics/agricultural-and-biological-sciences/kefir; doi: 10.1080/10408398.2010.540360 (accessed on 05 June 2023).

Sun, Z., Wang, X., Zhang, X., Wu, H., Zou, Y., Li, P., et al., (2018). Class III bacteriocin Helveticin-M causes sublethal damage on target cells through impairment of cell wall and membrane. *J. Ind. Microbiol. Biotechnol., 45*, 213–227. https://doi.org/10.1007/s10295-018-2008-6.

Tamang, J. P., Shin, D. H., Jung, S. J., & Charle, S. W., (2016). Functional properties of microorganisms in fermented foods. *Front Microbiol.* https://doi.org/10.3389/fmicb.2016.00578.

Yang, S. C., Lin, C. H., Sung, C. T., & Fang, J. Y., (2014). Antibacterial activities of bacteriocins: Application in foods and pharmaceuticals. *Front Microbiol., 5*, 241. doi: 10.3389/fmicb.2014.00241.

Zimina, M., Babich, O., Prosekov, A., Sukhikh, S., Ivanova, S., Shevchenko, M., et al., (2020). Overview of global trends in classification, methods of preparation and application of bacteriocins. *Antibiotics (Basel), 9*(9), 553.

CHAPTER 5

IMMUNITY-BOOSTING MACRONUTRIENTS AND MICRONUTRIENTS

DIKSHA KUMARI SHARMA

MSc in Applied Nutrition, All India Institute of Hygiene and Public Health, BN Campus, Saltlake, Kolkata, West Bengal, India

ABSTRACT

A healthy immune system is necessary for survival. Nutritional immunology is the condition, which altered the supply of nutrients to adapt the immune response. Several factors influence the system and its competency, together with nutrition. Nutrition, infection, and immunity all have a two-way relationship: changes in one aspect have an impact on the others. For instance, distinct immune options present throughout every life stage could have an effect on the sort, prevalence, and severity of infections, whereas poor nutrition will compromise the immune operation and increase infection risk. Numerous micronutrients are essential for immuno-competence, notably vitamins A, C, D, E, B2, B6, and B12, folic acid, iron, selenium, and zinc. The fundamental functions of the immune system are outlined in this prolog, immunity boosting macronutrients (carbohydrates, proteins, fats) and micronutrients (vitamins, minerals, and some functional foods). Continued study in this topic will eventually lead to a greater knowledge of the role of diet and nutrients in immune function, as well as the use of tailored nutrition to promote human health. And in this study, we looked at the evidence for the importance of nutrition in immune cell function, specifically dietary

Micronutrients and Macronutrients as Nutraceuticals. Prakash Chandra Gupta, PhD, Sayan Bhattacharyya, MBBS, MD, Nisha Sharma, PhD, MPharm, Rajesh K. Kesharwani, PhD, Raj K. Keservani, PhD, MPharm (Eds.)
© 2024 Apple Academic Press, Inc. Co-published with CRC Press (Taylor & Francis)

carbohydrates, protein, and fat, as well as how dietary intervention can help to enhance intestinal health and immunological state.

5.1 INTRODUCTION

In this 21st century, a strong immune system is a requisite for each and every individual to be alive. And this pandemic COVID-19 has proved it very well.

Immunity – in biological terms immunity can be described as the ability to resistances to pathogens; however, reactions to some harmless environmental molecules, tumors, and even unaltered host components are also considered forms of immunity (Allergy, tumor immunity, and autoimmunity).

And it is simply defined as the host's ability to fight disease-causing organisms or to resist the invasion of dangerous chemicals.

5.2 GOOD HEALTH IS A NEW WEALTH

The immune system is a magnificently complex network consisting of certain types of cells that have phagocytic, detective, and protective properties for foreign and cancerous cells. The immune system is Comprises of organs like the Thymus, spleen, and bone marrow, they produce and store specific immune cells (Sultan et al., 2014). Bone marrow is the main lymphoid organ where all the lymphoid cells and blood cells are produced. They trap microorganisms from different areas like blood, and tissue fluids. Lymphoid tissue also located in the lining of the major tracts (respiratory, gastrointestinal, urogenital tracts) called MALT mucosa-associated lymphoid tissue and it consists of 50% of lymphoid tissue.

Nonspecific substances such as macrophages and neutrophils boost the immune system, enhancing phagocyte defense capabilities. They cling to the surfaces of phagocytes and lymphocytes, stimulating the synthesis of interferon, interleukin, and other complex compounds, and thereby activating the immune system.

5.3 COMPONENTS OF THE IMMUNE SYSTEM

Immunity is comprised of two types: (a) innate and (b) adaptive.

1. **Innate Immunity:** It is a non-specific type of defense, that is present since birth. This is accomplished by providing different types of barriers to the entry of the foreign barrier.
   i. **Physical Barriers:** For example, the skin on our bodies that keep pathogens from entering. Mucous coating of the epithelial lining of respiratory, gastrointestinal, and urogenital tracts also helps in trapping microbes.
   ii. **Physiological Barriers:** Acid in the stomach, saliva present in the mouth, and tears in the eyes, all can prevent microbial growth.
   iii. **Cellular Barriers:** WBC, like polymorpho-nuclear leucocytes and monocyte, macrophages, and phagocytes destroy microbes.
   iv. **Cytokine Barriers:** Interferons (protein secreted by infected cells) that protect non-infected cells from further infection.
2. **Adaptive Immunity:** It is pathogen-specific and characterized by memory cells. When the body encounters pathogens for the first time that time it produces a primary response and appears to have a memory cell of that encounter. All these mechanisms are carried out by two lymphocytes B cells and T cells. B cells produce proteins in response to pathogens and T cells help them to produce them. Different types of antibodies are produced in our body IgA, IgM, IgG, and IgE are some of them. Antibodies that are produced in the host body are called active immunity, for example-vaccination. And when the antibodies are directly given to the body to protect from foreign agents is called passive immunity, example-colostrum which contains IgA to protect the child. These antigens are found in blood so it is called hormonal immune response. And the second type is a cell-mediated immune response, they are mediated by lymphocytes when vital organs (eyes, heart, kidney, liver) don't work satisfactorily.

5.4 MECHANISM OF IMMUNITY

When infectious agents invade our bodies or we are vaccinated, the immune system is activated. Macrophage catches, engulfs, and digests an antigen in specific instances, which might result in the demise of cells or tissues in our bodies. Macrophages transmit a piece of the antigen on their surface to antigen-presenting cells, and subsequently, proteins on the macrophage and helper T cells connect, triggering the helper T cell to proliferate into either TH 1 or TH 2 cells, which release distinct types of cytokines. The TH 1 cell's cytokines activate a cytotoxic T cell, which kills the infected target cell (Sultan et al., 2014; Nicholson et al., 2016; Anywar et al., 2020).

5.5 NUTRITION AND IMMUNITY

Nutrition and immunity are highly correlated. Nutrients play a major role in boosting immunity as deficiencies show attestation like infection, allergies, and inflammation. And diet is the only way that can affect the function of the immune system. Impairment of the immune system results in cancer, and chronic inflammation like atherosclerosis, cystic fibrosis, rheumatoid arthritis (RA), and bronchiole asthma all of these can be prevented and treated by the components of foods (Keservani et al., 2015; Areeba et al., 2020). The nutrients including macronutrients (protein, fats), micronutrients (vitamins and minerals), phytochemicals, and functional foods, elevate one's health if taken in adequate amounts. Inhibition of pro-inflammatory mediators, promotion of anti-inflammatory activities, modification of cell-mediated immunity, adjustment of APC function, and communication between the innate and adaptive immune systems are all examples of their immunomodulating actions (Areeba et al., 2020; Singh et al., 2017a, b).

Also, nutritional factors regulate metabolic processes which may include the activation or inhibition of key enzymes which result in a change in cellular immune function, particularly in the T lymphocytes lineage (Ibramin et al., 2016). Several micronutrients have an important role in key metabolic pathways and immune cell functions (Chandra et al., 1993). It is known that infection in children is the basic cause of child mortality and the reason behind it is malnutrition. It breaks down the immune system. Hence we must make sure to take nutritional adequate foods including both macro- and micronutrients to optimize better health.

The immune system is hampered by malnutrition. Cases of repressive immune systems have been on the rise recently (Palmar, 2011). The immune system can malfunction due to a lack of energy and macronutrients (CHO, protein, and lipids) or a lack of particular micronutrients. The nutrients which stimulate and support the immune system are called immune nutrition elements.

5.6 FACTORS AFFECTING IMMUNE RESPONSE

Effective defense against infectious organisms necessitates a well-functioning system, which is self-evident. Consequently, people with weakened immune systems are at inflated risk of turning infected and of infections being additional serious, even fatal. These embody some non-modifiable factors such as biological science, a stage in one's life (e.g., pregnancy, infancy, and previous age), and time of day, however, several modifiable factors additionally influence the response. These embody stress, fitness, frailty, body fat, and diet (Philip, 2021).

5.7 MACRONUTRIENT AND IMMUNITY

Carbohydrates are an important fuel needed by the cell for body functioning as well as the immune system. The fact that anaerobic glycolysis causes an increase in lymphocytes triggered by mitogens shows that glucose is being used as fuel. However, the usage of carbohydrates for energy reduces during lymphocyte multiplication. In this case, glycolytic mid-products are directly associated with purine and pyrimidine nucleotide synthesis for cell growth (Kursatkaracabey et al., 2012).

Carbohydrates are the nutrients seen in plant-based foods containing carbon, hydrogen, and oxygen molecule. The simplest sugar is found in fruits and fruit juices and complex carbohydrates are found in cereals, legumes, and vegetables. Carbohydrates are stored in the liver in the glycogen form. Very little are present in muscle and other organs. Carbohydrates are critical for recognizing immune recognition. In some earlier studies, sugars have long been described as T-cell-independent antigens that fail to induce immunologic memory and immunoglobulin class-switching. Studies of carbohydrate-based vaccines in mice have documented that IgM antibody production dominated the response (Beuvery et al., 1982), although some IgG production has been observed (Barrett et al., 1980).

In some studies, it was found that molecules containing carbohydrates like lipoprotein, can be recognized by TLR, which are "pattern recognition receptors" found on the surface of the immune cells and activate cytokine production.

Pure carbohydrates have poor immunogenicity as vaccines, and vaccines from them are developed by conjugation of a polysaccharide to a carrier protein, which strengths the adaptive immune system to generate a T-cell-dependent response that leads to protective IgG production and immunologic memory (antibody production) specific for the carbohydrate moiety (Kasper et al., 1996). This plan of action can produce protective antibodies against the bacteria from which the carbohydrates originated, as in the commercially available vaccines against *Haemophilusinfluenzae* type Band Streptococcus pneumoniae (Darkes et al., 2002). So carbohydrates are the most important fuel of the body for functioning, boosting immunity, and the prevention of diseases.

Hence, glucose is also an important fuel for cells of the immune system, including lymphocytes, neutrophils, and macrophages. Phagocytes utilize glucose at a rate 10-fold greater than they utilize glutamine when these substrates are both present in the culture medium at normal physiological concentrations (Blannin et al., 1998). Cells of the immune system have very high metabolic rates (Calder, 1995) and this consequently highlights the importance of adequate nutrition for the provision of fuels to maintain immune-competence.

5.8 PROTEIN AND IMMUNITY

Proteins make up the framework for Cells, defense systems, enzymes that govern physiological activities, and certain hormones (Kavacabey, 2012). The mostly immune system relies on the production of active protein compounds. And protein deficiency can lead to a decrease in the function of the immune system and may affect the regulator for some amino acids. Obtaining and maintaining the adequate requirement of amino acid is very crucial, as its deficiency or too much consumption can cause repression of the immune system. Latest studies are evident that protein plays an important role in the formation of natural as well as acquired immunity against foreign bodies (Kavacabey, 2012).

It is well known that protein is most important for boosting the immune system. PEM or protein energy malnutrition (PEM) is the major cause of secondary immune deficiency. Lessen caused by protein deficiency can alter the physical barrier. Skin is misplaced by collagen and connective tissue which makes them thinner (Saeed et al., 2016). It decreases the amount of antibodies in mucous and small intestine particularly in microvilli which leads to a more complementary atmosphere for the aggressor (Whitney et al., 1998).

Children suffering from severe PEM, can ultimately cause the degeneration of the thymus and hence lead to decrease cells and poor development of lymphoid organs. The loss of immune function in malnourished children is related to reducing leptin levels while re-feeding recovers immune function due to the increase in leptin concentration (Saeed et al., 2016; Matarese et al., 2005). Peripheral "T" lymphocytes from ill children suffering from PEM have less cell-mediated immunity and lack the defensive IgE antibodies (Saeed et al., 2016; Whitney et al., 1998; Ing et al., 2000).

5.8.1 ARGININE

Arginine is a conditionally essential amino acid. It mainly contains nitrogen for growth, illness, and metabolic stress. It also helps in wound healing and the progression of lymphocyte and phagocytic activity. It normalizes the T cell after a serious injury. It maintains nitrogen balance and stimulates the anabolic hormonal cycle (Marti et al., 2021).

Arginine availability, synthesis, and catabolism are all interconnected elements of immune responses, and fine-tuning them can result in pro-inflammatory or anti-inflammatory immune responses. It has emerged as a molecule having an effect on a variety of immune functions; arginine metabolism is implicated in conditions, from infectious diseases to autoimmunity and cancer (Marti et al., 2021).

5.8.2 GLUTAMINE

One of the non-essential amino acid-Glutamine, present abundantly in human muscle and utilized by lymphocytes to obtain energy and helps in nucleotide biosynthesis. It is important for lymphocytes, gut mucosa, stem cells, cell division, and phagocytic activity. It is an important nutrient for multiplying cells (like blood cells, intestinal cells, cancer cells, etc.).

Glutamine is the most common free amino acid in blood and plasma, and leucocytes use it at a rapid rate to provide energy and create the best circumstances for nucleotide production. Indeed, glutamine is considered important, if not vital, to lymphocytes and other rapidly dividing cells, including the gut mucosa and bone marrow stem cells. It is also needed for optimal macrophage phagocytic activity (Kavacabey, 2012; Evoy et al., 1998).

Glutamine belongs to the glutathione (GSH) pyrimidine nucleotides, gamma amino butyric acid syntheses category of antioxidants. It provides restorative effects for the digestive system in specific circumstances, such as cancer therapy and intestinal damage. Glutamate metabolism is regulated by organs such as the liver, intestines, and skeletal muscles, which also regulate glutamine circulation and secretion. Glutamate can be vital and play a key function in the body's metabolism in catabolic and hypercatabolic states. The disturbance of homeostasis in the inter-tissue metabolism of amino acids may jeopardize its supply. This is why glutamine supplementation is often utilized in therapeutic nutrition and is especially advised for immune-compromised individuals (China et al., 2020). It is a precursor of GSH, maintains acid-base balance, and is important for nitrogen transport within organs. Supplements can be given to the patients orally or through vascular access.

Because glutamine levels in plasma and bodily tissue are 10 to 100 times greater than any other amino acid, glutamine is considered the most abundant amino acid in the mammalian body. The glutamine level in an animal's body is determined by the synthesis and usage of body/uptake by various organs and tissues. The brain, liver, lungs, adipose tissues, and skeletal muscles produce the majority of glutamine. Various enzymes, including phosphate-dependent glutaminase (GLS) and glutamine synthetase, are involved in glutamine metabolism (GS). Glutamate availability is required for glutamine production by GSH synthetase. According to the current study, infections caused by bacteria like Escherichia coli may alter its harness and GS to reduce the effects of copper poisoning and acid stress (China et al., 2020). As a result, infections can adapt and endure by altering central metabolic pathways, which is critical for the host's defense against bacterial assault.

5.8.3 FAT

There are two necessary fatty acids that the body cannot produce, like linoleic acid (LA) and alpha-linolenic acid, which must be acquired through the diet additionally. These essential fatty acids can then be elongated and de-saturated in the body cells forming the –6 and –3 forms of PUFA. The major end products of the –3 pathway are eicosapentaenoic acid (EPA) and docosahexaenoic acid (DHA), but the major end product of the –6 pathway is arachidonic acid (AA). Omega-6 fatty acids account for the majority of PUFA in the food supply and the prime in all diets whereas AA is the forerunner of prostaglandins, leukotrienes, and other substances that play a vital part in immune control (Keservani et al., 2010a; Kavacabey, 2012).

5.8.4 OMEGA 3 FATTY ACID

Dietary fatty acid, especially long-chain omega 3 fatty acid is beneficial in boosting immunity and reducing the inflammatory response, when consumed it replaces the AA by increasing their fluids within the membrane and improving phagocytic activity and hence improving immunity. And also they are the precursor for cell signaling. Omega 3 fatty acid helps in boosting immunity in many ways like improving B cells, decreasing cytokines and inflammation, and increasing phagocytosis. DHA provides cell protection from oxidation of cells (BourBour et al., 2020; Keservani et al., 2010b).

A study reveals that a formula containing EPA, DHA, and GLA (gamma-linolenic acid) is very beneficial to critically ill patients (like sepsis, acute respiratory distress syndrome, and lung injury) it improves oxygenation and reduces the ventilation stay period (Gadek et al., 1997). It shows a positive effect on sepsis patients by improving the CD4–CD8 ratio and thus improving their immunity. It improves the partial pressure of the oxygen (BourBour et al., 2020).

5.8.5 FIBER

Foods rich in fiber promote a healthy gastrointestinal system. Beneficial bacteria in the colon convert them into short-chain fatty acids (SCFAs), which are utilized as fuel by gastrointestinal tract (GIT) cells. They also aid in the clearance of pollutants that might harm your GIT cells, as well as supporting general digestive health. Many studies reveal that fermentable dietary fiber can modulate the immune system which includes gut-associated lymphoid tissues (GALT). Changes in the intestinal microflora can mediate the immune system by LAB or by their production. The production of short-chain fatty acids (SCFAs) from fiber fermentation (Keservani et al., 2020; Saeed et al., 2016) (Table 5.1).

TABLE 5.1 Micronutrients Effects on Several Elements of Immunity

Micronutrients	Role in Barrier Function	Role in T-Cells-Mediated Immunity	Roles in B-Cells-Mediated Immunity
Vitamin A	Epithelial development, B-cells, T-cell homing in the gut, intestinal immunoglobulin A+ cells, and epithelial integrity are all aided by this compound.	Controls Th1 and Th2 cell proliferation and differentiation promote the conversion of naïve T cells to regulatory T cells and regulate the production of IL-2, IFN-, and TNF.	Immunoglobin A production is required for B cell function.

TABLE 5.1 *(Continued)*

Micronutrients	Role in Barrier Function	Role in T-Cells-Mediated Immunity	Roles in B-Cells-Mediated Immunity
Vitamin B6	Promoting gut homing of T cells.	T-cell differentiation, proliferation, and function are increased, particularly in Th1 cells, whereas IL2 production is reduced (promoted).	Antibody synthesis should be encouraged.
Vitamin B9	In the small intestine, a survival factor for regulatory T cells.	This chemical enhances T-cell proliferation and the Th1 response.	Encourage the synthesis of antibodies.
Vitamin B12	For the gut flora, this is an important cofactor.	Promotes T-cell development, proliferation, and function, particularly in cytotoxic T cells; maintains a balance between T helper and cytotoxic T cells.	Antibody manufacturing requires it.
Vitamin C	Promote collagen production, keratinocyte differentiation, wound healing, and anti-oxidant protection.	IFN production is controlled, and T cells, particularly cytotoxic T cells, are generated, differentiated, and proliferated.	Enhance antibody production.
Vitamin D	Promote the development of antimicrobial proteins, the tight connection between the stomach and the skin, and cell homing.	Increases antigen processing but inhibits antigen presentation; can inhibit T-cell proliferation, Th1-cell function, and cytotoxic T-cell function; can inhibit T-cell proliferation, Th1-cell function, and cytotoxic T-cell function; can inhibit T-cell proliferation, By encouraging the production of regulatory T cells, it inhibits dendritic cell development and maturation.	Antibody production may be reduced.

TABLE 5.1 *(Continued)*

Micronutrients	Role in Barrier Function	Role in T-Cells-Mediated Immunity	Roles in B-Cells-Mediated Immunity
Vitamin E	Defend against oxidative stress.	Promotes T-cell proliferation and function, particularly in Th1 cells, by promoting contact between dendritic cells and T cells.	Encourage the synthesis of antibodies.
Zinc	Maintain skin and mucosal membrane integrity while promoting complement activity.	Th1 cell response is promoted; cytotoxic T cell proliferation is promoted; regulatory T cell growth is promoted; IL-2 and IFN-production are controlled (promoted); Th9 and Th17 cell development is inhibited.	Encourage the synthesis of antibodies, notably immunoglobin G.
Iron	Essential for epithelial tissue growth and differentiation.	T cell growth and proliferation, as well as IFN production, are all controlled.	–

IFN: interferon, IL: interleukin, NK: natural killer; Th: T-helper; TNF: tumor necrosis factor.

Source: China et al. (2020).

5.8.6 MICRONUTRIENTS

Vitamins are micronutrients having physiological effects on various biological responses, like host immunity. Their deficiency leads to an increased risk of developing infectious, allergic, and inflammatory diseases.

5.8.7 VITAMIN A

Vitamin A supplementation is evident to reduce the rate of infection. Vitamin A helps in growth, reproduction, and blood cell formation, and improves the body's immune response. β-carotene is helpful as an antioxidant that enhances the immune response. The retinoic acid derivative of retinol increases the multiplication of lymphocytes and T-cells at the site of inflammation or the infection in gut. It is essential to competent innate responses. A deficiency of vitamin A leads to the destruction of natural killer (NK) cells, and eosinophils, and reduces the phagocytic and macrophage activity of cells. Supplementation of vitamins reduces the chances of infectious diseases like measles-related pneumonia, HIV, and malaria. The increased amount of supplement dose of vitamin A shows a negative correlation between vitamin A and immunity. A severe deficiency in children shows, the disappearance of goblet cells, and hence the production of mucus get hampered and it became a major cause of disease-causing agent. Vitamin A has phagocytic oxidative activities in macrophages. It is important for immunocompetence. They help in the structural and functional integrity of mucosal cells in the innate barrier, e.g., skin, respiratory tract, etc. (BourBour et al., 2020; Stephensen, 2001; Semba, 1999).

5.8.8 VITAMIN B1

Vitamin B could be a compound for many enzymes that are concerned in the TCA cycle. It is obvious that the TCA cycle is related to the useful management of the system, referred to as immunometabolism. Accumulating proof suggests that energy metabolism—particularly the balance between metastasis and therefore the TCA cycle—is related to the useful management of immune cells, in what's currently noted as immunometabolism (Methis et al., 2011).

In immunometabolism, regulative kind cells are used for energy generation, and activated or professionally activated used for metastasis (Buck et al., 2017).

Its deficiency causes lethargy, and hypovitaminosis sickness and might have an effect on the vas and system too. World Health Organization (WHO)/ Food and Agriculture Organization (FAO) advocates a daily thiamin intake of 1.1–1.2 mg for adults.

5.8.9 VITAMIN B2

In hepatoflavin (riboflavin) and its 2 co-factors ketone purine dinucleotide (FAD) and ketone mononucleotide (FMN)) are utilized in the TCA cycle and beta chemical reaction. Vitamin B2 deficiency suppresses the activity of acylCoA dehydrogenases concerned within the chemical reaction of fatty acids to get acetyl-CoA, which is employed by mitochondria to provide adenosine triphosphate via the TCA cycle. A carboxylic acid chemical reaction is concerned with the activation, differentiation, and proliferation of immune cells through the generation of acetyl-CoA and its entry into the TCA cycle. This means that hepatoflavin is related to the management of differentiation and performance of immune cells through the regulation of carboxylic acid chemical reactions. In addition to energy generation, hepatoflavin is related to reactive atomic number 8 species (ROS) generation in immune cells through the priming of NADPH enzyme two (Schramm et al., 2019); ROS are necessary effector and signaling molecules in inflammation and immunity (Yoshii et al., 2019). Green leafy vegetables, organ meats especially liver, and eggs are made in hepatoflavin. Dietaryhepatoflavin exists as rage or FMN and is regenerated to free B-complex vitamin by rage pyrophosphatase and FMN enzyme within the gut. Free B-complex vitamin is absorbed via B-complex vitamin transporter expressed on the animal tissue of the little gut and is then discharged into the blood. Within the blood, free B-complex vitamin is regenerated back to rage or FMN and distributed throughout the body.

5.8.10 VITAMIN B3

Vitamin B3 (niacin) is usually referred to as B vitamin and nicotinamide. These compounds are precursors of nicotinamide purine dinucleotide (NAD), a molecule that plays a key function in aerobic respiration as part of the cellular oxidation-reduction process. GPR109a, a G-protein coupled receptor expressed on a variety of cells, including immune cells, is also a substance of nourishment B3. Sustenance B3-GPR109a sign induces differentiation of regulative T cells and suppresses inflammation during a GPR109 a dependent manner. Sustenance B3 conjointly inhibits the assembly of the pro-inflammatory cytokines IL-1, IL-6, and growth mortification issue alpha (TNF-α) by macrophages and monocytes. Thus, sustenance B3 has medicine properties by modulating host immune cells and taking part in a vital role in the maintenance of immunologic equilibrium. Indeed, in humans, sustenance B3 deficiency causes mal rosso, which could be a sickness characterized by internal organ inflammation, diarrhea, dermatitis, and dementedness (Yoshii et al., 2019).

Vitamin B3 is found in animals foods like meat and fish within the type of nicotinamide and plant-based mostly foods like beans contain sustenance B3 as {nicotinic acid|niacin|B-complexsustenance|Bcomplex|vitamin B complex|vitamin B|B vitamin|B}. And B vitamin is regenerated into nicotinamide both are directly absorbed in the gut (Goa et al., 2018).

5.8.11 VITAMIN B5

Pantothenic acid or water-soluble vitamin five is a crucial compound for the TCA cycle and carboxylic acid and acts as a coenzyme A. Vitamin B5 generates energy for immune cells. Its deficiency causes sleep disorders and fatigue. CoAreswitches with sustenance B5 or cysteamine through {pantothenic acid|pantothen|B-complexsustenance|Bcomplex|vitamin B complex|vitamin B|B vitamin|B|acid} to take care of vitamin B5 levels throughout deficiency (Yoshii et al., 2019; Leonardi et al., 2005). However, cysteamine suppresses signal transduction of the peroxisome proliferator-activated gamma (PPARγ) and induces inflammation (Yoshii et al., 2019; Zhang et al., 2011). In fact, inflammation was improved in plantain-producing protein knockout mice. Therefore, sustenance B5 deficiency causes borders of nutrition. B vitamins and host immunity Ignition through the disfunction of animal tissue barriers and production of pro-inflammatory molecules (Zhang et al., 2011).

It promotes promote the protection action against mycobacterium by enhancing each innate and adjustive immunity. Vitamin B5 is found mainly in the liver, egg, chicken and broccoli members of the cabbage family, etc. CoA and pantothenic acid regenerate from free acid within the gut by endogenous enzymes. The free acid is absorbed and discharged into the blood via a sodium-dependent vitamin pill transporter (SMVT) that expresses within the animal tissue of the little gut (Yoshii et al., 2019; Quick, 2015). Finally, free acid is regenerated to CoA and distributed throughout the body, particularly within the liver and kidneys.

5.8.12 VITAMIN B6

Pyridoxine, pyridoxal, and pyridoxamine are some of the several forms of vitamin B6. These vitamin B6 forms are precursors to the coenzymes pyridoxal phosphate (PLP) and pyridoxamine phosphate (PMP), which are involved in a variety of cellular metabolic activities such as amino acid, lipid, and carbohydrate metabolism. Inflammatory illnesses including allergies and RA, as well as neurological dysfunction, are linked to vitamin B6 deficiency (Yoshii et al., 2019; Doke et al., 1997; Jannusch et al., 2017). A lack of vitamin B6 causes the Th1–Th2 balance to shift toward an overactive Th2 response, resulting in allergies (Qian et al., 2017). Low plasma vitamin B6 levels, as well as higher levels of pro-inflammatory cytokines including TNF- and IL-6, have been reported in RA patients.

Vitamin B6 is lavish in fish, chicken, tofu, sweet potato, and avocado. Dietary adermin exists as PLP or PMP; it is regenerated to free adermin by endogenous enzymes such as the adermin enzyme and is then absorbed by the small intestine. This absorption of aderminis acidic pH-dependent and carrier-mediated transport has also been shown. An intestinal pyridoxamine transporter is however to be known in any mammalian species (Said, 2011). When the absorption of free aderminoccurs, it enters the blood and is regenerated back to PLP or PMP.

5.8.13 VITAMIN B7

Vitamin B7 (biotin) may be a chemical compound for many carboxylases that area unit essential for aldohexose, amino acid, and carboxylic acid metabolism (Tong, 2013). As an example, victuals B7 is an important chemical compound for ethanoyl radical CoA carboxylase and carboxylic acid synthase, an area unit enzyme concerned with carboxylic acid biogenesis (Ikeda et al., 2013; Ikeda, 2017).

Vitamin B7 is lush in foods like haywire, beans, and seed. However, raw egg white contains an outsized quantity of avidin, that binds powerfully to victuals B7 and prevents its absorption within the gut (Zempleni et al., 2008). Therefore, victuals B7 deficiency is caused not solely by depleted victuals B7 intake, but additionally by excessive intake of raw egg white. Dietary B exists as a free protein-bound type or as a biocytin (Said, 2011). Within the bowel, Bidase releases free B from the sure macromolecule, and also the free B is absorbed via the biotin transporter (Zempleni et al., 2009).

5.8.14 VITAMIN B9

Vitamin B9 (folate), in its active kind as tetrahydrofolate (THF), is a chemical compound in many metabolic reactions, as well as polymer and organic compound synthesis. Attributable to the high demand of victuals B9 by red blood cells (RBC), vitamin B9 deficiency results in malignant anemia (Said, 2011). Vitamin B9 deficiency conjointly inhibits the proliferation of human CD8+ T cells *in vitro* by stunning the cell cycle within the S part and increasing the frequency of polymer injury (Courtemanche et al., 2004). Foods like beef liver, inexperienced bifoliate vegetables, and asparagus contain high levels of victuals B9. Victuals B9 exists as each mono- and polyglutamate B species within the diet (Melse-Boonstra et al., 2002). Folatepolyglutamate is deconjugated to the mono glutamate kind and then absorbed within the gut via B transporters such as proton-coupled B transporter (PCFT) (Said, 2011; Zhao et al., 2011). In the viscus animal tissue, B monoglutamate is reborn to THF, a lively kind, and chemical compound, prior to being transported to the blood (Said, 2011).

5.8.15 VITAMIN B12

Vitamin B12 (cobalamin) may be a cobalt-containing nourishment that, in its active styles of methylcobalamin and adenosylcobalamin, catalyzes essential amino acid synthesis. Together with B-complex vitamin and B9, the antipernicious anemia factor plays necessary roles in red vegetative cell formation and supermolecule synthesis, especially in neurons. Therefore, antipernicious anemia factor deficiency causes megaloblastic anemia and systema nervosum symptoms such as symptoms of the hands and feet (Bite et al., 2016). When it comes to hosting immunity. In mice, dietary antipernicious anemia factor deficiency reduces the number of CD8+ T cells and lowers NK T cell activity; methylcobalamin supplementation improves these circumstances. Tamura et al. (1999) suggest that the antipernicious anemia factor contributes to the reaction via CD8+ T cells and NK T cells. Antipernicious anemia factor is abundant in beef liver, bivalves, fish, poultry, and eggs. Dietary B {complex|vitamin B complex|vitamin B|B vitamin|B} exists in complex with dietary supermolecule and is rotten to free antipernicious anemia factor by an enzyme in the abdomen. Free antipernicious anemia factor is absorbed by the animal tissue cells of the tiny bowel via factor (IF), a gastric glycoprotein. Within the animal tissue cells, IF-vitamin B12 complicated is rotten to free antipernicious anemia factor by organelle so free into the blood, wherever it's born-again to the active kind and distributed throughout the body (Beedholm-Eblen et al., 2010; Green, 2017).

5.9 BCAA AND IMMUNITY

Although BCAAs (branched-chain amino acids) are important for the development of immunity *in vitro*, The mechanism behind their essentiality does not appear to be their role as energy-generating substrates. This is often a result of cells in culture having several different energy sources out there to them, as well as aldohexose and amino acid, and the relative involvement of BCAAs to energy generation is small. Hence, a complete lack of any single BCAA isn't seemingly to impair the power of cells to get energy. BCAAs conjointly act as donors of N and of C skeleton for the synthesis of other amino acids, for example – amino acid, that square measure necessary in supporting immune cell operation (Philip et al., 2006; Calder, 1994, 1999) without doubt, this is crucial at the whole-body level and should account for a few of the immunological edges of enhanced-BCAA offer to athletes (Bassit et al., 2002) and postsurgical and septic patients (Nuwev et al., 1983) who often show partial depletion of muscle and plasma glutamine (lund et al., 1986). However, cell culture studies conducted within the presence of the best amino acid concentrations reveal an associate degree of absolute demand for BCAAs (Chuang et al., 1990), suggesting that the BCAAs themselves square measure required, a minimum of by isolated immune cells. It looks presumably that the essentialness of BCAAs for immune cells is linked to supermolecule synthesis. Because generating an associate degree response necessitates the production of new cells and the synthesis of antigen-presenting apparatus, immunoglobulins, cytokines, protein receptors, acute section proteins, and other supermolecules, the immune system has a significant reliance on supermolecule synthesis. The lack of BCAAs makes it difficult for those proteins to be synthesized properly, which might prevent the best reaction from being generated. The failure to generate a supermolecule eliminates lymphocytes' capacity to respond to stimulation in cell culture situations where one BCAA is missing. Wherever the amount of BCAA available is restricted, cell culture and animal feeding experiments clearly reveal that there is a dose-response relationship between BCAA offer and response. This association, however, is not well defined.

5.9.1 VITAMIN C

Vitamin C, water-soluble vitamin gets higher constituents in the human system as antimicrobial and NK T cell actions, white corpuscle proliferation (Rayment et al., 2003), taxis, and delayed-type hypersensitivity.

Upon activation, phagocytes unleash a deposit of oxidizing substances with the power to exterminate viruses and bacterium, however, these oxidants also can be prejudicial to the host cells; equally, the oxidants generated throughout infective agent associated microorganism infections play an extensive role within the production of pathogens (Akaike et al., 1998). Water-soluble vitamins could be a competent soluble inhibitor that has protecting role for the host cells aboard the oxidants discharged by phagocytes and pathogens (Hemila, 1997). Water-soluble vitamin tends to uphold the chemical reaction and cut back the truthfulness of cells, so works against reactive element species (ROS) that are unit shaped within the method of respiration and the inflammatory reactions (Wintergerst et al., 2006). Higher amounts of ROS generated have roots in the aerophilic problem and ultimately escort to the dysfunction of immune rejoinder in conjunction with the failure of integrity and liquidity of cell wall. Whereas contact between cells is varied, all of the changes occurred because of aerophilic reactions lead to chronic diseases having chronic issues within the body. A lot of specific and necessary operations of vitamin C within the system are to unleash the aerophilic burden caused by the method of infection, so having physiologically imperative position through non-specific mechanisms (Hemilä, 1997). For the utilization of modified aliment, phagocytes have special channels through which it's transported into the cell to regenerate the reduced type (Nualart et al., 2003). Animate thing stress of ROS is considerably contributed by mitochondria and to safeguard this energy center from chronic changes because of oxidization. Vitamin C gets entered into it through an aldohexose transporter and protects it (Sagun et al., 2005).

Supplements may stimulate the system by enhancing the proliferation of T lymphocytes in response to infection increasing the synthesis of immunoglobulins and protein production. Most animals are ready to fill again their offer of nutrition C by manufacturing this immune booster in their livers. On the contrary, the Guinea pigs and this rodent's distant cousin—Homo sapiens—are ineffective to create their ascorbic acid because they lack the actual catalyst that synthesizes this vitamin. Owing to the lack of synthesized ascorbic acid, it is important that people consume adequate amounts of this nutrition from foods made of ascorbic acid. Once supplementation of ascorbic acid is consumed, the tissue level first-rate off at a pair of to 3 g whereas the surplus is excreted into body waste.

5.9.2 VITAMIN D

Vitamin D receptors are expressed on immune cells, particularly B cells, T cells, and antigen-presenting cells, which are all capable of synthesizing the active vitamin D substance. However, the very best concentration of vitamin D receptors is found within the immature immune cells of the thymus. Vitamin D, especially in its active form 1, 25, dihydroxycholecalciferol [1, 25 (OH) 2 D3], absorbs calcium and helps in bone mineralization.

Cholecalciferol (D3) causes a modulation in the system by golf stroke a stop to the unnecessary look of inflammatory cytokines and boosting the aerobic activity of macrophages. At the utmost likelihood, it causes the looks of powerful antimicrobial peptides (AMPs), that area unit gift in neutrophils, monocytes, and in animal tissue cells lining the respiratory tract wherever they resist the lungs from infections (Meydani et al., 2005). The studies on reaction diseases have incontestable a positive relation with vitamin D standing particularly for rheumatism, degenerative disorder, etc.; it acts as a sovereign shielding unit manipulating Th-1 mediate reaction operation (Hypoponen et al., 2001).

Inadequate and reduced responses from innate and antibody-specific immunity result in infection vulnerability due to vitamin D deficiency (Wintergerst et al., 2006). Vitamin D works as a modulator in the T-cell antigen receptor and plays a significant role in the regulation of immune responses via vitamin D receptors found in the nuclei of antigen-presenting cells and activated T-cells, as well as vitamin D responsive factors found in the promoter region of cytokine genes (Vanetten et al., 2005).

The deficiency of vitamin D causes a higher receptiveness to activate Mycobacterium tuberculosis infection (Bikle, 2008). Studies have demonstrated that macrophages also stimulate the expression of vitamin D and stimulate the genes of an enzyme that converts cholecalciferol to its most active form, and this process results in the antimicrobial protein's appearance, ultimately leading to destroy the *M. tuberculosis* present within the cell (Liu et al., 2006).

5.9.3 VITAMIN E

Vitamin E inhibits the synthesis of immune modulators and protoplasm since it is a fat-soluble inhibitor, but also improves the immunological response. Taking antioxidants stimulates the sweetening of white corpuscle, aids in the death of foreign cells, aids in the eradication of germs by special cells, and improves resistance to infectious agents. It conjointly enhances immune body productivity. Despite the fact that completely different results are obtained, multiple infections have been observed to be reduced, however, there has been no effect on tract infections. An antioxidant is effective in strengthening the system. It improves the body's resilience against infection and alternative infections and prevents vitamin A oxidation from obtaining change. It's vital to consume sensible sources of antioxidants like vegetables, hazelnut, walnut, and legumes adequately (Barba et al., 2012). Its inhibitor functions build antioxidants primarily effective against infections.

5.9.4 VITAMIN K

Vitamin K is incredibly crucial for traditional activity and the modification of supermolecules that are dependent on it is supposed to be linked with some immune cells (Shand, 1984). Protein S one of the dependent supermolecules to vitamin K (Griener, 2011), is thought to be associated with C4B binding supermolecule (C4BP) and this relation indicates the United States of America positive ends up in favor of B cells (Blom et al., 2004).

5.9.5 MINERALS

Minerals are colorless chemicals present in foods that are essential for biological processes. Minerals are substances that the body cannot produce on its own. Minerals usually collaborate with vitamins and supply vitamin transportation to the foremost required areas. At the same time, they are involved in blood pressure, heart rhythm, muscle functioning, fluid balance, fertility, and a variety of other tasks. Mineral loss and deficiency have been shown in scientific research to have a direct impact on human health. There are at least 13 vitamins that are known to be poorly distributed in tissues and are all present in fish. Fish, in addition to being a good source of polyunsaturated fatty acids (omega-3), metal, phosphor, chemical elements, and iodine needed for brain development, also helps to boost the immune system (Barba et al., 2012). Aquatic merchandise area unit splendid metallic element and phosphor course, containing between 15 to 200 mg metallic element in each a 100 g and 100 to 400 mg in each a 100 gm in some species. Fish is usually recommended to people who need Na died with the containable 60 mg Na in a 100 g muscle. Within the magnitude, potassium is present in fish flesh, which serves as a catalyzer in the supermolecule and supermolecule metabolism, eurhythmy, nerve transmission, and muscle contraction.

5.9.6 TRACE ELEMENT

1. **Iron (Fe):** It is projected to possess a task in innate immune responses. It enhances immunity because of its presence as metalloenzymes or supermolecules like iron enzyme and lactoferrin (iron transporting protein in blood serum). It reduces the danger of diarrheal issues and morbidity. The genes/proteins that square measure concerned in the iron physiological state also have connected immunologic functions (Ward et al., 2011): each deficiency and copiousness have an effect on the system. Viruses and bacteria want iron for proliferation. As a result, administering iron during acute infectious infections should be avoided. The movement of white blood cells (WBCs) into infectious sites diminishes with iron shortage, and the killing of germs that have entered cells decreases. The fertility of leucocytes, responsible for defense, is affected negatively within the proliferative part of leucocyte activation is the Fe-requiring step, in the main as a result of Fe is essential for enzymes like ribonucleotide reductase concerned in deoxyribonucleic acid (DNA) synthesis (Brock et al., 2000). Iron is joined with the regulation of protein production and mechanism of action and within the activation of supermolecule enzyme C that is indispensable for phosphorylation of things that control cell proliferation deficiency of iron. However, cells, responsible for the production of protein, and what is known as body substance immunity aren't littered with iron deficiency. It can be emphatic that, in areas wide unfold of malaria; iron support to kids will increase the complications.
2. **Zinc (Zn):** It is needed for the best functioning of each innate and bought immunity. Its deficiency ends up in a rise in fungal, viral, and microorganism infections, a decrease within the endocrine and lymph cell variety, and additionally ends up in diarrhea, malabsorption, and physical impairment (Blom et al., 2011). Atomic number 30 deficiency causes diarrhea that is joined to the deteriorated immune system and internal organ membrane cell transport. It's illustrious to point out each medicinal drug and inhibitor activity (Kavacabey, 2012). It acts as a catalyst and regulates enzymes, proteins, and transcription factors and is, therefore, an important element in immune responses. It will increase the secretion of IL-1 And has an antagonizing result on toxic metals like nickel that suppress the system (Contora et al., 2012). It is required for several catalyst activations together with

DNA and RNA synthesis. It's additionally an inhibitor result. In deficiency, cellular immunity fails, acrodermatitis and enteropatica deteriorate, fungus, virus, and microorganism infections increase, endocrine, and lymphocytes decline, and rations alter diarrheal, absorption, and growth delay.

3. **Copper (Cu):** It plays a role in leukocyte differentiation, maturation, and proliferation. Studies show that Cu reduces the individual's susceptibility to bacterial agents (Djoko et al., 2015). Mild to moderate deficiency of Cu causes a reduction in Interleukin 2 and T cell proliferation (Alpert et al., 2017). In severe copper shortage, the number of neutrophils in human peripheral blood is decreased. Not solely area units they reduced in variety however their ability to kill ingested microorganisms is additionally reduced in each obvious and marginal copper deficiency (Percival, 1998).
4. **Selenium (Se):** It is concerned with biological functions because of its integration into selenoprotein which is concerned with the activation, proliferation, and differentiation of cells that create innate and adaptive immune responses. Dietary consumption of Se and selenoproteins is vital for initiating or enhancing immunity, and also are concerned with immunoregulation. This helps in preventing pathology or chronic inflammation (Huang et al., 2012). It plays a crucial role in reduction-oxidation regulation and inhibitor activities. It additionally protects people from the heart-damaging effects of herpes (Alpert et al., 2017).

5.9.7 PROBIOTIC AND PREBIOTIC

Beneficial microorganisms are additional nutrients in probiotic products. The intestines are the body's largest immunological organ. Intestinal bacteria are always in communication with the immune system. This connection is critical for the development of the immune system. Probiotics improve the immune system's overall response as well as the immune system's gastrointestinal system.

They have enhanced humoral immune responses, particularly intestinal IgA responses, and thereby promote the intestine's immunologic barrier. They assist immune clearance by stimulating nonspecific host resistance to microbial infections. Probiotics are non-pathogenic bacteria that control mucosal and systemic immunity when consumed with food and manage the nutritional and microbial balance (Kavacabey, 2012). Immunological system stimulation, modulation, protection against enteric infections, and immune inflammatory illnesses are some of the advantages of probiotic ingestion (inflammatory bowel diseases) Lactose intolerance symptoms are reduced, blood cholesterol levels are reduced, and cancer is prevented (Khan et al., 2012). It has been said that probiotics strengthen the mucosal defense systems and affect immune stimulation. Probiotic microorganisms boost the host's immune system by influencing both specific and natural immunity. The presence of probiotic bacteria in the intestines of viable cells is also thought to boost and improve the immune system (Kavacabey, 2012).

Probiotic products are beneficiary being additional nutrients. The intestine is called the largest immune organ in the body. Intestines microorganism invariably interactivity with the system microorganism in the intestines. This interactivity is incredibly vital for system development, they need increased body substance immune responses, notably enteral immune gamma globulin responses, and thereby promote the intestine's immunological barrier. They stimulate the non-specific host resistance to pathogenic microorganisms. Non-pathogenic microorganisms, that regulate tissue layer and general immunity by taking foods and that regulate the biological process and microorganism balance, are referred to as probiotics (Kavacabey, 2012). Probiotic consumption has some immune-inflammatory diseases benefits; system stimulation, regulation, protection against enteric infections, and immune-inflammatory diseases (inflammatory gut diseases) mitigation of genetic disorder symptoms, reduction in blood cholesterol level, and prevention of cancer (Khan et al., 2012). It is now clear that probiotics strengthen the membrane defense systems and affect immune stimulation. Probiotic microorganisms stimulate the system of the host, by effecting specific immunity and resistance. It's additionally been explicit that, the presence of probiotic microorganisms in viable cell's intestines, stimulates, and strengthens the system (Kavacabey, 2012).

Probiotics and prebiotics are present within the lamiapropria and among Peyer's patches, surveying the contents of the gut lumen to envision what has been eaten for potential pathogens and providing protection (Rescigno et al., 2001). Microorganism prevents the colonization of infective bacterium by, increasing the production of metabolites like SCFAs and dairy product acid, that additional inhibit the expansion of infective bacteria, and enhance host immunity domestically and systemically through a range of mechanisms (Rook et al., 2016). Due to the connection between the gut microbiota and the system, it looks logical that dietary components that are incontestible to favorably modification the composition of the gut microbiota, such as prebiotics (So, et al., 2018) and probiotics which briefly colonize the gut, providing opportunity for interactions with commensal bacterium, pathogens, and host tissues through direct contact or released substances (Calder et al., 2020), might influence immunity (Table 5.2).

TABLE 5.2 Role of Nutrients in Supporting Immune System and Their Sources

Micronutrients	Role in Innate Immunity	Role in Adaptive Immunity	Food Sources
Vitamin C	When pathogens are eliminated by immune cells, this antioxidant protects against the development of reactive oxygen and nitrogen species (ROS and RNS). Other important antioxidants, such as glutathione and vitamin E, are reactivated. Collagen synthesis is enhanced, which aids in the maintenance of epithelial barrier integrity. The generation, function, and mobility of leukocytes (such as neutrophils, lymphocytes, and phagocytes) are all stimulated. The amount of complement protein in the blood has risen. Has antimicrobial, NK cell, and chemotaxis properties. Apoptosis and macrophage clearance of waste neutrophils from infection sites are also involved.	Antibody levels in the blood might be raised. Has a function in lymphocyte proliferation and differentiation.	Green vegetables and fruits are rich in vitamin C.

TABLE 5.2 *(Continued)*

Micronutrients	Role in Innate Immunity	Role in Adaptive Immunity	Food Sources
Vitamin D	The vitamin D receptor is expressed by innate immune cells such as monocytes, macrophages, and dendritic cells. Increases the conversion of monocytes to macrophages. Immune cell proliferation and cytokine production are boosted, which aids in pathogen defense. Vitamin D's active form, 1,25-dihydroxy vitamin D3, regulates the antimicrobial proteins cathelicidin and defensin, which may fight bacteria and other diseases directly.	1,25-dihydroxy vitamin D3 suppresses adaptive immunity by lowering antibody formation in B cells and inhibiting T cell proliferation, for example.	Oily fish, eggs, fortified cereals, and dairy products are all good sources of omega-3 fatty acids.
Vitamin A	Maintains the structural and functional integrity of mucosal cells in innate barriers (such as the skin and respiratory tract) and is required for innate immune cells to function appropriately (e.g., NK cells, macrophages, neutrophils).	Required for proper T and B cell activity and, as a result, the generation of antigen-specific antibody responses. Growth and differentiation of Th1 and Th2 cells, as well as the anti-inflammatory response of Th2.	liver, eggs, and cheese dark green leafy vegetables, and orange-colored fruits and vegetables.
Vitamin E	It's a fat-soluble antioxidant with a lot of strength. Protects cell membrane integrity against free radical damage. Increases the production of interleukin-2 (IL-2) as well as the cytotoxic activity of natural killer cells (NK cells).	The functions of T cells and lymphocyte proliferation have improved. The th1 response is amplified and optimized, whereas the Th2 response is repressed.	Plant oils such as vegetables, sunflower, soya, and olive oils. Nuts and seeds.

TABLE 5.2 *(Continued)*

Micronutrients	Role in Innate Immunity	Role in Adaptive Immunity	Food Sources
Vitamin B6	Aids in inflammation control. Plays a part in the generation of cytokines and the activity of natural killer (NK) cells.	Amino acid metabolism and synthesis, which are the building blocks of cytokines and antibodies, are both necessary. Proliferation, differentiation, and maturation of lymphocytes are all functions of this protein. Th1 immune response is maintained. Has a function in the synthesis of antibodies.	–
Vitamin B12	Play a vital part in the functioning of NK cells.	It's possible that it works as a cellular immunity immunomodulator, affecting cytotoxic cells (NK cells, CD8+ T cells) in particular. The generation of T lymphocytes is enhanced. Immunity at the cellular and humoral levels, as well as one-carbon metabolism, are all implicated (interactions with folate).	–

TABLE 5.2 *(Continued)*

Micronutrients	Role in Innate Immunity	Role in Adaptive Immunity	Food Sources
Folate	It keeps the body's natural defenses in place (NK cells).	It plays a part in cell-mediated immunity. Antibody response to antigens must be adequate. Supports the immune response mediated by Th1.	Green vegetables, pulses, fruits, nuts, some cheeses, brown, and whole meal bread, and fortified breakfast cereals.
Zinc	Antioxidant qualities protect against reactive oxygen and nitrogen species (ROS and RNS). Modulating cytokine release aids CD8+ T cell proliferation. Maintains the skin's and mucosal membranes' integrity.	In cellular development and differentiation, immune cells with a high rate of differentiation and turnover play a crucial role. The intracellular binding of tyrosine kinase to T cell receptors is required for T lymphocyte growth and activation. The Th1 response is encouraged.	Protein may be found in meat, poultry, cheese, shellfish (such as crab, cockles, and mussels), nuts and seeds (especially pumpkin seeds and pine nuts), wholegrain breakfast cereals, and wholegrain and seeded bread.
Iron	Involved in the production of cytokines and the regulation of their activity. Forms highly toxic hydroxyl radicals, which are necessary for neutrophils to kill germs. Reactive oxygen species (ROS) are important in the destruction of infections.	T-lymphocyte differentiation and proliferation are aided by this protein. Essential for cell differentiation and proliferation, and a component of enzymes necessary for immune cell function (e.g., ribonucleotide reductase involved in DNA synthesis).	Green vegetables, lentils, tofu, beans, etc.

Source: Adapted from Silvia et al. (2018).

5.10 SOME HERBS ARE EFFECTIVE MEDIATORS

5.10.1 ASHWAGANDHA

It is used as an Antioxidant. It notebly an increase in neutrophil adhesion and delayed-type hypersensitivity (DTH) response. It significantly boosts cellular immunity (Evoy et al., 1998). It provides various health benefits, including boosting brain function, lower blood sugar and cortisol levels, anti-inflammatory, immune stimulant energy endurance, adoptogen, antistress, and helps fight symptoms of anxiety and depression.

It Can also treat different diseases like cold and cough, diabetes mellitus, conjunctivitis, epilepsy, dementia, insomnia, leprosy, Parkinson's disease, rheumatism, arthritis, intestinal infections, bronchitis, asthma, impotence, and a suppressant in HIV/AIDS patients. And also been shown effective in the hematopoietic system acting as an immunoregulatory and chemoprotective agent.

5.10.2 GINGSENG

Gingseng contains triterpene glycosides, or saponins, commonly referred to as ginsenosides. Many active compounds can be obtained from all parts of the plant, including amino acids, alkaloids, phenols, proteins, polypeptides, and vitamins B1 and B2. It is said to be beneficial in many conditions like anemia, diabetes, gastritis, neurasthenia, erectile dysfunction, impotence, male infertility, fever, hangover, and asthma.

It is also used for treating bleeding disorders, loss of appetite, vomiting, colitis, dysentery, cancer, insomnia, neuralgia, rheumatism, dizziness, headache, convulsions, disorders of pregnancy and childbirth, hot flashes occurring due to menopause, and slow the aging process (Leonardi et al., 2005; Quick, 2015). Researchers have shown the immune system is stimulated by ginseng extract, and the standardized extract is more efficient than the liquid ginseng extract, according to a number of consistent researchers that studied the effects of *P. ginseng* extract on vaccine response.

5.10.3 AMLA

Amla also includes punicafolin and phyllanemblinin A, as well as flavonoids and other polyphenols (Keservani & Sharma, 2014), kaempferol, ellagic acid, and gallic acid. Amla is water-soluble vitamin-rich fruit that boosts the assembly of WBCs within the body which facilitates fighting many infections and diseases. Amla is additionally rich in Iron, calcium other minerals that build the whole fruits.

5.10.4 LIQUORICE

It is a root extract of *Glycyrrhizaglabra*. It holds anti-inflammatory as well as antiallergic activities that are mainly attributed to its bioactive constituents, like glycyrrhizin, aglycone, and beta-glycyrrhetinic acid (Kroes et al., 1997). Glycyrrhizin has anti-inflammatory and anti-allergic properties (Shibata, 2000), and its mode of action actually revolves around the production of IL-10 and IL-12 in large amounts. Some scientific evidence supports its use as an immune-stimulating agent that is involved in the proper metabolisms and functioning of various important mediators in adaptive or acquired immunity.

5.10.5 GARLIC

Garlic is thought to help the immune system maintain homeostasis. The garlic protein fraction has been discovered to have a stimulatory impact on lymphocytes, NK cells, and macrophage cytotoxic capacity.

The immunomodulatory properties of garlic and its bioactive profile include immune stimulation, reduce platelet aggregation, and chemoprevention of cancer (Gomboaleon et al., 2007). Garlic can scavenge free radicals and can serve as a successful antioxidant. The consumption of garlic results in an improvement in hematological parameters like total WBC count as well as homeostasis characteristics (Wilson et al., 2007). Consumption of garlic may also increase the production and release of nitric oxide (NO) which is responsible to reduce blood pressure and prevent proliferative diseases (Bhattacharya et al., 2007).

5.10.6 TURMERIC

Turmeric powder is about 60 to 70% carbohydrates, 6 to 13% water, 6 to 8% protein, 5 to 10% fat, 3 to 7% dietary minerals, 3 to 7% essential oils, 2 to 7% dietary fiber, and 1 to 6% curcuminoids. Turmeric contains 67.8 mg of iron per 100 g of turmeric powder, making it one of the most iron-rich foods. One teaspoon (3 g) of turmeric powder provides 2 mg of Iron. Iron is important for improving immunities power and turmeric has riches Iron (Vasudevan, 1994).

5.11 CONCLUSION

For healthy growth and development of the human body, appropriate, and balanced nourishment is essential. Nutrients have an important influence on immune system function.

Innate immunity and adaptive immune systems are two essential components of the human immune system. There are many nutrients that are linked with destruction of the cell-mediated immunity or phagocytic function, secretory immunoglobulin A, antibody concentrations, or cytokine production. Some nutrients have good immune responses like vitamin C, zinc, folic acid, selenium, copper, iron, and vitamins A, C, and E. Many studies are proven that weak immune systems are special because of nutrient deficiency of either macronutrients or micronutrients.

Immune functions vary in different age groups (low during infancy, old age) disease conditions having co-morbidities like hypertension, CVD, Diabetes mellitus), unhealthy lifestyle (pollution, smoking, alcohol consumption, less sleep, psychological stress, no physical activity), eating habits (junk foods, oily, and fried foods, salty food, processed foods, unhealthy fats), pathogen infection (viral or bacterial infection). Somehow, nutritional inadequacy leads to a nutrient deficiency which can lead to a weak immune response that invites various severe infections and ultimately death. Therefore, attention should be directed to addressing the nutritional inadequacies leading to weak immune systems for a better and healthy future for the county. So, proper knowledge, physical fitness, adequate nutrition, positive approach toward health can boost the immune system.

KEYWORDS

- **glutamine**
- **immune system**
- **immunity**
- **macronutrients**
- **micronutrients**
- **omega 3 fatty acid**
- **protein**

REFERENCES

Akaike, T., Suga, M., & Maeda, H., (1998). Free radicals in viral pathogenesis: Molecular mechanisms involving superoxide and NO. Experiment. *Reviews in Medical, Virology*. PMID: 9421208. doi: 10.3181/00379727-217-44206.

Alpert, P. T., (2017). The role of vitamins and minerals on the immune system. *Home Healthcare Management & Practice,* 199–202. Available form: Research Article https://doi.org/10.1177/1084822317713300.

Anywar, G., Kakudidi, E., Byamukama, R., Mukonzo, J., Schubert, A., & Oryem-Origa, H., (2020). Medicinal plants used by traditional medicine practitioners to boost the immune system in people living with HIV/AIDS in Uganda. *Eur. J. Integr Med., 35*, 101011. https://doi.org/10.1016/j.eujim.2019.101011.

Areeba, F., Harshitha, C., & Priyanka, A., (2020). Food and nutrition as natural immune-boosters: An elaborative review. *IJISET – International Journal of Innovative Science, Engineering & Technology, 7*(6). ISSN (Online) 2348 – 7968 | Impact Factor – 6.72.

Barba, F. J., Esteve, M. J., & Frigola, A., (2012). Impact of high-pressure processing on vitamin E (α-, γ-, and δ-tocopherol), vitamin D (cholecalciferol and ergocalciferol), and fatty acid profiles in liquid foods. *J. Agric. Food Chem., 60*(14), 3763–3768. doi: 10.1021/jf205355h. Epub 2012 Apr 2.

Barrett, D. J., Ammann, A. J., Stenmark, S., & Wara, D. W., (1980). Immunoglobulin G and M antibodies to pneumococcal polysaccharides detected by enzyme-linked immunosorbent assay. *Infect. Immun., 27*, 411–417. doi: 10.1128/iai.27.2.411-417.1980.

Bassit, R. A., Sawada, L. A., Bacurau, R. F. P., Navarro, F., Martins, E. Jr., Santos, R. V. T., Caperuto, E. C., et al., (2002). Branched-chain amino acid supplementation and the immune response of long-distance athletes. *Nutrition, 18*, 376–379. doi: 10.1016/s0899-9007(02)00753-0.

Beedholm-Ebsen, R., Van De, W. K., Hardlei, T., Nexø, E., Borst, P., & Moestrup, S. K., (2010). Identification of multidrug resistance protein 1 (MRP1/ABCC1) as a molecular gate for cellular export of cobalamin. *Blood, 115*, 1632–1639. doi: 10.1182/blood-2009-07- 232587.

Beuvery, E. C., Van, R. F., & Nagel, J., (1982). Comparison of the induction of immunoglobulin M and G antibodies in mice with purified pneumococcal type 3 and meningococcal group C polysaccharides and their protein conjugates. *Infect. Immun., 37*, 15–22. doi: 10.1128/iai.37.1.15-22.1982.

Bhattacharyya, M., Girish, G. V., Karmohapatra, S. K., Samad, S. A., & Sinha, A. K., (2007). Systemic production of IFN-alpha by garlic (*Allium sativum*) in *humans. J. Interfer. Cytok. Res., 27*, 377–382. doi: 10.1089/jir.2006.0124.

Bikle, D. D., (2008). Vitamin D and the immune system: Role in protection against bacterial infection. *Current Opinion in Nephrology and Hypertension, 17*, 348–352. doi: 10.1097/MNH.0b013e3282ff64a3.

Bito, T., & Watanabe, F., (2016). Biochemistry, function, and deficiency of vitamin B12 in *Caenorhabditis elegans. Exp. Biol. Med., 241*, 1663–1668. doi: 10.1177/1535370216662713.

Blannin, A. K., Walsh, N. P., Clark, A. M., & Gleeson, M., (1998). Rates of glutamine and glucose utilization by quiescent and stimulated human neutrophils *in vitro. J. Physiol., 506*, 121. doi: 10.1074/jbc.M309520200. Epub 2004 Jan 13.

Blom, A. M., Villoutreix, B. O., & Dahlbäck, B., (2004). Complement inhibitor C4b-binding protein-friend or foe in the innate immune system? *Mol. Immunol., 40*(18), 1333–1346. doi: 10.1016/j.molimm.2003.12.002.

BourBour, F., MirzaeiDahka, S., Gholamalizadeh, M., Akbari, M. E., Shadnoush, M., Haghighi, M., et al., (2020). Nutrients in prevention, treatment, and management of viral infections; special focus on coronavirus. *Arch. Physiol. Biochem.*, 1–10. doi: 10.1080/13813455.2020.1791188.

Brock, J. H., & Mulero, V., (2000). Cellular and molecular aspects of iron and immune function. *Proceedings of the Nutrition Society, 59*, 537–540. doi: 10.1017/s002966510000077x.

Buck M. D., Sowell, R. T., Kaech, S. M., & Pearce, E. L., (2017). Metabolic instruction of immunity. *Cell, 169*(4), 570–586. doi: 10.1016/j.cell.2017.04.004.

Calder, P. C., & Yaqoob, P., (1999). Glutamine and the immune system. *Amino Acids, 17*, 227–241. doi: 10.1007/BF01366922.

Calder, P. C., (1994). Glutamine and the immune system. *Clin. Nutr., 13*, 2–8. doi: 10.1016/0261-5614(*94*)90003.

Calder, P. C., (1995). Fuel utilization by cells of the immune system. *Proc. Nutr. Soc., 54*, 65–82. doi: 10.1079/pns19950038.

Calder, P. C., Carr, A. C., Gombart, A. F., et al., (2020). Optimal nutritional status for a well-functioning immune system is an important factor to protect against viral infections. *Nutrients, 12*, 1181. doi: 10.3390/nu12041181.

Cantorna, M. T., Zhao, J., & Yang, L., (2012). Vitamin D, invariant natural killer T-cells and experimental autoimmune disease. *Proc. Nutr. Soc., 71*(1), 62–66. doi: 10.1017/S0029665111003193.

Chandra, R., (1993). Nutrition and the immune system. *Proceedings of the Nutrition Society, 52*(1), 77–84.

Chuang, J. C., Yu, C. L., & Wang, S. R., (1990). Modulation of human lymphocyte proliferation by amino acids. *Clin. Exp. Immunology, 81*, 173–176. doi: 10.1111/j.1365-2249.1990.tb05310.x.

Courtemanche, C., Elson-Schwab, I., Mashiyama, S. T., Kerry, N., & Ames, B. N., (2004). Folate deficiency inhibits the proliferation of primary human CD8+ T lymphocytes *in vitro. J Immunol. Baltim. Md., 173*, 3186–3192. doi: 10.4049/jimmunol.173.5.3186.

Darkes, M. J. M., & Plosker, G. L., (2002). Pneumococcal conjugate vaccine (Prevnar; PNCRM7): A review of its use in the prevention of *Streptococcus pneumoniae* infection. *Paediatr. Drugs, 4*(9), 609–630.

Djoko, K. Y., et al., (2015). The role of copper and zinc toxicity in innate immune defense against bacterial pathogens. *Journal of Biological Chemistry, 290*(31), 18954–18961. doi: 10.1074/jbc.R115.647099.

Doke, S., Inagaki, N., Hayakawa, T., & Tsuge, H., (1997). Effect of vitamin B6 deficiency on an antibody production in mice. *Biosci. Biotechnol. Biochem., 61*, 1331–1336. doi: 10.1271/bbb.61.1331.

Evoy, D., Lieberman, M. D., Fahey, T. J. 3rd., & Daly, J. M., (1998). Immunonutrition: The role of arginine. *Nutrition, 14*, 611–617. doi: 10.1016/s0899-9007(98)00005-7.

Gadek, J. E., DeMichele, S. J., Karlstad, M. D., Pacht, E. R., Donahoe, M., Albertson, T. E., et al., (1999). Effect of enteral feeding with eicosapentaenoic acid, gamma-linolenic acid, and antioxidants in patients with acute respiratory distress syndrome. Enteral nutrition in ARDS study group. *Crit Care Med., 27*(8), 1409–1420. doi: 10.1097/00003246-199908000-00001.

Gamboa-Leon, M. R., Aranda-Gonzalez, I., Mut-Martın, M., Garcia-Miss, M. R., & Dumonteil, E., (2007). *In vivo* and *in vitro* control of *Leishmania mexicana* due to garlic-induced NO production. *Scand. J. Immunol., 66*, 508–514. doi: 10.1111/j.1365-3083.2007.02000.x.

Gao, J., Xu, K., Liu, H., Liu, G., Bai, M., Peng, C., et al., (2018). Impact of the gut microbiota on intestinal immunity mediated by tryptophan metabolism. *Front Cell Infect. Microbiol., 8*, 13. doi: 10.3389/fcimb.2018.00013.

Green, R., (2017). Vitamin B12 deficiency from the perspective of a practicing hematologist. *Blood, 129*, 2603–2611. doi: 10.1182/blood-2016-10-56918.

Hemilä, H., (1997). Vitamin C intake and susceptibility to the common cold. *British Journal of Nutrition, 77*, 59–72. doi: 10.1017/s0007114500002889.

Huang, Z., Aaron, H. R., & Peter, R. H., (2012). The role of selenium in inflammation and immunity: From molecular mechanisms to therapeutic opportunities. *Antioxidants & Redox Signaling, 16*(7), 705–743. doi: 10.1089/ars.2011.4145.

Hyppönen, E., Läärä, E., Reunanen, A., Järvelin, M. R., & Virtanen, S. M., (2001). Intake of vitamin D and risk of type 1 diabetes: A birth-cohort study. *The Lancet, 358*, 1500–1503. doi: 10.1016/S0140-6736(01)06580-1.

Ibrahim, K. S., & El-Sayed, E. M., (2016). Potential role of nutrients on immunity. *International Food Research Journal, 23*(2).

Ikeda, M., Miyamoto, A., Mutoh, S., Kitano, Y., Tajima, M., Shirakura, D., et al., (2013). Development of biotin prototrophic and hyper-auxotrophic *Corynebacterium glutamicum* strains toward biotin production. *Appl. Env. Microbiol., 2013*, 828. doi: 10.1128/AEM.00828-13.

Ikeda, M., Nagashima, T., Nakamura, E., Kato, R., Ohshita, M., Hayashi, M., et al., (2017). *In vivo* roles of fatty acid biosynthesis enzymes in biosynthesis of biotin and αlipoic acid in *Corynebacterium glutamicum*. *Appl. Environ. Microbiol., 83*, 17. doi: 10.1128/AEM.01322-17.

Ing, R., Su, Z., Scott, M. E., & Koski, K. G., (2000). Suppressed T helper 2 immunity and prolonged survival of a nematode parasite in protein-malnourished mice. Proceedings of the *National Academy of Sciences, 97*, 7078–7083. doi: 10.1073/pnas.97.13.7078.

Jannusch, K., Jockwitz, C., Bidmon, H. J., Moebus, S., Amunts, K., & Caspers, S., (2017). A complex interplay of vitamin B1 and B6 metabolism with cognition, brain structure, and functional connectivity in older adults. *Front Neurosci., 11*, 596. doi: 10.3389/fnins.2017.00596 85.

Karacabey, K., (2012). The effect of nutritional elements on the immune system. *J. Obes. Weight Loss Ther. [Internet], 2*(9). Available from: http://dx.doi.org/10.4172/2165-7904.1000152.

Kasper, D. L., Paoletti, L. C., Wessels, M. R., Guttormsen, H. K., Carey, V. J., Jennings, H. J., et al., (1996). Immune response to type III group B streptococcal polysaccharide- tetanus toxoid conjugate vaccine. *J. Clin. Invest., 98*, 2308–2314. doi: 10.1172/JCI119042.

Keservani, R. K., Kesharwani, R. K., Sharma, A. K., & Jarouliya, U., (2015). Dietary supplements, nutraceutical and functional foods in immune response (immunomodulators). In: Debasis, B., Harry, G. P., & Anand, S., (eds.), *Nutraceutical and Functional Foods in Human Life and Disease Prevention* (pp. 343–358). Chapter 20, Boca Raton, Fl. CRC Press, Taylor and Francis. ISBN: 9781482237214.

Keservani, R. K., Kesharwani, R. K., Sharma, A. K., Vyas, N., & Chadoker, A., (2010b). Nutritional supplements: An overview. *International Journal of Current Pharmaceutical Review and Research, 1*(1), 59–75.

Keservani, R. K., Kesharwani, R. K., Vyas, N., Jain, S., Raghuvanshi, R., & Sharma, A. K., (2010a). Nutraceutical and functional food as future food: A review. *Der Pharmacia Lettre, 2*(1), 106–116.

Keservani, R. K., Sharma, A. K., & Kesharwani, R. K., (2020). *Nutraceuticals and Dietary Supplements: Applications in Health Improvement and Disease Management*. CRC Press. ISBN: 9781771888738.

Khan, R. U., Rahman, Z. U., Javed, I., & Muhammad, F., (2012). Effect of vitamins, probiotics and protein on semen traits in post-molt male broiler breeders. *Anim. Reprod. Sci., 135*(1–4), 85–90. doi: 10.1016/j.anireprosci.2012.09.005.

Kroes, B. H., Beukeman, C. J., Van, D. B. A. J., Wolbink, G. J., Van, D. H., & Labadie, R. P., (1997). Inhibition of human complement by beta-glycyrrhetinic acid. *Immunology, 90*, 115–120. doi: 10.1046/j.1365-2567.1997.00131.x.

Kursat, K., & Nurfer, O., (2012). The effect of nutritional elements on the immune system. Physical education and sports, Gaziantep University, Gaziantep, Turkey. *Journal of Obesity & Weight Loss Therapy*. doi: 10.4172/2165-7904.1000152.

Leonardi, R., Zhang, Y. M., Rock, C. O., & Jackowski, S., (2005). Coenzyme A: Back in action. *Prog. Lipid Res., 44*, 125–153. doi: 10.1016/j.plipres.2005.04.001.

Liu, P. T., Stenger, S., Li, H., Wenzel, L., Tan, B. H., Krutzik, S. R., & Meinken, C., (2006). Toll-like receptor triggering of a vitamin D-mediated human antimicrobial response. *Science, 311*, 1770–1773. doi: 10.1126/science.1123933.

Lund, J., Stjernstrom, H., Bergholm, U., Jorfeldt, L., Vinnars, E., & Wiklund, L., (1986). The exchange of blood-borne amino acids in the leg during abdominal surgical trauma: Effects of glucose infusion. *Clin. Sci., 71*, 487–496. doi: 10.1111/j.1475-097x.1981.tb00918.x.

Martí, I. L. A. A., & Reith, W., (2021). Arginine-dependent immune responses. *Cell Mol. Life Sci., 78*(13), 5303–5324. PMCID: PMC8257534. doi: 10.1007/s00018-021-03828-4.

Matarese, G., Moschos, S., & Mantzoros, C. S., (2005). Leptin in immunology. *The Journal of Immunology, 174*, 3137–3142.

Mathis, D., & Shoelson, S. E., (2011). Immunometabolism: An emerging frontier. *Nat. Rev. Immunol., 11*, 81–83. doi: 10.1038/nri2922.

Melse-Boonstra, A., De Bree, A., Verhoef, P., Bjørke-Monsen, A. L., & Verschuren, W. M. M., (2002). Dietary monoglutamate and polyglutamatefolate are associated with plasma folate concentrations in Dutch men and women aged 20–65 years. *J. Nutr., 132*, 1307–1312. doi: 10.1093/jn/132.6.1307.

Meydani, S. N., Han, S. N., & Wu, D., (2005). Vitamin E and immune response in the aged: Molecular mechanisms and clinical implications. *Immunological Reviews, 2*(5), 269–284. PMCID: PMC7158858. doi: 10.1111/j.0105-2896.2005.00274.x.

Nualart, F. J., Rivas, C. I., Montecinos, V. P., Godoy, A. S., Guaiquil, V. H., Golde, D. W., & Vera, J. C., (2003). Recycling of vitamin C by a bystander effect. *Journal of Biological Chemistry, 278*, 10128–10133. doi: 10.1074/jbc.M210686200.

Nuwer, N., Cerra, F. B., Shronts, E. P., Lysne, J., Teasley, K. M., & Konstantinides, F. N., (1983). Does modified amino acid total parenteral nutrition alter immune-response in high level surgical stress. *J. Parenter. Enter. Nutr., 7*, 521–524. doi: 10.1177/0148607183007006521.

Palmer, A. C., (2011). Nutritionally mediated programming of the developing immune system. *Adv Nutr., 2*, 377–395.

Percival, S. S., (1998). Copper and immunity. *The American Journal of Clinical Nutrition, 67*, 1064S–1068S. https://doi.org/10.1093/ajcn/67.5.*1064S.*

Philip, C. C., (2021). Nutrition and immunity: Lesson for COVID-19. *European Journal of Clinical Nutrition.* https://doi.org/10.1038/s41430-021-00949-8.

Qian, B., Shen, S., Zhang, J., & Jing, P., (2017). Effects of vitamin B6 deficiency on the composition and functional potential of T cell populations. *J. Immunol. Res., 2017*, 2197975. doi: 10.1155/2017/2197975.

Quick, M., & Shi, L., (2015). The sodium/multivitamin transporter: A multipotent system with therapeutic implications. *Vitam. Horm., 98*, 63–100. doi: 10.1016/bs.vh.2014.12.003.

Rayment, S. J., Shaw, J., Woollard, K. J., Lunec, J., & Griffiths, H. R., (2003). Vitamin C supplementation in normal subjects reduces constitutive ICAM-1 expression. *Biochemical and Biophysical Research Communications, 308*, 339–345. doi: 10.1016/s0006-291x(03)01383-4.

Rescigno, M., Urbano, M., Valzasina, B., et al., (2001). Dendritic cells express tight junction proteins and penetrate gut epithelial monolayers to sample bacteria. *Nature Immunology, 2*, 361. doi: 10.1038/86373.

Rooks, M. G., & Garrett, W. S., (2016). Gut microbiota, metabolites and host immunity. *Nature Reviews Immunology, 16*, 341–352.

Saeed, F., Nadeem, M., Ahmed, R. S., TahirNadeem, M., Arshad, M. S., & Ullah, A., (2016). Studying the impact of nutritional immunology underlying the modulation of immune responses by nutritional compounds a review. *Food Agric. Immunol., 27*(2), 205–29https://doi.org/10.1080/09540105.2015.1079600.

Sagun, K., Cárcamo, J. M., & Golde, D. W., (2005). Vitamin C enters mitochondria via facilitative glucose transporter 1 (Glut1) and confers mitochondrial protection against oxidative injury. *The Federation of American Societies for Experimental Biology, 19*, 1657–1667. doi: 10.1096/fj.05-4107com.

Said, H. M., (2011). Intestinal absorption of water-soluble vitamins in health and disease. *Biochem. J., 437*, 357–372. doi: 10.1042/BJ20110326.

Samorodov, V. N., & Pospelov, S. V., (1996). Phytochemical composition of plants *Echinacea* (*Echinaceae moench*) and their pharmacological properties. *Pharm. Chem. J., 30*(4), 32–37. doi: 10.3390/ph13060125.

Schramm, M., Wiegmann, K., Schramm, S., Gluschko, A., Herb, M., Utermöhlen, O., et al., (2014). Riboflavin (vitamin B2) deficiency impairs NADPH oxidase 2 (Nox2) priming and defense against *Listeria monocytogenes*. *Eur. J. Immunol., 44*, 728–741. doi: 10.1002/eji.201343940.

Semba, R. D., (1999). Vitamin A and immunity to viral, bacterial and protozoan infections. *The Proceedings of the Nutrition Society, 58*(3), 719–727.

Shands J. W. Jr., (1984). Macrophage procoagulants. *Haemostasis, 14*(5), 373 doi: 10.1017/s0029665199000944.

Shibata, S., (2000). A drug over the millennia: Pharmacognosy, chemistry, and pharmacology of licorice. *Yakugaku Zasshi, 120*, 849–862. doi: 10.1248/yakushi1947.120.10_849.

Silvia, M., Adeline, P., & Philip, C. C., (2018). Immune function and micronutrient requirements change over the life course. *Nutrients, 10*(10), 1531. doi: 10.3390/nu10101531.

Singh, P., Kesharwani, R. K., & Keservani, R. K., (2017a). Antioxidants and vitamins: Roles in cellular function and metabolism. In: *Sustained Energy for Enhanced Human Functions and Activity* (pp. 385–407). Academic Press. ISBN:9780128054130.

Singh, P., Kesharwani, R. K., & Keservani, R. K., (2017b). Protein, carbohydrates, and fats: Energy metabolism. In: *Sustained Energy for Enhanced Human Functions and Activity* (pp. 103–115). Academic Press. ISBN:9780128054130.

So, D., Whelan, K., Rossi, M., et al., (2018). Dietary fiber intervention on gut microbiota composition in healthy adults: A systematic review and meta-analysis. *The American Journal of Clinical Nutrition, 107*, 965–983. doi: 10.1093/ajcn/nqy041.

Stephensen, C. B., (2001). Vitamin A, infection, and immune function. *Annual Review of Nutrition, 21*(1), 167–192. doi: 10.1146/annurev.nutr.21.1.167.

Sultan, M. T., Butt, M. S., Qayyum, M. M. N., & Suleria, H. A. R., (2014). Immunity: Plants as effective mediators. *Crit. Rev. Food Sci. Nutr., 54*(10), 1298–1308. doi: 10.1080/10408398.2011.633249.

Tamura, J., Kubota, K., Murakami, H., Sawamura, M., Matsushima, T., Tamura, T., et al., (1999). Immunomodulation by vitamin B12: Augmentation of CD8+ T lymphocytes and natural killer (NK) cell activity in vitamin B12-deficient patients by methyl-B12 treatment. *Clin. Exp. Immunol., 116*, 28–32. doi: 10.1046/j.1365-2249.1999. 00870.x.

Tong, L., (2013). Structure and function of biotin-dependent carboxylases. *Cell Mol. Life Sci., 70*, 863–891. doi: 10.1007/s00018-012-1096-0.

Van, E. E., & Mathieu, C. (2005). Immunoregulation by 1, 25-dihydroxyvitamin D 3: Basic concepts. *The Journal of Steroid Biochemistry and Molecular Biology, 97*, 93–101. doi: 10.1016/j.jsbmb.2005.06.002.

Vasudevan, S. K., & Suresh, K., (1994). Augmentation of murine natural killer cell and antibody dependent cellular cytotoxicity activities by *Phylanthus emblica*, a new immunomodulator. *J. Ethnopharmacol., 44*, 55–60. doi: 10.1016/0378-8741(94)90099-x.

Ward, R. J., et al., (2011). Iron and the immune system. *Journal of Neural Transmission, 118*(3), 315–328.

Whitney, E. N., Cataldo, C. B., & Rolfes, S. R., (1998). *Understanding Normal and Clinical Nutrition*. Belmont, CA: Wadsworth.

Wilson, E. A., & Demmig-Adams, B., (2007). Antioxidant, anti-inflammatory, and antimicrobial properties of garlic and onions. *Nutr. Food Sci., 37*, 178–183. doi: 10.1108/00346650710749071.

Wintergerst, E. S., Maggini, S., & Hornig, D. H., (2006). Contribution of selected vitamins and trace elements to immune function. *Annals of Nutrition and Metabolism, 51*, 301–323. doi: 10.1159/000107673.

Yoshii, K., Hosomi, K., Sawane, K., & Kunisawa, J., (2019). Metabolism of dietary and microbial vitamin B family in the regulation of host immunity. *Front Nutr., 6*, 48. doi: 10.3389/fnut.2019.00048.

Zempleni, J., Hassan, Y. I., & Wijeratne, S. S., (2008). Biotin and biotinidase deficiency. *Expert Rev. Endocrinol. Metab., 3*, 715–724. doi: 10.1586/17446651.3.6.715.

Zempleni, J., Wijeratne, S. S. K., & Hassan, Y. I., (2009). Biotin. *Bio Factors Oxf. Engl., 35*, 36–46. doi: 10.1002/biof.8.

Zhang, B., Lo, C., Shen, L., Sood, R., Jones, C., Cusmano-Ozog, K., et al., (2011). The role of vanin-1 and oxidative stress–related pathways in distinguishing acute and chronic pediatric ITP. *Blood, 117*, 4569–4579. doi: 10.1182/blood-2010-09-304931.

Zhao, R., Diop-Bove, N., Visentin, M., & Goldman, I. D., (2011). Mechanisms of membrane transport of folates into cells and across epithelia. *Annu. Rev. Nutr., 31*, 177–201. doi: 10.1146/annurev-nutr-072610-145133.

CHAPTER 6

GUT MICROBIOTA AND PATHOGENS

SAYAN BHATTACHARYYA

Associate Professor, Microbiology, All India Institute of Hygiene and Public Health (AIIH&PH), Kolkata, West Bengal, India

ABSTRACT

Gut microbiota is very important for normal digestion and metabolism in man and can also prevent the development of immune-mediated diseases. They can be affected to a great extent by diet and its various components. Some dietary elements stimulate the growth of healthy bacteria in the gut and suppress the growth of pathogens. Hence some micronutrients, macronutrients, and nutraceuticals can be incorporated into the human diet for optimum health, good growth of beneficial bacteria in the gut, and inhibition of pathogens.

6.1 INTRODUCTION AND SCOPE

Microbiota generally refers to the whole population of microorganisms which can usually colonize a particular site in the human body. It includes not only bacteria, but also other types of microorganisms like fungi, archaea, viruses, and protozoa (Jandhyala et al., 2015). The normal gut of a human being is sterile at birth but is colonized soon after. Normal gut flora consists of Gram-positive aerobic and anaerobic bacteria (cocci and bacilli) and Gram-negative facultative and strict anaerobes. They exist in the gut as normalized commensal microorganisms. Each part of the gut has its unique

Micronutrients and Macronutrients as Nutraceuticals. Prakash Chandra Gupta, PhD,
Sayan Bhattacharyya, MBBS, MD, Nisha Sharma, PhD, MPharm, Rajesh K. Kesharwani, PhD,
Raj K. Keservani, PhD, MPharm (Eds.)
© 2024 Apple Academic Press, Inc. Co-published with CRC Press (Taylor & Francis)

set of microbiota. The intestinal tract is big enough and is often called the 'second brain.' This is because it has a bearing on our physical and mental well being this microbiota varies with the dietary pattern of a person. Dietary elements contain various micronutrients. Micronutrients and nutraceutical (Keservani et al., 2010a, b) like compounds have a prominent effect on the composition of this gut microbiome. In the following sections, the author will try to elaborate upon them.

6.2 COMPOSITION OF THE MICROBIOME OF THE HUMAN GASTROINTESTINAL TRACT (GIT)

Although it was earlier considered that the gut microbial flora has about 500 to 1,000 different species of microbes, the latest large-scale studies have documented that the total mean human gut microbial flora is composed of over 35,000 different bacterial species (Jandhyala et al., 2015). Moreover, if defined from the viewpoint of the total genes of bacteria, studies like the Microbiome Project of Humans, and also other studies like the 'Metagenome of the Human Intestine (also called MetaHIT) studies have shown conclusively, that there can exist more than 10 million different non-redundant genes that could be present in the microbiome of man. Overall, the healthy gut microbiota in man is predominantly made up of bacteria that belong to the bacterial Phyla Bacteroidetes and also Firmicutes. This is closely followed by bacteria that fall under the two phyla, namely Verrucomicrobia and Actinobacteria. Although this general pattern is fairly consistent with time, gut microbiota can show both temporal and spatial differences with respect to distribution at the genus and species levels. As one moves downwards from the esophagus towards the rectal area, there is a profound observable difference in the diversity and the number of bacteria. This ranges right from 10^1 bacteria in each gram of luminal contents in the esophagus and gastric lining, to up to 10^{12} per gram of contents, as seen in the large gut (Jandhyala et al., 2015).

6.3 FUNCTIONS OF GUT MICROBIOTA

The normal gut microbiota has got some specified functions in the human host, like nutrient metabolism of xenobiotics and drugs, maintaining the structural oneness of the mucosal and epithelial barrier of the gut, as well as immunomodulation. It also confers protection against a number of different bacterial pathogens (Jandhyala et al., 2015). It has been postulated for a long time that the gut microbiota has a very important role to play in maintaining the gut functions and integrity in the normal host, and is beneficial for the health of man as a whole. A lot of such evidence is now coming out from several studies carried out on human subjects as well as animal models like germ-free mice. These data second these findings. There is a lot of similar high-quality data from studies like the Human Microbiome Project carried out in the U.S. (also called HMP), European MetaHIT, and several other important studies. These have now shown very clearly the beneficial effects of the normal microbial population of the human gut on several health parameters, right down to the genetic level. For example, studies have identified a number of genes of the gut microbes, like the HMO-related gene cluster 1, which is responsible for digesting the human milk oligosaccharides (HMOs) (Jandhyala et al., 2015; Human Microbiome Project Consortium, 2012).

Other notable functions of gut microflora are as follows:

- The gut microbes have been shown to have a positive influence on lipid metabolism. It does so by suppressing the inhibition of the lipoprotein lipase (LPL) enzyme in the fat cells or adipocytes. Moreover, the Gram-negative bacterium *Bacteroidesthetaiotaomicron, present* in the gut has been shown to improve the efficacy of hydrolysis of lipids by upregulating the expression of a co-lipase enzyme required by pancreatic lipase enzyme, for the digestion of lipids (Hooper et al., 2001).
- Many amino acid transporters present on the cell wall of these bacteria can also facilitate the entry of amino acids from the lumen of the intestine into the bacterial cell. After this, many gene products in these bacteria convert these amino acids into smaller signaling molecules as well as many peptides having antimicrobial effects, also called bacteriocins (Jandhyala et al., 2015).
- Members of the genus *Bacteroides* can produce conjugated linoleic acid (also called CLA) which is known for its antidiabetic, antiatherogenic, anti-obesogenic, and hypolipidemic properties. It also has potent immunomodulatory effects.
- Recent studies have also documented that human gut microorganisms can also break down various polyphenols (or phenolic compounds) that are found in foods. Polyphenol-containing secondary metabolites are also found in many plant, fruit, or plant-derived products, namely tea, cocoa, and wine. These include flavan-3-ols, isoflavones,

flavanols, flavanones, flavones, and also other compounds like lignans, anthocyanidins, tannins, and chlorogenic acids.

- These gut microorganisms are also imperative for the fermentation of much non-digestible matter, like dietary fibers and also the host's intestinal mucus. This fermentation can then support the growth of specialist microorganisms, which lead to the production of short chain-bearing fatty acids (or SCFAs) and many gases. The most common SCFAs thus produced are propionate, acetate, and butyrate.
- Butyrate is the major energy source for human intestinal cells, and it can effectively stimulate apoptosis of the cancer cells of the colon. It can also activate intestinal gluconeogenesis, thus exerting beneficial effects on glucose and energy homeostasis. Butyrate is required by the gut epithelial cells to utilize large amounts of oxygen via β-oxidation. In this way, it generates a state of hypoxia in the gut. This helps maintain Oxygen balance in the gastrointestinal tract (GIT). This also helps in keeping at bay dysbiosis of the microbiome of the gut (Valdes et al., 2018).
- The gut microbial flora possesses various structural components and metabolites. They may also induce the synthesis of numerous antimicrobial proteins (AMP) like, C-type lectins, cathelicidins, and prodefensins by the Paneth cells of man.
- Lactobacilli and other bacteria present in the gut help in the production of vitamins such as Vitamin K and Vitamin B12. Vitamin K is pivotal for the blood clotting pathway in the human body. Vitamin K is available to humans in either of two forms: (i) vitamin K1, or phylloquinone, which is mainly found in green leafy vegetables; and (ii) vitamin K2, also known as menaquinone, which is synthesized by intestinal bacterial flora and is present in many fermented foods (Paiva et al., 1998). Vitamin B12 is required for optimum bone marrow and nerve functions. These bacteria thus exist as symbionts because they help the host in many ways.
- Another important function of the gut microbiota is to regulate the overgrowth of pathogenic microbial strains by induction of local immunoglobulins. The gut microbes, particularly Gram-negative bacteria like *Bacteroides* spp. have been shown to activate the dendritic cells (DCs) in the intestine. DCs stimulate the plasma cells that are present in the intestinal mucosal lining, which then expresses secretory IgA (sIgA). The sIgA can then, in turn, form a covering over the gut microbiota. The sIgA that coats these microbes belong

predominantly to the sIgA2 subclass, which is also more refractory to lysis by the bacterial proteases (Jandhyala et al., 2015).

6.4 FACTORS AFFECTING GUT MICROBIOTA

6.4.1 ROLE OF FOOD AND DIET IN THE DEVELOPMENT OF GUT FLORA

Mother's breast milk is passed onto the baby's gut and it helps in preventing the colonization by pathogenic bacteria. Breast milk not only meets the nutritional and physiological demands of the infant but also has many bioactive compounds which are not typically found in commercially available formula feeds. These compounds have a very important role to play in the digestion and absorption of nutrients, immune-mediated protection, and development of anti-microbial defenses (Human Microbiome Project Consortium, 2012). HMOs nourish the colonic bacteria of the infant. Hence they impart a selective growth advantage to *Bifidobacterium* sp. (Zivkovic et al., 2011).

Diet is indeed a very important factor that can determine the composition, diversity, and richness of gut microbial flora. This is seen even in adulthood. Generally, intake of a diet that is rich in fruits, fiber, and vegetables, ultimately leads to a higher richness and variety of the gut microbiota. Individuals who subsist on this type of diet have a higher count of insoluble carbohydrate-metabolizing microbes which belong to the phylum Firmicutes in their gut, namely *Ruminococcusspp., Eubacterium rectale* and *Roseburia* spp. (Walker et al., 2011). Several studies in humans and rats have shown a pronounced change in the microbial population of the GIT when people use seaweeds as part of their food. In human beings, supplying the seaweed *Gelidium* in the diet has been found to increase significantly the population of *Bifidobacteriumgenus* in the gut, without inducing any appreciable changes in the other bacteria (Jandhyala et al., 2015). Dietary fiber, found in fruits and vegetables is also good for the gut. This is because it is broken down by the healthy microorganisms present in the gut to produce short-chain fatty acids (SCFAs).

Gut bacteria can break down polyphenols present in different food matters. Vegetables and fruits like onions, broccoli, apples, capers, grapes, and plums are indeed very rich in polyphenols (which are also called Flavanols) which are broken down by gut bacteria like *Bacteroidesdistasonis*, *Bacteroidesuniformis*, *Enterococcus casseliflavus* and the Gram-positive anaerobic bacterium *Eubacteriumramulus* (Jandhyala et al., 2015). Citrus fruits and tomatoes are rich in Flavanones that are degraded by gut bacteria like *Clostridium* spp. and *Eubacterium ramulus*. Tannins are found in high quantity in foods like Raspberries, cranberries, strawberries, walnuts, pomegranates, and also in grapes. They can be broken down by *Butyrivibrio* spp. present in the GIT. Hence, a healthy gut microbiome is deemed to be very essential for food degradation, digestion, and absorption.

Animal protein, however, when consumed in excess amounts, is bad for the gut and can lead to Inflammatory bowel disease. The composition and function of commensal bacteria in the gut are highlighted in Table 6.1.

TABLE 6.1 Bacterial Flora of Healthy Gut and its Functions

Bacterial Species	Functions
Lactobacillus spp.	Competition with pathogens, formation of Vitamins B12 and K.
Akkermansia spp.	Prevention of obesity.
Blautia spp.	Manufacturing of short-chain bearing fatty acids (or SCFAs).
Roseburia spp.	Metabolism of insoluble carbohydrates.
Bifidobacterium spp.	Helps in digestion.

6.4.2 *ALTERATION OF GUT FLORA BY PROBIOTICS*

The World Health Organization (WHO) has defined probiotics to be "live microorganisms which can impart some benefits to human health when administered in small or adequate amounts." Many strains of beneficial bacteria, like *Lactobacillus casei*, *L. bulgaricus*, *Lactobacillus plantarum*, *L. acidophilus*, *Streptococcus thermophilus*, *Bifidobacterium longum*, *B. infantis*, and *E. coli* strain Nissle 1917, have been documented to possess many novel immunomodulatory and gut barrier-maintaining properties. These compounds and several other molecules akin to these have been used for commercial purposes for managing human illnesses, *like* IBD and antibiotic-associated diarrhea (AAD). The fundamental idea behind the inclusion of these microorganisms in the treatment armamentarium is to simulate the normal health-augmenting properties of the "good" or healthy gut bacteria (Human Microbiome Project Consortium, 2012).

6.5 PREBIOTICS

The addition of a prebiotic substance can also possibly augment the functioning of the probiotic microbes. Prebiotics can be formally defined to be food components that contain some non-digestible oligosaccharides (NDOs) (*like* galactooligosaccharides and inulin). They help in the growth of the probiotic bacteria in the gut. The gut commensal bacteria can also selectively ferment these types of fibers. Thus they lead to the production of volatile SCFAs. SCFAs in turn exert their pro-health effects on the host (Jandhyala et al., 2015). If probiotics and prebiotics are used together, it is termed a "Synbiotic."

6.6 EFFECT OF THE MICRONUTRIENTS AND MACRONUTRIENTS ON THE GUT MICROFLORA

As discussed earlier, dietary fiber and complex carbohydrates, like galactooligosaccharides (GOS), arabinoxylans, oligofructose, and inulin (fructan), can promote the growth of beneficial bacteria in the gut and simultaneously suppress the potentially harmful ones (Yang et al., 2020). Preclinical studies also indicate that both the amount and types of fat present in the diet can modulate effectively both the beneficial and potentially pathogenic microorganisms. This also affects the Firmicutes/Bacteroides ratio in the gut, which is crucial for gut health. Some clinical as also preclinical research has shown that the type and quantity of proteins in the diet have substantial but variable effects on the nature of microbiota present in the gut (Yang et al., 2020). Polyphenols, consisting of phenolic acids, stilbenes, flavonoids, and lignans (Keservani & Sharma, 2014) can be obtained from natural foods like fruits, vegetables, cereals, plant-based hot beverages (Kesharwani et al., 2016), and wine, have attracted the interest of scientists owing to their potent antioxidant effect, anti-inflammatory action, and anti-carcinogenic properties (Yang et al., 2020). *In vitro*, studies have also revealed that these types of polyphenols present in diet can modulate the gut microbiome of man by inhibiting potential pathogens, such as *H. pylori* and *Staphylococcus* spp. They also favor the growth of potentially beneficial bacteria, notably *Lactobacillus* and *Bifidobacteria* spp. (Yang et al., 2020). *In vitro*, studies have also demonstrated that flavonoids, namely anthocyanins, phenolic acids (epicatechins, *p*-coumaric acid, and *o*-coumaric acid), as well as other polyphenolic compounds like rutin, chlorogenic acid, quercetin, and caffeic acid can increase the counts of beneficial gut bacteria, like *Bifidobacterium* and *Lactobacillus spp.*, and simultaneously decrease the number of potentially pathogenic bacteria which colonize the human gut (Yang et al., 2020). Catechin compounds can also enhance the numbers of the *Clostridium coccoides-Eubacterium rectale* group of bacteria, *Bifidobacterium* sp., and *E. coli* and also concomitantly suppress the growth of potentially pathogenic microorganisms (like the *Clostridium histolyticum* group). By means of agar plate methods, it has been also demonstrated that Naringenin, a flavone found in abundance in citrus fruits, can modulate the multiplication and gene-mediated control of the gut commensals (Yang et al., 2020). Ethyl acetate or watery-polyphenolic extract of a variety of tea, which is also called Yunnan Chinese is able to suppress unhealthy gut bacteria like *Clostridium* and *Bacteroides in vitro*.

Preclinical studies have shown persistent evidence that a multitude of various polyphenols in diet (Keservani & Sharma, 2014) can enhance various aspects of human health by way of changes induced in the gut microbiome. In animals who have been administered a diet rich in fats, or even a control or normal diet, polyphenol compounds found in grapes, pomegranate peel, red wine, and green tea can increase the numbers of *Bifidobacteria* and *Lactobacilli* in the stool. A study published by Axling et al. has also found that powdered green tea if administered along with the bacterium *Lactobacillus plantarum*, might increase significantly the bacterial richness and variety present in the gut. It could also increase the counts of genera like *Akkermansia* and *Lactobacillus* in the murine colon as per reports (Jandhyala et al., 2015). Some clinically-done trials have also reported such profound changes in the constituents of the gut microbiological flora after the administration of polyphenol-rich foods like red wine, green tea, cocoa, green leafy veggies, and fruits (Keservani et al., 2016a, b). Polyphenols derived from cocoa fruit, at an amount of 494 mg/day, when taken orally for 4 weeks, could significantly increase the fecal load of *Bifidobacterium* and *Lactobacillus spp.*, which is exactly opposite to the effects of a low polyphenol diet (23 mg/day). This prebiotic-like effect has been seconded also by some *in-vitro* fermentation studies on stool samples from the same individuals. In another study, polyphenols extracted from red wine could elevate the number of *Bifidobacteria* and *Lactobacilli* in the stool of obese individuals. This was also accompanied by a marked improvement in the extant indicators of metabolic syndromes, such as plasma glucose concentrations, blood pressure, and plasma lipid profile (Yang et al., 2020). In other studies, the modulation of the beneficial microbiota of the gut, like *Bifidobacterium* sp., *B. uniformis*, *Eubacteriumrectale*, and *Eggerthellalenta* has also been reported in people who tend to consume red wine polyphenols on a regular basis. Subsisting on a vegetable or fruit juice-based diet for three successive days can also bring down the F/B ratio of healthy adults.

Dietary (Keservani et al., 2020; Rane & Keservani, 2020a–c) polysaccharides can be intertwined in many complex mechanisms via an intricate array of bonds present between the monosaccharide units. This is shown also by the mere number of carbohydrate-activating enzymes which can be seen in the human intestinal microbiome (Oliphant et al., 2019). For example, the anaerobic rod *B. thetaiotaomicron* alone has as many as 260 varieties of glycoside hydrolase enzymes in its genome. This highlights the evolutionary need for adaptation. By this, one can fully carry out the utilization of resistant starch and also the choice of fibers available as a component of our staple diet. On the other hand, human cells produce very few copies of these enzymes. However, we do produce the amylase enzyme to remove α-linked sugar units from starch. Human cells can also utilize sugars like sucrose, glucose, fructose, and lactose which are present in the small intestine. Thus, humans have to depend heavily on the gut microbiota in order to acquire energy from the remaining complex sugars (Oliphant et al., 2019). Dietary proteins (Jarouliya & Keservani, 2019a) characteristically have a number of conserved peptide bonds which can be well degraded by the protease enzymes. The gut commensal bacteria can also help produce cysteine-, serine-, aspartic-, and metalloprotease enzymes. However, normally in any such stool sample, these bacterial enzymes are far lesser in number than the protease enzymes in human cells (Oliphant et al., 2019). However, the 20 amino acids which are the building blocks for protein synthesis, need some more processing steps for getting assimilated into the biochemical pathways, as compared to the monosaccharide units. Hence, it is unusual for a given healthy microbe of the gut, to be able to ferment all the amino acids for producing energy (Oliphant et al., 2019). Also, assimilation of amino acids by microbes, from the environmental sources into the anabolic processes of the body might indeed conserve more energy when compared to their catabolic use. This is accomplished by omitting any need for amino acid biosynthesis. It is for this very reason that amino acids are usually not considered as efficient an energy source as carbohydrates, particularly for the human gut microbes. Thus, it is not at all astonishing to find that the gut microbiota preferentially break down carbohydrates more swiftly than proteins, depending on the relative proportions that are made available to them. Once pyruvate is formed, mainly derived from dietary carbohydrates but at the same time from other food-based substrates, too, the human gut microbes develop many fermentation mechanisms to produce more energy. Pyruvate can be broken down next, into end products like lactate, succinate, or acetyl-Coenzyme A. However, these intermediary products do not achieve very good concentrations in maximum stool samples. This is because they can be further metabolized by cross-feeding microbes, producing short-chain-bearing fatty acids (SCFAs) like acetate, propionate, and butyrate in the process (Oliphant et al., 2019). These fecal metabolites are this way, actually the most plentiful and well-researched microbial end-products. This is because their effects are deemed to be very significant for good health. For example, the cells of the intestinal epithelium of the host (IECs) can use them well as a source for deriving energy (Oliphant et al., 2019).

6.6.1 ROLE OF VITAMINS PRESENT IN DIET AND THEIR EFFECTS ON MICROBIOTA

Vitamins are organic compounds that are needed in very small or minute quantities for supporting the different normal physiological functions in our body. They often perform a number of functions in the body, like their role as cofactors for many enzymes. Our diet is by default, the most important source of vitamins for man. This is because our bodies cannot produce enough vitamins to meet the daily dietary requirements. However, certain vitamins, like vitamin K and B complex groups of vitamins (Singh et al., 2017a), are produced very efficiently by the gut microbiota, as told earlier (Yang et al., 2020). Vitamin A in moderate amounts, increases *Bifidobacterium* spp. and *Akkermansia* spp. in the gut but in high doses, it also suppresses the counts of *Bifidobacterium* spp. In a study in mice models, supplementation done by Retinoic acid, which is a precursor molecule of Vitamin A, could reduce significantly the rates of Norovirus infection by increasing counts of lactobacilli. In an *in vitro* model, *Lactobacillus* spp. showed antiviral effects against Norovirus. Hence, based on these data, it can be safely postulated that high levels of *Lactobacillus* in the gut seen after Retinoic acid supplementation could be in part, responsible for this inhibition of Norovirus. In addition to this, supplementation by vitamin A has even been found to bring down both the mortality and morbidity associated with many infectious and non-infectious gastrointestinal ailments. Vitamin D found in the diet, can, however, suppress the load of *Bifidobacterium* spp. present in gut (Yang et al., 2020).

The B-complex group of vitamins (Singh et al., 2017a, b) is a family of 8 distinct water-soluble vitamins which are imperative for a number of metabolic processes. It is well known that these B-complex groups of vitamins are to be seen in many different types of foods like animal-based foods, green leafy vegetables, peas, and beans. However, they are easily destroyed, especially by the presence of alcohol or by cooking. Some B complex vitamins may also help in bacterial colonization, modulate bacterial virulence factors, and partake in host-pathogen interactions by modulating the host defenses (Yang et al., 2020). For example, adding vitamin B12 to the diet has been noted to increase the colonization of the gut by the anaerobic gram-negative bacterium *Bacteroidesthetaiotaomicron*. This has been found in a gnotobiotic mice model in experiments. Furthermore, Vitamin B12 is also used by some pathogens in the gut so that they can take up ethanolamine. The latter can enhance *Salmonella Typhimurium* growth and expression of its virulence gene in *in-vitro* and *in vivo* models (Yang et al., 2020). Additionally, all Fusobacteria and more than 90% of Bacteroidetes in the gut have been predicted to be producers of cyanocobalamin or Vitamin B12 by biological synthesis. Similarly, vitamin B6 or Pyridoxine can also be produced by the gut microbiota, just like Bacteroidetes. Vitamin B6 is used mostly as a co-factor by various enzymes for many normal physiological reactions of our body, which are associated with the immune response of the host. Vitamin B6 deficiency can induce atrophy of the lymphoid tissue, cause a marked reduction of the lymphocyte counts, and also lead to deranged antibody responses as well as IL-2 formation (Yang et al., 2020). Animal experiments and human clinical data show furthermore that Vitamin B6 in the intestine promotes the growth of the genus *Bacteroides*. This is mediated either by modification of the immune system of the host or by impairing the growth or expression of the virulence factors of pathogens like *Salmonella Typhimurium*.

Amino acids also affect the gut microbiota. Tryptophan is one example of such an essential amino acid molecule. In the human gut, Tryptophan metabolism has a tremendous impact on the host and also on immunity and metabolism, gut microflora, the intestinal barrier, and the process of transit of the gut contents. Tryptophan can be degraded in the gut by bacteria into Indole. Indole helps in the formation of IL 22. IL-22 cytokine has been documented to regulate the proliferation of epithelial cells and antimicrobial peptide (AMP) production, preventing the ability of commensal bacteria to cause inflammation.

6.7 EFFECT OF OTHER MICRO- AND MACRO-NUTRIENTS ON GUT MICROBIOTA

6.7.1 *EFFECT OF FATS, METALS, AND TRACE ELEMENTS*

Human diet may have a direct influence on the microbiome of the gut. This can finally result in changes in the patterns of the biochemical reactions which occur in the lumen of the gut (Hemarajata et al., 2013). In experiments with germ-free mice that have been transplanted with human stool microbiota, animals that were fed a high-fat, high-sugar Western diet showed rapid changes in the intestinal microbial community. There was an increased load of bacteria belonging to phylum *Firmicutes* and a concomitant decreased burden of members of the phylum *Bacteroidetes* (Hemarajata et al., 2013). Western diet is loaded with meat and fats. The Mediterranean diet usually refers to a diet that involves a daily intake of fruits, whole grains, green vegetables, and healthy fats. It promotes regular intake of fish, poultry products, eggs, and beans on a weekly basis, consumption of moderate amounts of dairy products, and limited eating of red meat. The Paleo diet, on the other hand, offers a lot of lean meat, fish, fruits, vegetables, nuts, and seeds. It limits the use of legumes, dairy products, and food grains. This is in contrast with what is often called the "Western diet" which involves a high intake of meat and fats. The less FODMAP (Fermentable oligo, di-, and monosaccharides and polyol) diet has now gained some foothold for the management of disorders like IBS (or irritable bowel syndrome) and promotes overall good health. However, this low FODMAP diet can also lead to a reduction in healthy bacteria in the gut, like *Bifidobacter* spp. Similarly, a Gluten free diet for Coeliac disease patients not only leads to a reduction in *Escherichia coli* and *Staphylococcus aureus* in the gut. It results in low levels of beneficial bacteria in the gut also. Dietary non-digestible complex carbohydrates can be subjected to fermentation in the lumen of the intestine, leading to the production of SCFAs like acetate, propionate, and butyrate. Metabolically active volatile SCFAs, as said before, are involved in a number of biological processes and provide metabolic energy sources for human colonic epithelial cells. Furthermore, the fermentation of prebiotic carbohydrates like inulin and fructooligosaccharides helps in the proliferation of beneficial microbes (mainly *Bifidobacterium* spp. and *Lactobacillus* spp.) in the GIT. A high-fat diet increases the levels of Endotoxin in the gut as well as in the blood. Copper is a very essential trace element for animals and also man. It has not only been used as an adjunct at pharmacological doses in piggeries in swine production to uplift the growth parameters but also has an effect on the intestinal microbes by increasing the host bacterial resistance. In piglets, Copper treatment in particular increases the richness of *E. coli* in the distal gut and also the *E. coli* resistance to chloramphenicol and ciprofloxacin (Zhang et al., 2019). In other studies, Copper along with Fructose in the diet increased the load of Firmicutes and led to the depletion of *Akkermansia* spp. in the gut, as seen in rat models.

Unabsorbed iron in the gut can kill the gut microbiota and can also cause marked changes in the ratio of protective and pathogenic bacteria. The gut microbiota of man faces a lot of unabsorbed luminal iron which is ingested via a diet having lots of red meat as well as fortified cereals. Iron is deemed to be very important for the multiplication and survival of almost all bacteria. However, there are a few notable exceptions. This latter group of bacteria has alternative metabolic pathways, right from their evolution. *Lactobacillus plantarum* is one such example. It is the first accurately described iron-independent bacterial species. It has just one or two iron atoms, which is usually too insufficient to provide iron with any concrete biological functions. This feature can explain their relative abundance in natural gut microbial flora and milk, which has a very iron-restricted environment owing to the presence of lactoferrin. Siderophores are small moieties with high iron affinity, which are released by bacteria and can also chelate Iron effectively. They are the most common iron-utilizing techniques available in the aerobes as well as facultative anaerobic bacterial families like Enterobacteriaceae, Streptomycetaceae, and Bacillaceae, for scavenging inorganic iron from the GIT (Yilmaz et al., 2018). Many experimental animal model studies which have employed transgenic animals or chemically induced colitis, also show that oral Iron administration might actually worsen intestinal inflammation. This occurs firstly by the formation of reactive oxygen species or ROS, and also secondly by means of promoting the growth of bacteria that love Iron. Iron can also enhance the replication and virulence of enteropathogens like *Shigella*, *Salmonella*, and *Campylobacter* spp. An *in-vitro* study has also documented that a moderate degree of iron in the extracellular milieu may help in cellular invasion by *Salmonella* spp. when co-cultured with the intestinal epithelial cells. It has also been seen that infants who are fed cow's milk preparations fortified with Iron have lesser *Bifidobacterium* but higher counts of *Bacteroides* and *E. coli* in the gut, as compared to infants receiving an unfortified cow's milk preparations. Dietary iron in high levels can also increase the chances of acquiring Colorectal cancer by generating Hydroxyl (OH) free radicals. Plant-based diets containing Phytic acid can bind well to this Iron and help reduce the incidence of Colorectal cancer.

Magnesium is a mineral that plays a very crucial role in many cellular reactions, like DNA repair and replication, mineral transport (particularly Calcium and Potassium), and metabolism. Dietary Magnesium supplementation, in a situation with no mineral deficiency, can produce intestinal dysbiosis. On the contrary, less magnesium consumption in the diet is linked with gut microbial flora with a greater capacity to capture energy from the diet (Garcia-Legorreta et al., 2020).

Zinc is also a very vital metal or micronutrient that can determine the versatility of the gut microflora. Zinc enhances the growth and survival of SCFA-generating bacteria in the human gut. This, in turn, reduces the pH of the contents of the gut lumen. This leads to further enhanced absorption of Zinc in blood from the gut. Zinc deficiency in piglet models can be associated causally with a lower burden of beneficial *Clostridiales* and *Verrucomicrobia* and a higher load of *Enterobacteriaceae* and *Enterococcus spp. in the gut (*Lopez et al., 2018*). Long-term deficit of Zinc in the feed or diet can even deplete Clostridium spp. in the chicken gut. This leads to lesser production of* SCFAs *in the intestine. Consequently, there is then abundant growth of E. coli and Proteus spp. in the gut.*

Tea and coffee contain caffeic acid. Caffeic acid can downregulate the growth of many pathogenic bacteria in the gut, like *E. coli*, *Salmonella*, *Pseudomonas*, *Clostridium* spp., and *Bacteroides* sp. (Vamanu et al., 2019).

6.8 EFFECT OF NUTRACEUTICAL SUBSTANCES ON THE GUT FLORA

The nomenclature "nutraceuticals" was first postulated by the renowned scientist, Dr. Stephen L. DeFelice in 1989. It implies a vital link between nutrition and the pharmaceutical or drug realm. After 1994, when the true value of food supplements for the upkeep of good health was strongly and firmly put forth, the category "nutraceuticals" was also included in this group. Compared to classical drugs, a nutraceutical moiety generally consists of a cluster of various bioactive substances, whose actions are determined by their biotransformation occurring in the human large gut.

Nutraceuticals thus fall into these three categories:

- Nutrients;
- Herbal products; and
- Dietary supplements.

The healthy microbes of the gut produce SCFAs. SCFAs play many important roles in the prevention and treatment of major illnesses like metabolic syndrome, and gastrointestinal ailments like Ulcerative colitis, Crohn's disease, or antibiotic-associated diarrhea. They are even helpful in several types of malignancies. Nutraceuticals can promote the growth of healthy bacteria in the gut and thus further increase SCFA formation.

Resveratrol is a well-known dietary polyphenol. It is used widely as a food supplement, in most cases by patients with cardiovascular disorders (Catinean et al., 2018). It is to be seen in those species of berries which belong to the genus *Vaccinium* and is also found in appreciable amounts in grapes (Vamanu et al., 2019). It can also be found in the skin of grapes and peanuts, and can also be found in quite high amounts in red wine. It facilitates the luxuriant growth of *Bacteroides*, *Lactobacillus*, and *Clostridium* in the gut, and simultaneously inhibits *Enterococcus fecalis*. It is directly related to the health-promoting effects which are observed after consumption of red wine.

Oregano vulgare is a popular aromatic herb. It has many culinary and therapeutic applications. Carvacrol and thymol are the two commonest phenolic compounds that are derived from oregano. They have quite potent antimicrobial effects on *E. coli* (Catinean et al., 2018).

1. **Omega-3 and Omega-6 Fatty Acids:** These are the most widely used supplements that are very useful in dyslipidemia, particularly in health conditions like hypertriglyceridemia. Fish oil, having both of these things, has been found to suppress the growth of *Helicobacter sp., Clostridium, Sphingomonas, Firmicutes, Pseudomonas* sp., and many other such non-cultivable bacterial species. Thus, it may be helpful for managing different conditions like gastritis, gastric ulcer, and a lot of other illnesses.
2. **Berberine:** It is an alkaloid with a quaternary ammonium structure. It is isolated from plants like *Rhizomacoptidis* or *Berberis sp.* It can effectively modulate the gut microbial flora. It has good antimicrobial activity on the phyla *Firmicutes* and *Bacteroidetesin gut*, as well.
3. **Spirulina (*Arthrospira platensis*):** This which is a type of cyano-bacteria (or blue–green algae). It is very commonly used as a food supplement. It is also a nutraceutical moiety. It has the unique ability to inhibit several Gram-positive and Gram-negative bacterial pathogens like *Staphylococcus aureus, Bacillus subtilis, E. coli*, and *Pseudomonas aeruginosa*. This antibacterial effect is believed to occur due to an extracellular metabolite liberated by Spirulina.
4. **Curcumin:** It is a bioactive compound that is extracted from the rhizome of the *Curcuma longa* plant. It is commonly known in India as Turmeric or Haldi. In an experimental murine model of inflammatory bowel disease, dietary supplementation with curcumin produced a significant alteration in the gut microbial flora by increasing the

growth of *Lactobacillus* spp. and decreasing the abundance of *Coriobacterium* spp. in the gut. Curcumin also produces appreciable antimicrobial effects in the GIT. This is amply clear in its actions against pathogens like *Staphylococcus aureus, Candida* spp., and *Salmonella* spp. (Vamanu et al., 2019). We ourselves had also seen the same observations and saw that Turmeric can also inhibit many virulence factors of many pathogenic bacteria.

5. **Arginine:** It is an Amino acid that is very useful when included in the diet. Highly reactive polyamine compounds, like spermidine, putrescine, and spermine, are produced from arginine by the action of the microbes of the gut. These amines are utilized in the development and upkeep of the epithelial lining of the intestine. They also help in the development of the resident immune cells.
6. **Fish Oil:** It is quite rich in omega-3 fatty acids. Many clinical studies using n–3 PUFAs have shown their power of (i) Diminishing the *Firmicutes/Bacteroidetes* ratio; (ii) lowering the levels of *Coprococcus* and *Facecalibacterium*; (iii) increasing the abundance of butyrate-producing bacterial strains, like *Bifidobacterium spp.*, *Lachnospira* spp., *Roseburia* spp., and *Lactobacillus* spp.
7. **Aloe Vera (*Aloe barbadensis* Miller):** This plant has several active compounds like anthraquinones, carbohydrates, minerals, vitamins, enzymes, and amino acids which have many beneficial effects. Aloe vera juice in the concentration of 5–25% has been found to help increase counts of the bacterium *Lactobacillus plantarumin gut. Aloin is said to be the bioactive molecule present in this juice.* Aloin shows antimicrobial and antimalarial effects against many disease-producing microbes. It can inhibit Gram-negative bacteria quite well, and to a lesser degree gram-positive *bacteria also.* Five *Lactobacillus brevis* strains have been isolated so far from *Aloe vera* pulp which is fermented naturally. They have all properties of a good probiotic agent, like high tolerance to acid, with the capability of survival for 4 hours at a pH of 2.5. These strains can lead to inhibition of the growth and multiplication of many pathogens of the gut like *Campylobacter jejuni*, *C. perfringens*, *Staphylococcus aureus*, *Salmonella* spp., and *E. coli*. However, it can also sometimes inhibit the normal commensal flora of the gut.
8. **Isoflavones or isoflavonoids:** These compounds are found in abundance in many dietary elements like Soy, soy protein, and Miso. These isoflavone compounds are inactive in their primary

form, in which they exist as conjugated glucosides. The gut bacteria and intestinal mucosa help in the conversion of these glucosides into compounds which are subsequently well absorbed and further converted by the intestinal microbial flora into other metabolites. Isoflavonoids contain a number of glycoside molecules, such as daidzein, genistein, and glycitein. Daidzein can be metabolized by the bacteria of the gut into a molecule called equol. The latter plays a very important function in conferring the health benefits of soy. It also shows potent estrogen-like and antioxidant effects (Catinean et al., 2018).

9. **Polyphenols:** These have been studied extensively in the last few years for their purported beneficial effects in improving both cardio-vascular diseases (CVDs) as well as cancers (Keservani & Sharma, 2014). They are to be seen in the human diet in quite high amounts. The major sources of polyphenols in food are dark chocolate, cocoa powder, nuts, berries, black pepper, and beans. They are also found in high amounts in vegetables like red onion, soy, spinach, black, green tea, and also red wine. There is often a bidirectional relationship between polyphenols present in food and the gut microbes. The bioavailability of these polyphenols is enhanced by the microbial flora in the gut of man. However, the unabsorbed polyphenols maintain the equilibrium of healthy microorganisms in the gut. Quercetin is one of the most thoroughly studied polyphenols with respect to the gut microbiota. It is a polyphenol compound that can be obtained from plant sources and belongs to the group of flavonols, a subclass of flavonoid compounds. It can be seen not only in fruits like grapes, berries, apples, onions, tea, tomatoes, seed, and nuts but also in many medicinal plants like *Ginkgo biloba, H. perforatum*, and *Sambucuscanadensis.* Quercetin improves the *Firmicutes/ Bacteroidetes* ratio in rat models who are fed a diet rich in Sucrose and fat. It does so by increasing the levels of beneficial gut bacteria like *Bacteroidesvulgatus* and *Akkermansiamuciniphila.* They also have a reverse link with obesity (Jarouliya & Keservani, 2019). Moreover, it can also lower the burden of*Bilophilawadsworthia and Eubacteriumcylindroides in the gut.* These bacteria are known to be associated with diet-induced obesity. Tea, obtained from the young leaves of the *Camellia sinensis* plant, has got high concentrations of polyphenols, like flavanols, catechin, and epicatechin and their various esters. High amounts of unabsorbed residues of tea are also

found in the gut. They play a tremendous function in the overall upkeep of gut health. Caffeic acid, for instance, can bring down the numbers of many intestinal bacterial pathogens, like *E. coli, Pseudomonas aeruginosa, Salmonella, Clostridium* spp., and *Bacteroides* spp.

10. **Anthocyanins:** These are grouped in the flavonoid class of compounds, and can be found in berries, grapes, and apples. These compounds have a stellar role to play in the prevention of many cardiovascular (Keservani et al., 2016a) and neurological disorders (Keservani et al., 2015, 2016b), metabolic disorders, and even cancer, particularly owing to their antioxidant property (Catinean et al., 2018). Berries are also reputed for their high levels of anthocyanin compounds. The supplementation of the diet of mice with blueberries or black currants has shown a marked alteration in the gut microbial make-up, in different experiments. There is an increase in anaerobic bacteria like *Bacteroidetes* and *Actinobacteria*, possibly via some antioxidant-like mode of action.

6.9 FERMENTED FOODS AS IN DIET AND AS NUTRACEUTICALS

Fermented foods have a firm place in the cuisine of almost every culture in the world. In the West, there is a recent uptick in the popularity of various fermented foods in recent times. The major reasons for this are the purported health benefits of fermented foods and man's growing interest in gastrointestinal health.

Kefir is a popular fermented milk product. It can suppress the growth of many harmful bacteria like *E. coli* and *Staphylococcus aureus*. It concomitantly helps in the growth of healthy gut bacteria. It has got appreciable antioxidant activities also. Kefir has its origins in the countries of Eastern Europe and Southwest Asia. The name may be traced back to the Turkish word "*keyfir*," which translates to "feeling good after eating." Fermented dairy products have since antiquity, been considered to impart a number of health benefits to those individuals who regularly eat or drink them. Kefir is a fermented milk product containing tons of Lactobacilli. It can inhibit pathogenic microorganisms like *Candida* spp., *Staphylococcus*, and *Escherichia coli.* Kefir grains contain as many as 61 different categories of bacteria as well as yeasts. This makes Kefir a very rich and diverse source of probiotics. Kefiran, a type of carbohydrate that is commonly found in kefir, has good antibacterial properties, too (Bourrie et al., 2016).

Ellie Metchnikoff first postulated that the effect of kefir on the bacteria in the gut can contribute to a healthy and prosperous life (Bourrie et al., 2016). The most common bacterial genera found in kefir using culture are *Lactobacillus* spp., *Streptococcus spp., Lactococcus* spp., and *Leuconostoc spp.* In addition to bacteria, Kefir also consists of many strains of yeasts like *Candida kefyr*, *Saccharomyces* spp., and *Kluyveromyces* spp. Eating kefir or the carbohydrate kefiran in the animal models has been accompanied by an increase in the counts of microbes considered beneficial, like *Lactobacillus* and *Bifidobacterium*. It also decreases the harmful microbes like *Clostridium perfringens* at the same time (Bourrie et al., 2016). Lactobacilli isolated from kefir also have the unique ability to shield Vero cells from the effect of Shiga toxin type II, produced by *E. coli* O157:H7.

There are a multitude of ways by which fermented foods may actuate their beneficial effects. Lactic acid bacteria (LAB) are indeed quite relevant to both dairy and non-dairy fermented foods. They help generate a number of bioactive peptides and polyamine molecules. These compounds hence have profound effects on the heart, blood vessels, immunity, and also metabolic health. Thirdly, fermentation may help convert some of these compounds into several biologically active metabolites. For example, lactic acid-producing bacteria are able to convert many phenolic compounds like flavonoids into more biologically active metabolites. Fourthly, food components that are found in fermented foods, like prebiotics and vitamins, may also exert many health benefits. Lastly, fermentation may lower the levels of toxins and anti-nutrients. For example, soybean fermentation may reduce the phytic acid levels. Fermented vegetable products like Sauerkraut also have *Lactobacillus sakei* and other bacterial strains, which may be very helpful in this regard.

Many preservatives also have some notable impact on the microbial flora of the intestine. Carboxymethyl cellulose and Polysorbate 80, two very commonly used food additives, can also reduce *Lactobacillus rhamnosus* levels in the gut. It also increases anxiety-like behavior in mice models. Epsilon-polylysine has some transient inhibitory effects on the gut flora, but the bacteria can overcome these effects with time (Figure 6.1).

6.9.1 EFFECT OF SOME COMMON SPICES, HERBS, AND THEIR OILS ON GUT FLORA

Many herbs and spices have appreciable quantities of micronutrients and have nutraceutical-like effects and hence be beneficial. It has been observed that the essential oil of Cinnamon can induce the proliferation of *Lactobacillus* spp. and *Bacteroides* spp. Hence it is said to be useful in IBD (inflammatory bowel disease) where it acts by restoring the dysbiotic gut microbiota. In similar studies, Clove oil can also inhibit pathogens in the gut and also commensals to a lesser extent. Cardamom (*Elletariacardamomum*) oil has an inhibitory effect on *Helicobacter pylori* and can help in the synthesis of SCFA in the gut. It can also inhibit *Campylobacter* spp. in the intestine. Black pepper powder, in the Drosophila model, has been found to increase the counts of *Lactobacillus* spp. and *Acetobacter* spp. in the gut. This is possibly due to its high polyphenol and carotenoid content.

FIGURE 6.1 Green tea promotes good bacteria in the gut.
Source: Image: Author.

Fennel seeds (*Foeniculum vulgare*) are commonly used along with or after meals in India for a very long time as digestive and mouth fresheners. It has many chemicals which have a probiotic effect in the gut and also help build probiotic bacteria.

Ajwain or bishop's weed (*Trachyspermumammi*) has many antibacterial properties and can also inhibit *Staphylococcus aureus* virulence factors, as observed by the author of this chapter himself. Many of the essential oils found in ajwain, notably thymol, and carvacrol, can help in preventing the growth of bacteria and fungi. They may also help eliminate pathogens like *Salmonella* and *E. coli*, which can lead to food poisoning and stomach ailments in man.

Neem (*Azadirachtaindica*) leaves have been consumed for many disorders in India for ages. It contains many essential oils which have potent antipyretic and anthelmintic effects. It exerts its functions via many alkaloids like nimbolide, Nimbin, nimbidinand limonoids, quercetin, and beta-sitosterol. Ethanolic extract of neem leaves can prevent the growth of Escherichia coli. Neem can also inhibit other bacterial pathogens like *Salmonella gallinarum* and *S. pullorum*. It also has many hepatoprotective and gastric ulcer-healing effects.

Curry leaves (*Murraya koenigii*) are an integral constituent of the average Indian diet. They also have many well-documented antibacterial and laxative properties. They are also very good for the liver and skin and are quite good for treating anemia as well. It is rich in many micronutrients, Vitamins, Iron, and Carbazole alkaloids. These Carbazole alkaloids can inhibit appreciably the growth of disease-producing *E. coli* in the gut (Figure 6.2).

FIGURE 6.2 Curry leaves are beneficial for gut microflora.
Source: Image: Author.

Onion (*Allium cepa*) is a staple component of the Indian diet. It is a very good source of important nutrients like carbohydrates, and also vitamins and minerals. The health-promoting effects of this plant may be attributed to the presence of various chemical components like carbohydrates (fermentable oligosaccharides and polysaccharides) having notable prebiotic effects, sulfur compounds (like thiosulfinates), and various phenolic compounds like flavonoids and quercetin derivates (Catinean et al., 2018).

Nigella sativa, also called black cumin seeds is very popular in Indian cuisine. They can inhibit *Escherichia coli* and other bacteria present in the GIT. It also protects against peptic ulcers. Thymoquinone is the active component that is present in black cumin seeds.

Green chili is an essential part of the diet in almost all nooks and corners of the world. Green chili consists of Capsaicin, a nutraceutical-like chemical that helps in growth of the healthy bacteria in the gut, like *Faecalibacterium* and *Roseburia* spp. As of now, prebiotics, as well as probiotics, are also considered nutraceuticals.

Ginger (*Zingiberofficinale*) is also good for gut bacteria. Ginger supplementation in diet can have a notable effect on the constituents of the microbiome of our gut. It increases the counts of the species which belong to the *Bifidobacterium* genus and also increases SCFA-producing bacteria (like *Alloprevotella* and *Allobaculum*). Along with this, there is an increase in fecal SCFA concentrations. *Faecalibacterium* levels also go up after exposure to ginger juice. The molecule 6-gingerol (6G) is the major polyphenol present in Ginger extract, which increases the bio-load of *Bifidobacterium* significantly in the intestine. Ginger is actually the rhizome of the plant, *Zingiberofficinale* Roscoe (*Zingiberaceae*). It is commonly eaten as a spice or also used as a good flavoring agent the world over.

Garlic extract (*Allium sativum*) inhibits all the bacteria in the gut other than *Lactobacillus* spp. and is thus deemed to be very good for health. Garlic also demonstrates potent bactericidal action against a number of pathogens like *Staphylococcus aureus*, *E. coli*, *Salmonella Typhimurium*, *Neisseria gonorrhoeae,* and *Enterococcus faecalis*. *Clostridium nexile* is the bacterial strain that is reported to be most vulnerable to Garlic. Fructans are the prebiotic compounds that are seen in garlic, while allicin is the putative antibacterial compound. Allicin also induces the proliferation of the bacterium *Blautia* spp. in the gut. The latter is a major producer of SCFAs.

Lemon is also proven to be very beneficial for gut health. It contains Vitamin C, Citric acid, and polyphenol molecules such as Eriocitrin. Eriocitrin is a strong antioxidant molecule. It helps prevent the overaging of the microbiota of the gut; it also maintains a good Bacteroides: Firmicutes ratio in the gut. Naringenin is a flavonoid compound found in lemon and also grapes, which can help abrogate the adherence of *Salmonella* Typhimurium in cell lines.

Soybeans and lactose obtained from cow milk are major contributors to dietary GOS or galactooligosaccharides. These are included particularly in the infant food formulae. The salient advantages of GOS are their high solubility, neutral taste, thermostability, acidity, and low glycemic index. It was also documented that if the preterm infants are fed 90% GOS along with 10% FOS, the make-up of the intestinal microbiological flora shows an increase in beneficial bacteria like *Bifidobacterium* spp. and *Lactobacilli*. This resembles a breastfed infant's microbial flora of the GIT (Catinean et al., 2018).

Food additives can also disturb the overall homeostasis of the gut and thereby contribute to the hosts' tissue-damaging inflammatory responses. For example, mice that have been fed emulsifiers like carboxymethylcellulose and Tween 80 (polysorbate 80) can often develop gut dysbiosis along with concomitant overgrowth of the bacteria capable of degrading mucus. This can, in turn, trigger colitis in animals that are deficient in interleukin-10 (IL-10), a cytokine that shows powerful anti-inflammatory and immunoregulatory actions. Monosodium glutamate (MSG) or Azinomoto, which is used quite commonly as an additive in various varieties of fast foods, has been found to induce such dysbiosis or derangements in the microbial population of the gut. It can also sometimes induce obesity in mice models. High consumption of MSG decreases the load of *Faecalibacterium*, *Megamonas*, and *Blautia* in the gut, and also leads to an increase of *Collinsella* spp. at the same time. Quercetin found in onion and other herbs is able to reverse this (Figure 6.3).

Sometimes fungi or macro-fungi are themselves used as good nutraceuticals, like the various varieties of Mushrooms. They contain lots of protein, Vitamin B5, moisture, beta Glucan and other carbohydrate substances that can have many health benefits. They are also rich in minerals and trace elements like Phosphorus and Potassium. They are believed to have many antitumor and antioxidant activities through the production of many secondary metabolites. They also possess potent antibacterial and antiviral effects.

FIGURE 6.3 Black pepper powder enhances good microbial growth in the gut via polyphenols.
Source: Image: Author.

6.10 ALTERATION OF GUT MICROBIOTA BY ANTIBIOTIC USE AND AAD (ANTIBIOTIC-ASSOCIATED DIARRHEA)

Antibiotics can alter the composition of the microbial flora of the gut to a large degree. Major alterations in the gut microbiome, seen as a response to antibiotic usage are diminished taxonomic diversity and the persistence of these changes in a major chunk of the population. It has been documented that the effects of even short-term use (for 7 days) of broad-spectrum antibiotics having predominantly anaerobic spectra of coverage like Clindamycin could persist for a period of up to 2 years. The diversity of *Bacteroides* spp. cannot be then recovered at all (Jernberg et al., 2007). Similarly, a short course *H. pylori* eradication regime with Clarithromycin triple therapy tactic can lead to a dramatic reduction in the diversity of *Actinobacteria* present in the gut. There is then also a 1,000-fold elevation in the expression of the *ermB* resistance gene (Jakobsson et al., 2010).

6.11 EFFECT OF AGE ON GUT MICROBIOTA

Even though it is frequently held that the GIT tends to get colonized by microbes just after the birth of the child, there is enough mounting evidence now that the gut of the infant might be colonized by microorganisms even in utero. New genetic sequencing based on 16S rRNA, has shown very clearly that the mother's initial meconium fluid is quite rich in bacteria like*Escherichia* spp., *Shigella*, *Enterococcus*, Leuconostoc, Lactococcus, and *Streptococcus*. Still, it is now evident that the first microbiome profile of the newborn is mostly shaped by the mode of delivery. The intestines of infants who are born vaginally are initially colonized by microorganisms arising from the mother's vagina. This is best shown by the bacteria *Lactobacillus* spp. and *Prevotella* spp. (Jandhyala et al., 2015). On the other hand, in cesarean delivery mostly the mother's skin flora colonizes the infant's gut. This is evident by the predominance of *Streptococcus* spp., *Corynebacterium* spp., and *Propionibacterium* spp. In pre-term infants, bacteria colonizing the gut are mainly *Bifidobacterium* and *Lactobacillus*. Basically, these things may differ, depending on the different types of feeding practices. In formula-fed infants, Enterobacteria, *Bacteroides* spp., *Enterococcus*, Clostridia, and other anaerobic *Streptococci* tend to dominate the gut niche. In breastfed infants, however, *Bifidobacterium* and *Lactobacillus* predominate the gut microbiome. Breast milk contains many indigestible glycan moieties which are now called oligosaccharides of human milk (HMO). They are then easily decomposed by these types of bacteria (Jandhyala et al., 2015).

6.12 MICROBES OTHER THAN BACTERIA AS GUT FLORA

6.12.1 FUNGI AND THEIR ROLE IN GUT

Fungi are also very important members of the gut microbiome. However, their role as members of the gut microbial flora is not so well researched yet (Chin et al., 2020). Recent tests; like shotgun metagenomics sequencing analysis have shown that fungi make up about 0.1% of the total microorganisms in the gut. In terms of phyla, so far, most studies show clearly that Ascomycota is the commonest fungal phylum encountered in the gut. This is followed by fungi of phyla Zygomycota and Basidiomycota (Chin et al., 2020). However, as regards fungal genera, a recent work by Hallen-Adams and other coworkers has identified potential fungal pathogens that inhabit the gut. These belong to the genera *Cryptococcus*, *Malassezia*, *Aspergillus*, *Candida*, *Galactomyces*, *Saccharomyces*, *Trichosporon*, and *Cladosporium*. The yeast-like fungus, *Candida albicans*, is a member of the normal mycobiome of the intestine. It can also become a pathogen in rare cases if the other competing bacterial flora of the gut is reduced by oral antibiotics (Kumamoto et al., 2020). Specific conditions like a dysbalanced microbiome, suppression of man's immunity, and also a deranged gut barrier can make one prone to the development of invasive *C. albicans* infections. These can be nosocomial in most cases. Fungal members of the gut microbiome, which can also be termed the mycobiome, have got much lesser attention when compared to the gut bacteriome. Nevertheless, what is known now about this emerging arena and how the mycobiome of man can affect health or disease is in fact very interesting to know (Forbes et al., 2019). To this day, the phylum *Ascomycota* (which includes fungi like *Cladosporium* spp., *Candida* spp., and *Saccharomyces cerevisiae*) and also the Basidiomycota phylum (having genera like *Cryptococcus* spp., *Filobasidium* spp., and *Malassezzia* spp.) have been found to dominate most human anatomic sites as colonizers. Whilst most mycobiome research till now has focussed only on small cohorts, the Microbiome Project of Humans has newly analyzed the mycobiome of 317 healthy stool specimens by a study of the ITS2 region, the 18S rRNA gene, and also by a novel shotgun metagenomics approach (Garcia-Legorreta et al., 2020; Lopez et al., 2018). Fungal diversity has been shown to be significantly lesser as compared to bacterial diversity in the gut. Yeasts usually dominate the gut mycobiome. In fact, here the genera like *Saccharomyces* (*S. cerevisiae*), *Malassezia* (*M. restricta*), and *Candida* (*C. albicans*) are the commonest. They are present in about 97, 88, and 81% of all human fecal specimens, respectively. A lot of inter- and intra-individual variability in this regard has also been found. *S. cerevisiae, M. restricta*, and *C. albicans*, however, have been detected in 92, 78, and 64% of individuals, respectively. These species are really the resident commensal fungi of the gut; they form an important part of the core gut mycobiome. Recent findings support this idea that a competitive association is present between bacteria and fungi in the human gut. For instance, studies have proven that long-term antibiotic usage can be directly linked to fungal infection and overgrowth, particularly in the gut. Studies have also shown that germ-free mice are highly prone to infection caused by fungi like *Candida* spp. Fungal overgrowth in the mouse gut, induced by antibiotics, can also facilitate the subsequent occurrence of allergic airway reactions to *Aspergillus fumigatus* spores (Kumamoto et al., 2020). Also, commensal bacteria of the gut, like

Bacteroidesthetaiotaomicron and *Blautiaproducta* can incite the secretion of peptides with antifungal action, by the colonic epithelial cells (Forbes et al., 2019).

Gutmycobiome: Some newly discovered factors considered to be associated with the community composition of the fungal flora of the gut are the host genotype, host physiology like age, gender, and also the presence of some comorbidities, lifestyle parameters like diet, hygiene, and occupation, and also the immune system (Nash et al., 2017; Cui et al., 2013). The gut fungal composition can also be affected by a number of other factors, like age, diet, host immunity, medication, host genetics, and also bacterial microbiome. The latter occurs through interkingdom interactions (Chin et al., 2020). Diet in particular represents a major thing that shapes the composition of the normal gut fungal flora (Cui et al., 2013; Huseyin et al., 2017). Many fungi are actually foodborne microorganisms that are seen in common animal and plant-based food products. Hence upon ingestion, they become transient colonizers and are considered to affect the mycobiome of our gut. In a study using unadjusted analyzes via Spearman's correlation, the bioburden of *Candida* spp. correlated positively with the recent consumption of carbohydrates and negatively correlated with the consumption of proteins, amino acids, and fatty acids. In the same study, *Aspergillus* spp. was found to correlate inversely with the recent ingestion of SCFAs (Forbes et al., 2019). A fatty diet also influences gut mycobiome greatly. Fat deposition in the liver, a transcription-mediated adaptation of metabolically active sites of the body, and serum metabolic biomarker levels can be causally linked with changes in diversity as well as the composition of the mycobiome in the GIT. Particularly, variation or alteration in fungi of the genera *Thermomyces* and *Saccharomyces* have been seen to be strongly linked with a lot of metabolic abnormalities as well as weight gain. These data can show that the metabolic interactions of the host with microbes can also be influenced by the variability of the host mycobiome (Mims et al., 2021). Exposure to a processed diet can downregulate these fungi-associated networks. Also, metabolic activity is strongly associated with variability in fungal flora of the intestine. The gut fungal flora of healthy mice, as found in scientific experiments, is decided by the environment as well as diet. It correlates significantly with the metabolic changes in the human host. For example, increased triglyceride levels and metabolic biomarkers, and the deposition of fats in the liver, correlate also with the abundance of *Thermomyces* and decreased count of *Saccharomyces*. These results thus highlight the significance of the mycobiome of the gut in good health. They have hence got a lot of importance for human and experimental metabolic research (Mims et al., 2021).

Another very important function of gut microorganisms, which is often attributed only to gut bacteria, is the formation of bile acids that are secondary (or BAs). Our notion of the role of BAs has moved forward from basically detergent molecules to hormones that are intricately related to many metabolic processes. This shows the important functions of Bile Acids in conditions like altered blood lipid levels and also Type 2 DiabetesMellitus (Mims et al., 2021). BAs also help in host–microbiome interactions. This can occur by cellular receptors like TGR5 (*Gpbar1*) as well as FXR (*Nrlh4*). So, BA composition is intertwined with levels of metabolic hormones like resistin, ghrelin, leptin, GLP-1 (glucose-like peptide-1), and peptide YY. However, fungi also generate Bile Acids. They possibly do also take part in the gut BA pool to a large extent. *Fusarium* spp. can help shape the structure of processed diet-induced weight gain-resistant mic. It can metabolize deoxycholic acid very fast. Other important fungi, like *Aspergillus* and *Penicillium*, can also help synthesize secondary BAs in the gut.

Alzheimer's disease (AD) is a progressive, fatal degenerative CNS disease. It can affect more than 40 million people across the world. It is indeed the major cause of dementia nowadays, as well as a decline in cognitive ability, especially in the elderly. Although so much research is now done regarding the risk factors of AD pathogenesis at genetic and molecular levels, no treatment option is currently there still to prevent or slow AD progression. This also stresses the complexity of AD and our insufficient knowledge of its pathogenesis. This points out the need for finding out new therapeutic targets, as also some non-conventional measures, which may help control this devastating condition. Recent evidence gives the hint that the gut microbiome does play a very important function in the pathogenesis of many nerve-related disorders (Nagpal et al., 2020). Studies have shown the importance of this gut-brain axis, which is an integrated network. In this, the gut microbiome and the central nervous system can cross-talk among themselves via immune, endocrine, or nerve signaling pathways, in various aspects of host health and diseases. Recent studies by other workers have also quoted an altered microbiome of the gut, in patients with MCI, dementia, or AD. Of late, it has been found that patients having a mild cognitive impairment (MCI) nurture a specified set of bacteria in their intestines. Also, the modified Mediterranean ketogenic diet (also called MMKD) can improve the markers of AD in cerebrospinal fluid (CSF) and also the pattern of different bacteria in the gut. However, the precise role of the fungi of the gut (mycobiome) with respect to MCI/AD pathology, bacteria of the gut, and food, is yet to be known accurately. Compared to cognitively normal

(CN) counterparts, in MCI there is a higher load of the fungi of families Sclerotiniaceae, Cystofilobasidiaceae, Togniniaceae, Phaffomyceteceae, Trichocomaceae, and the genera *Botrytis*, *Phaeoacremonium*, *Kazachstania*, and *Cladosporium*. They also have a lower gut load of fungi like *Heteronym* spp. (Jakobsson et al., 2010). Many fungi commonly found in our gut, like *Botrytis, Penicillium, Agaricus,* and *Fusarium* do not remain viable for long in the gut, either due to unfriendlysurvival conditions, or ecological niche that is unsuitable. So, they may be frequently as well as transiently introduced through food. Nonetheless, the inability to colonize the gut in a stable manner may not prevent these fungi from having a deep impact on the host or interacting with gut bacteria (Nagpal et al., 2020). Mycobiome analysis has been done in a few studies to assess the true role of microbiota in the pathogenesis of Diabetes mellitus. In the relevance of fungal involvement, some workers have found out that besides *E. coli*, *C. albicans*, and *Enterobacteriaceae* colonization increases in the patients with T1DM (Type 1 diabetes), whereas *Bifidobacterium* load gets diminished. Food also has a great influence on the gut mycobiome. A high prevalence of *Saccharomyces* in the gut can occur, due to the consumption of yeast-rich foods like beer and bread. On the other hand, high levels of *Candida* can be strongly linked to recent consumption of sugary foods (Yilmaz et al., 2018). Furthermore, diet therapy and antibiotics can help lower the burden of fungi in the gut in patients with Crohn's disease. This has been shown by Lewis et al. (2015).

6.12.2 ROLE OF FUNGI AS PROBIOTICS

Fungi have also been used for a long time as probiotics. *Saccharomyces boulardii* var. *cerevisiae* is one such commonly used probiotic yeast. It has got many health benefits. People have used *S. cerevisiae* to allay *C. difficile* infection of the gut. However, rarely they may also enter the bloodstream and cause fungemia (Stone et al., 2021). Probiotic yeasts like *Saccharomyces* spp. and *Issatchenkiaoccidentalis* when applied to surfaces, may prevent biofilm formation and also adhesion of *C. auris* (https://www.wpi.edu/news/probiotic-yeast-may-offer-effective-treatment-drug-resistant-fungal-infections, 2019). Sacc*haromyces boulardii* is a commercialized probiotic yeast. It is widely applied to control the virulence factors of *C. albicans. In vitro,* experiments have shown clearly that *S. boulardii* and its cellular extracts can significantly inhibit the adhesion, morphological transition as well as biofilm formation in *C. albicans* (Lohith et al., 2018). Yeast-based probiotics have phytase enzymes that can break down the phytate present in cereal-based food. Thus it can better iron absorption from the gut and thus also minimize Iron deficiency. Esters produced by *Pichia kudriavzevii* produce a fruity odor in foods and render them attractive to taste (Ogunremi, 2020). Yeasts like *Starmeracaribaea* as well as *Hanseniasporaguilliermondii* have been reported to produce very high levels of acetate esters. Thereby they impart a highly desirable flavor to food.

6.12.3 FUNGAL EXTRACTS AND EFFECT ON GUT MICROBIOTA

Kombucha tea is a popular beverage that is produced by the fermentation of black tea sucrose by fungi and also some bacteria. In many studies in mice, feeding Kombucha tea is found to increase Bacteroidetes in the gut and decrease the levels of bacteria of Erysipelotrichia class. The latter can also cause nonalcoholic fatty liver disease (NAFLD) which is now emerging as the most common factor behind chronic liver disease in man. NAFLD occurs due to obesity and absorption of LPS of harmful bacteria like in class Erysipelotrichia along with lipid-rich food, as found in animal models. Kombucha tea can prevent this process (Jung et al., 2019). Kombucha tea also has many significant antibacterial effects, perhaps due to Acetic acid generation.

6.13 DISCUSSION

Hence, as we approach the very last part of the chapter, it can be summarized that various dietary elements have some effect or influence on both beneficial bacteria of the gut as well as gut pathogens. *Lactobacillus* spp., *Akkermansia* spp., and *Bifidobacterium* sp. Are considered among the good bacteria in the gut. Many nutrients like probiotics and prebiotics have a lot of impact on the bacteria and fungi in the gut. These effects need to be studied more and are of great practical significance. Micronutrients, macronutrients, and minerals are hence very important for promoting the growth of healthy bacteria in the gut and inhibiting the proliferation of pathogens. They do so by various molecules and mechanisms. Spices, herbs, plant-based products, and other fermented products also have almost the same effects. Hence they should form components of a healthy and wholesome diet. Microbiologists and nutritionists should hence work in tandem for formulating a balanced diet pattern that has a good effect on gut microbiota.

6.14 CONCLUSION

Gut microbiota, also called microbiome, can be influenced greatly by diet and many trace elements found in one's diet. Some nutrients enhance the load and diversity and the beneficial gut microbes while others can suppress them. This is a very interesting area of research and can be called one of the upcoming topics of the future. This is a new avenue of health and is very important from public health viewpoint also. However, more studies are needed in this very interesting area.

ACKNOWLEDGMENTS

I heartily acknowledge the help I got from my beloved wife, Dr. Satarupa for being my support, and also Dr. Rajesh Kumar Kesharwani in assisting me in bringing out this chapter in the treatise on micronutrients, macronutrients, and nutraceuticals.

KEYWORDS

- **gut microbiota**
- **macronutrients**
- **micronutrients**
- **nutraceuticals**
- **prebiotics**
- **trace elements**
- **vitamins**

REFERENCES

Bourrie, B. C. T., Willing, B. P., & Cotter, P. D., (2016). The microbiota and health-promoting characteristics of the fermented beverage Kefir. *Front Microbiol.* https://doi.org/10.3389/fmicb.2016.00647.

Catinean, A., Neag, M. A., Muntean, D. M., Bocsan, I. C., & Buzoianu, A. D., (2018). Overview on the interplay between nutraceuticals and gut microbiota. *Peer J, 6*, e4465. doi: 10.7717/peerj.4465.

Chin, V. K., Yong, V. C., Chong, P. P., Amin, N. S., Basir, R., & Abdullah, M., (2020). Mycobiome in the gut: A multiperspective review. *Mediators Inflamm.*, 9560684. doi: 10.1155/2020/9560684.

Cui, L., Morris, A., & Ghedin, E., (2013). The human mycobiome in health and disease. *Genome Med., 5*, 63. doi: 10.1186/gm467.

Forbes, J. D., Bernstein, C. N., Tremlett, H., Domselaar, G. V., & Knox, N. C., (2019). A fungal world: Could the gut mycobiome be involved in neurological disease? *Frontiers Microbiol., 9*. https://doi.org/10.3389/fmicb.2018.03249. https://www.frontiersin.org/article/10.3389/fmicb.2018.03249 (accessed on 05 June 2023).

Garcia- Legorreta, A., Soriabo- Perez, L. A., Flores-Buendia, A. M., Medina- Campos, A. M., Noriega, L. G., Granados- Portillo, O., et al., (2020). Effect of dietary magnesium content on intestinal microbiota of rats. *Nutrients, 12*(9), 2889.

Hemarajata, P., & Versalovic, J., (2013). Effects of probiotics on gut microbiota: Mechanisms of intestinal immunomodulation and neuromodulation. *Therap. Adv. Gastroenterol., 6*(1), 39–51.

Hooper, L. V., Wong, M. H., Thelin, A., Hansson, L., Falk, P. G., & Gordon, J. I., (2001). Molecular analysis of commensal host-microbial relationships in the intestine. *Science, 291*, 881–884. doi: 10.1177/1756283X12459294.

Human Microbiome Project Consortium, (2012). Structure, function and diversity of the healthy human microbiome. *Nature, 48*(6), 207–214.

Huseyin, C. E., O'Toole, P. W., Cotter, P. D., & Scanlan, P. D., (2017). Forgotten fungi—The gut mycobiome in human health and disease. *FEMS Microbiol. Rev., 41*, 479–511. doi: 10.1093/femsre/fuw047.

Jakobsson, H. E., Jernberg, C., Andersson, A. F., Sjölund-Karlsson, M., Jansson, J. K., & Engstrand, L., (2010). Short-term antibiotic treatment has differing long-term impacts on the human throat and gut microbiome. *PLoS One, 5*, e9836.

Jandhyala, S. M., Talukdar, R., Subramanyam, C., Vuyyuru, H., Sasikala, M., & Nageshwar, R. D., (2015). Role of the normal gut microbiota. *World J. Gastroenterol., 21*(29), 8787–8803. doi: 10.3748/wjg.v21.i29.8787.

Jarouliya, U., & Keservani, R. K., (2019). Pathways leading to child obesity. In: Debasis, B., (ed.), *Global Perspectives on Childhood Obesity: Current Status, Consequences and Prevention* (2nd edn., pp. 137–146). Chapter 12, Academic Press, Elsevier. ISBN: 9780128128404.

Jarouliya, U., & Keservani, R. K., (2019a). Protein function as cell surface and nuclear receptor in human diseases. In: Sanjay, B., & Debarshikar, M., (eds.), *Medicinal Chemistry with Pharmaceutical Product Development* (pp. 1–32). Chapter 1, Apple Academic Press, CRC Press, Taylor& Francis Group. ISBN: 9781771887106.

Jernberg, C., Löfmark, S., Edlund, C., & Jansson, J. K., (2007). Long-term ecological impacts of antibiotic administration on the human intestinal microbiota. *ISMEJ, 1*, 56–66.

Jung, Y., Kim, I., Mannaa, M., Kim, J., Wang, S., Park, I., et al., (2019). *Food Sci. Biotechnol., 28*(1), 261–267. doi: 10.1007/s10068-018-0433-y.

Keservani, R. K., & Sharma, A. K., (2014). Flavonoids: Emerging trends and potential health benefits. *Journal of Chinese Pharmaceutical Sciences, 23*(12), 815.

Keservani, R. K., Kesharwani, R. K., Sharma, A. K., Vyas, N., & Chadoker, A., (2010b). Nutritional supplements: An overview. *International Journal of Current Pharmaceutical Review and Research, 1*(1), 59–75.

Keservani, R. K., Kesharwani, R. K., Vyas, N., Jain, S., Raghuvanshi, R., & Sharma, A. K., (2010a). Nutraceutical and functional food as future food: A review. *Der Pharmacia Lettre, 2*(1), 106–116.

Keservani, R. K., Sharma, A. K., & Kesharwani, R. K., (2016a). Nutraceutical and functional foods for cardiovascular health. In: *Food Process Engineering* (pp. 257–278, 291–312). Apple Academic Press, CRC Press. ISBN: 9781771884020.

Keservani, R. K., Sharma, A. K., & Kesharwani, R. K., (2016b). Medicinal effect of nutraceutical fruits for the cognition and brain health. *Scientifica,* 10. Article ID 3109254. doi: 10.1155/2016/3109254.

Keservani, R. K., Sharma, A. K., & Kesharwani, R. K., (2020). *Nutraceuticals and Dietary Supplements: Applications in Health Improvement and Disease Management*. CRC Press. ISBN: 9781771888738.

Keservani, R. K., Singh, S., Singh, V., Kesharwani, R. K., & Sharma, A. K., (2015). Nutraceuticals and functional foods in the prevention of mental disorder. In: Debasis, B., Harry, G. P., & Anand, S., (eds.), *Nutraceutical and Functional Foods in Human Life and Disease Prevention* (pp. 255–268). Chapter 15, Boca Raton, Fl. CRC Press, Taylor and Francis. ISBN: 9781482237214.

Kesharwani, R. K., Singh, P., & Keservani, R. K., (2016). Green coffee bean extract and chlorogenic acids: Chemistry and novel antioxidant benefits. In: Bagchi, D., Moriyama, H., & Swaroop, A., (eds.), *Green Coffee Bean Extract in Human Health* (pp. 1–17). Chapter 1, Boca Raton, Taylor and Francis, USA. ISBN: 9781498716376.

Kumamoto, C. A., Gresnigt, M. S., & Hube, B., (2020). The gut, the bad and the harmless: *Candida albicans* as a commensal and opportunistic pathogen in the intestine. *Curr. Opin. Microbiol., 56*, 7–15. doi: 10.1016/j.mib.2020.05.006. Epub 2020 Jun 27. PMID: 32604030; PMCID: PMC7744392.

Lewis, J. D., Chen, E. Z., Baldassano, R. N., et al., (2015). Inflammation, antibiotics, and diet as environmental stressors of the gut microbiome in pediatric Crohn's disease. *Cell Host & Microbe., 18*(4), 489–500. doi: 10.1016/j.chom.2015.09.008.

Lohith, K., & Anu-Appaiah, K. A., (2018). Antagonistic effect of *Saccharomyces cerevisiae* KTP and *Issatchenkia occidentalis* ApC on hyphal development and adhesion of *Candida albicans*. *Medical Mycology, 56*(8), 1023–1032. https://doi.org/10.1093/mmy/myx156.

Lopez, C. A., & Skaar, E. P., (2018). The impact of dietary transition metals on host-bacterial interactions. *Cell Host Microbe, 23*, 737–748. doi: 10.1016/j.chom.2018.05.008.

Mims, T. S., Abdallah, Q. A., Stewart, J. D., et al., (2021). The gut mycobiome of healthy mice is shaped by the environment and correlates with metabolic outcomes in response to diet. *Commun. Biol., 4*, 281. https://doi.org/10.1038/s42003-021-01820-z.

Nagpal, R., Neth, B. J., Wang, S., Mishra, S. P., Craft, S., & Yadav, H., (2020). Gut mycobiome and its interaction with diet, gut bacteria and Alzheimer's disease markers in subjects with mild cognitive impairment: A pilot study. *EBio. Medicine, 59*, 102950. doi: 10.1016/j.ebiom.2020.102950.

Nash, A. K., Stewart, C. J., Smith, D. P., Muzny, D. M., Metcalf, G. A., Gesell, J. R., et al., (2017). The gut mycobiome of the human microbiome project healthy cohort. *Microbiome, 5*, 153. doi: 10.1186/s40168-017-0373-4.

Ogunremi, O. R., Agrawal, R., & Sanni, A., (2020). Production and characterization of volatile compounds and phytase from potentially probiotic yeasts isolated from traditional fermented cereal foods in Nigeria. *J. Genet. Eng. Biotechnol., 18*(16). https://doi.org/10.1186/s43141-020-00031-z.

Oliphant, K., & Allen-Vercoe, E., (2019). Macronutrient metabolism by the human gut microbiome: Major fermentation by-products and their impact on host health. *Microbiome, 7*, 91. https://doi.org/10.1186/s40168-019-0704-8.a.

Paiva, S. A. R., Sepe, T. E., Booth, S. L., Camilo, M. E., O'Brien, M. E., Davidson, K. W., et al., (1998). Interaction between vitamin K nutriture and bacterial overgrowth in hypochlorhydria induced by omeprazole. *Am. J. Clin. Nutr., 68*, 699–704.

Probiotic Yeast May Offer an Effective Treatment for Drug-Resistant Fungal Infections, (2019). https://www.wpi.edu/news/probiotic-yeast-may-offer-effective-treatment-drug-resistant-fungal-infections (accessed on 05 June 2023).

Rane, B. R., Bharath, M. S., Patil, R. R., Keservani, R. K., & Jain, A. S., (2020b). Novel approaches in nutraceuticals. In: Rajesh, K. K., Raj, K. K., & Anil, K. S., (eds.), *Enhancing the Therapeutic Efficacy of Herbal Formulations Through Novel Drug Delivery Systems* (pp. 241–266). Chapter 11, IGI Global International Publisher, Pennsylvania, USA. ISBN13: 9781799844532.

Rane, B. R., Patil, A. S., Keservani, R. K., & Jain, A. S., (2020a). Novel approaches in herbal formulation. In: Rajesh, K. K., Raj, K. K., & Anil, K. S., (eds.), *Enhancing the Therapeutic Efficacy of Herbal Formulations Through Novel Drug Delivery Systems* (pp. 43–68). Chapter 2, IGI Global International Publisher, Pennsylvania, USA. ISBN13: 9781799844532.

Rane, B. R., Tadavi, S. A., & Keservani, R. K., (2020c). Naturopathy. In: Anil, K. S., Raj, K. K., & Surya, P. G., (eds.), *Herbal Product Development* (pp. 321–347). Chapter 12, Apple Academic Press, CRC Press, Taylor & Francis Group. ISBN: 9781771888776.

Singh, P., Kesharwani, R. K., & Keservani, R. K., (2017a). Antioxidants and vitamins: Roles in cellular function and metabolism. In: *Sustained Energy for Enhanced Human Functions and Activity* (pp. 385–407). Academic Press. ISBN:9780128054130.

Singh, P., Kesharwani, R. K., & Keservani, R. K., (2017b). Protein, carbohydrates, and fats: Energy metabolism. In: *Sustained Energy for Enhanced Human Functions and Activity* (pp. 103–115). Academic Press. ISBN:9780128054130.

Stone, J., (2021). *Fungemia, Other Fungal Infections Associated with S. Boulardii Probiotics*. https://www.medscape.com/viewarticle/954942 (accessed on 05 June 2023).

Valdes, A. M., Walter, J., Segal, E., & Spector, T. D., (2018). Role of the gut microbiota in nutrition and health *BMJ, 361*, k2179 doi: 10.1136/bmj.k2179.

Vamanu, E., (2019). Polyphenolic nutraceuticals to combat oxidative stress through microbiota modulation. *Frontiers Microbiol.* https://doi.org/10.3389/fphar.2019.00492.

Walker, A. W., Ince, J., Duncan, S. H., Webster, L. M., Holtrop, G., Ze, X., Brown, D., Stares, M. D., Scott, P., Bergerat, A., et al., (2011). Dominant and diet-responsive groups of bacteria within the human colonic microbiota. *ISMEJ, 5*, 220–230.

Yang, Q., Liang, Q., Balakrishnan, B., Belobrajdic, D. P., Feng, Q. J., & Zhang, W., (2020). Role of dietary nutrients in the modulation of gut microbiota: A narrative review. *Nutrients, 12*(2), 381. doi: 10.3390/nu12020381.

Yilmaz, B., & Li, H., (2018). Gut microbiota and iron: The crucial actors in health and disease. *Pharmaceuticals (Basel), 11*(4), 98. doi: 10.3390/ph11040098.

Zhang, Y., Zhou, Z., & Dong, Z., (2019). Effect of dietary copper on intestinal microbiota and antimicrobial resistance profiles of *Escherichia coli* in weaned piglets. *Frontiers Microbiol.* https://doi.org/10.3389/fmicb.2019.02808.

Zivkovic, A. M., German, J. B., Lebrilla, C. B., & Mills, D. A., (2011). Human milk glycobiome and its impact on the infant gastrointestinal microbiota. *Proc Natl. Acad. Sci. USA, 108*(Suppl 1), 4653–4658.

CHAPTER 7

NUTRIENT DELIVERY OF MICRONUTRIENTS AND MACRONUTRIENTS

AMIT BANIK

Associate Professor, Microbiology, All India Institute of Hygiene and Public Health (AIIH&PH), BN Campus, Kolkata, West Bengal, India

ABSTRACT

With the development of the food industry in the last century, numerous collateral developments have also occurred. Nutrition, as derived from food materials, needs to be delivered effectively to derive the maximum benefit. Nutrition drives our body and also strengthens our immune system. The delivery of these very nutrients has attracted a lot of recognition in the last two decades. With the availability of a wide array of food items to choose from and the emerging knowledge that regulating our food choices can positively impact our health, renewed interest has been striking in this sector.

7.1 INTRODUCTION

These days whenever you walk into a supermarket, you are flooded with food having met varied dietary requirements satiating every consumer's preferences. There is an explosion of knowledge available from popular sources which drives this frenzy. However, we must admit that we have more choices available today and hence more responsibility to choose wisely. All these developments have revolutionized the food industry.

Micronutrients and Macronutrients as Nutraceuticals. Prakash Chandra Gupta, PhD, Sayan Bhattacharyya, MBBS, MD, Nisha Sharma, PhD, MPharm, Rajesh K. Kesharwani, PhD, Raj K. Keservani, PhD, MPharm (Eds.)
© 2024 Apple Academic Press, Inc. Co-published with CRC Press (Taylor & Francis)

The main impetus of diet is to deliver appropriate nutrients in appropriate quantities to each individual to meet their body's requirements. A little knowledge about food definitions is pertinent here. According to Kalra (2003), when food is being cooked or prepared by using "scientific intelligence" with or without knowledge of how or why it is being used, that food is called "functional food." Thus, functional food provides the body with the required amount of vitamins (Singh et al., 2017a), fats, proteins, carbohydrates (Singh et al., 2017b), etc., needed for its healthy survival. When functional food aids in the prevention and/or treatment of disease(s) and provide other health benefits beyond just providing nutrition it is called a 'nutraceutical.' Nutraceuticals have the properties of both foods and medicinal products. If the substance leads only to the good maintenance of healthy tissues and organs, then it may be considered a food ingredient (Keservani et al., 2010a, b). On the other hand, if it can be shown to have a modifying effect on the body's natural physiological functions and is supported by favorable scientific knowledge, it is considered a nutraceutical (European Commission Food Safety, 2015). Over the last 10–12 years evidence has been accumulating that nutraceuticals reduce the risk for Chronic diseases. This fact has further fueled the market for food supplementation.

7.2 FOOD AND BIOACTIVES

Nutraceuticals with established health benefits are commonly seen with fruits, dairy, vegetables, soya, whole grains, etc. Nutraceuticals can be present in food in two different ways. It may be locked inside the food matrix or it may have been extracted prior to the manufacture of processed foods and subsequently added to the processed food at a later stage. Apart from traditional food items, certain newer items are also becoming popular in which biologically active substances are mixed with food matrix to enhance physiological effects. The presence of dietary (Keservani et al., 2020) compounds with health benefits is an excellent tool to improve public health. A number of such functional food components are being investigated for their potential beneficial roles which include vitamins, probiotics, oxidants, and bioactive peptides.

7.3 CHALLENGES IN NUTRACEUTICAL FORMULATION

Various physiochemical and physiological attributes need to be pondered to derive the effectiveness of nutraceuticals. Preservation and effective bioavailability maintained through proper dosage and formulation are important to derive maximum benefit. Any intervention which compromises the bioavailability of the food items needs to be considered seriously to maintain the efficacy of these bioactive compounds. Whatever method is tried to achieve this; the ultimate goal remains to protect, shield, control, and target the release of bioactive to the destined location. Most approaches share one key methodology: entrapment, encapsulation, and coating (Augustin & Hemar, 2009). Nutraceuticals can exist in different shapes and sizes and can be released in tablets, capsules, granules, and micro- and macro-particle formulations (Lee et al., 2015). However, different approaches to delivering nutraceuticals are followed without compromising their health benefits during consumption or processing.

7.4 ORAL DELIVERY SYSTEMS

Delivery through the Oral route is considered to be the best methodology based on patient acceptance, compliance, and ease of administration. Besides this route of administration helps maintain the inherent virtues of natural food consumption. But while introducing nutraceuticals through the oral route, a lot of problems need to be nullified (Chen et al., 2006). Challenges (Keservani & Kesharwani, 2016) include reduced gastric time, low solubility in the gut, and instability while under food processing conditions (light, temperature, oxygen) or during transit in GIT (presence of other organic matter, pH, enzymes). All or some of these factors could hinder the activity or health benefits of nutraceuticals (Agrawal et al., 2014). To circumvent these problems, researchers have found the use of nanoparticles (NPs) very efficient.

Usual oral delivery systems include:

- Phospholipids;
- Emulsions;
- Chemically modified;
- Chitosan based;
- Nano-dispersions;
- Mesosporous silica particles.

Using NPs as a drug delivery mode, we can increase the stability of the molecules, including hydrophobic and hydrophilic substances, increase carrier capacity and make it stable for various routes of administration, e.g., oral, pulmonary, and topical (Helson, 2013). Since biological interactions are largely dependent on size, by virtue of the larger surface area of NPs the bioavailability of these molecules is far higher compared to larger-sized particles. Hence smaller nano-sized particles have a high propensity to be disseminated faster vis-a-vis larger particles which tend to have a small initial burst release followed by slow longer sustained release.

NPs help improve bioavailability by enabling sustained controlled release of drugs besides providing stability to the molecules. This helps reduce the frequency and dosage of drug administration, besides protecting the liver and kidney from the deleterious effects of excess metabolites. Since most of the NPs are derived from biological materials which are compatible with our body, their degradation is not much of a plight and hence regulatory approvals are much easier to get than alternative compounds. NPs are also useful considering that they help retain taste, enhance flavor, prevent oxidation and enzymatic degradation, effective gastric residence time, and permeation ability which improves the overall usability experience. Several NPs have already got approval for clinical use while a multitude are in preclinical development. Most of these have found use in cancer treatment, tumor angiogenesis, or photodynamic therapy. A number of nutraceuticals have been investigated as bioactive compounds in nanoparticle formulations which include vitamins, minerals, probiotics, prebiotics, omega fatty acids, etc.

Targeted drug delivery through NPs can be achieved by two methods. One is Passive targeting which happens when NPs manifest localization to specific organs or sites of disease through biological mechanisms, such as in the reticuloendothelial system (RES) or the enhanced permeability and retention (EPR) effect (Maeda et al., 2000). The other method is active targeting which translates to a combination of peptides, antibodies, or other ligands onto the nanocarrier surface, which in turn perceive and attach to target cells through ligand–receptor interactions evident by the expression of receptors or epitopes on the cell surface (Nobs et al., 2004).

Variations in the sizes and shapes of NPs can have immense ramifications. An NP can be harmless at 40 nm size but useful at 2 nm and vice versa. As they are heavily influenced according to the micro-environment they work in, NPs can change shape, and size inside an organism. A big particle may disintegrate into a 1 nm particle or similarly smaller NPs may coalesce into a 150 nm NP.

7.5 FOOD-GRADE DELIVERY SYSTEMS

NPs have evinced huge interest as drug delivery modes when formulated into different delivery systems owing to their tremendous therapeutic potential.

The commonest carrier systems are magnetic, dendritic, polymeric, and soli-lipid liposomal. NPs can be fabricated in two main approaches: Top-down and Bottom-up (McClements, 2015). In a "bottom-up" approach, materials, and devices are built from molecular components which bind together chemically by principles of molecular recognition (Chan & Kwok, 2011). On the opposite hand, the "top-down" approach warrants the fabrication of nanoproducts from larger entities without atomic-level control (Acosta, 2009). Size and the main nutritional component are the main characteristics based on which food-grade delivery systems can be classified. The efficacy of the systems can be gauged based on the availability of the bioactive compound to the body according to size. The rate of release, retention at the site, and solubility can be altered in conformity to the size of the encapsulating material. According to the main nutritional component, these systems are conveniently classified into 3 groups: Lipid-based, protein-based, and carbohydrate-based delivery (Benshitrit et al., 2012).

7.6 PROTEIN-BASED DELIVERY SYSTEMS

Requirements like low cytotoxicity, biodegradability, generous renewable sources, substantive drug binding capacity, nonantigenicity, high nutritional value, and remarkable assimilation into the targeted cells are met with protein-based nanocarriers. They represent promising candidates for efficient drug and gene delivery. Proteins have the option of less opsonization by the RES through an aqueous steric barrier in addition to their excellent functional properties including emulsification, gelation, foaming, and water binding capacity. Furthermore, protein NPs can be easily prepared and scaled up while manufacturing. Also, protein NPs have desirability as metabolizable naturally occurring components. Protein hydrolysis by digestive enzymes creates many bioactive peptides that may make use of a number of physiological effects *in vivo* (Elzoghby et al., 2012). The distinct structure of proteins offers the benefit of drug conjugation according to site specifications and targeting using various ligands modifying protein nanocarrier surfaces besides being cost-efficient (Benshitrit et al., 2012). Proteins derived from various sources like those of plant and animal origin are already in high usage and provide cost-efficient delivery of bioactives. Major preparation methods for protein-based delivery include techniques like gelation, emulsification, extrusion, spray drying, and coacervation commonly used to produce delivery systems such as hydrogels, hydrocolloids, micelles, nano assemblies, films, and micro- and NPs (Vozza et al., 2017). The commonest protein carriers utilized are gelatin, casein, collagen, albumin, wheat proteins, elastin, silk proteins of animal origin and soy proteins, zein, gliadin, and lectins of plant origin.

In comparison to hydrophilic animal proteins, hydrophobic plant proteins have the capability of yielding sustained drug release, NPs may not need any further chemical or physical treatment to harden them, is less expensive as well as reduce the risk of spreading diseases such as bovine spongiform encephalitis.

Of particular mention among plant-origin proteins, zein has gathered much interest because of its unique beneficial properties. Zein is a protein that is water-insoluble but alcohol-soluble, having a molecular weight of about 40 kDa which is predominantly present in the endosperm of corn kernels. It has 75% of lipophilic and 25% of hydrophilic amino acid residues. Zein has been used as an edible coat for pharmaceuticals and food items because it shows good biodegradability *in vivo*, low water uptake values, high thermal resistance, and good mechanical properties. Owing to its significant hydrophobicity, zein has been a successful application as a promising carrier for the coating and release of fat-soluble compounds (e.g., gitoxin) at a controlled pace. Zein NPs have been used to encapsulate fish oil in solid as an alternative to emulsions and also delivery other fat-soluble compounds easily (Elzoghby et al., 2012).

7.7 LIPID-BASED DELIVERY SYSTEMS

Lipophilic components encompass a diverse range of substances such as bioactive lipids, vaccines, flavors, antimicrobials, antioxidants, drugs, omega fatty acids like linoleic acid (LA), and butyric acid which have been found beneficial for preventing coronary heart disease (CHD).

Lipid bioactive when supplied in a liquid edible form has been shown to increase their palatability, desirability, and bioactivity due to their differing molecular, physicochemical, and psychological properties. The most frequently-used lipophilic nutraceuticals scrutinized of late are fatty acids ('Ω-3 fatty acids, butyric acid, and CLA), carotenoids (β-carotene, lycopene, lutein, zeaxanthin), antioxidants (tocopherols, flavonoids, polyphenols) and phytosterols (stigmasterol, β-sitosterol, campesterol) (Keservani & Sharma, 2014; Vozza et al., 2017).

Emulsification is the primary method to produce lipid-based formulations such as oil-in-water (O/W) microemulsions, liposomes, multilayer emulsions, filled hydrogel particles, or solid lipid NPs. The formation of multiphase colloidal dispersions into nanoscale droplets leads to the production of nanoemulsions (NE). NE (also known as submicron emulsions) can be produced by the physical shear induced by rupturing of one liquid within another immiscible liquid (Mukherjee et al., 2009). Nano-emulsions are optically transparent due to less scattering due to their smaller size, whereas microemulsions appear opaque since they show abundant scattering in visible light.

Another important lipid-based delivery system is solid-lipid NPs (SLNs) which exist as spherical particles of 10–1,000 nm and can solubilize lipid molecules very easily. They have a solid/semisolid lipid core stabilized by surfactants. The formulation in SLN steers to better protein stability, proteolytic degradation is avoided, and also leads to the continuous release of the assimilated molecules. Keypeptides like Cyclosporine A, calcitonin, somatostatin, and insulin have been already incorporated into solid lipid NPs and are presently being investigated for other virtues. Many local or systemic medicinal applications may be prognosticated, like vaccination with protein antigens, cancer therapy, and treatment of infectious diseases and chronic diseases (Almeida & Souto, 2007).

SLNs have the ability to control the release profiles of a given bioactive due to the intrinsically slow degradation rate of the crystalline lipid core. These advantages have been shown to improve the pharmacokinetic profiles of bioactive delivered via SLNs, typically resulting in the slow evolution of peak therapeutic responses and more stable plasma concentration trends in comparison to lipid-lipid emulsion systems (Ting et al., 2014). SLNs can be simply prepared by the emulsification-evaporation and subsequent sonication methods which involves the addition of the hot Pre-Emulsion to cold surfactant-containing water, under constant mixing, in order to allow for solvent evaporation. Other methods include Lyophilization, Spray Drying but they come with inherent problems in SLNs structure integrity maintenance.

7.8 CARBOHYDRATE-BASED DELIVERY SYSTEMS

They are broadly divided into mono, oligo, and polysaccharides. These entities find widespread application in nutraceuticals known for their targeted approach. Mono and oligosaccharides are often used as constituents of the encapsulating matrix while polysaccharides most often being used as the carrier component due to their massive molecular structure and thus ability for greater bioactive entrapment. Polysaccharides are derived from animal, algal, microbial, and protein sources and are typically subdivided into three groups: neutral, anionic, and cationic. Polysaccharide examples include starch and cellulose (neutral), alginate, pectin, and xanthan gum (anionic), and chitosan and dextran-spermine (cationic).

Numerous approaches for the production of carbohydrate-based delivery systems are extant including coacervation, electrospinning, electrospray, spray drying, emulsion diffusion, reverse micelle, emulsion-droplet coalescence, supercritical fluid (SCF), emulsification/solvent evaporation, salting-out, ultrasonication, and high-pressure homogenization (Fathi et al., 2014). Anionic polysaccharides are neutral at pH values below their p*K*a value but negative above, whereas cationic polysaccharides tend to be neutral at pH values above their p*K*a value but positive below (Kumari et al., 2010). Alginate is an anionically charged polysaccharide that has been used for NP production. When being developed as a nanocarrier system, alginate is conjoined with divalent cations such as magnesium chloride and calcium chloride for complexing alginate-based nanoparticles (Alg-NPs). Alg-NPs have been applied for the delivery of drugs and nutraceuticals like insulin and vitamin D3.

Environmental and solution conditions may modify the effects expected from such nutraceuticals. Hence an ingredient or food manufacturer must therefore select the most appropriate combination of compounds and procedures for the development of a delivery system for a particular application. According to new research carbohydrate-based delivery systems can also be used for other routes of administration apart from only oral such as topical, nasal, and pulmonary as nanometric carriers (Fathi et al., 2014).

The presence of a multitude of functional groups in polysaccharide-based systems allows them to form complexes with bioactives through hydrophobic and ionic interactions which provides them with the ability to combine with a wide variety of both hydrophilic and hydrophobic bioactives. So unlike lipids, extensive modification of these bioactives is possible which makes them very viable and popular delivery systems besides being biodegradable and biocompatible in nature. Even under high temperatures they retain their structure and do not melt away. The stability provided by these is unmatched besides being natural and safe.

7.9 CHITOSAN DELIVERY SYSTEMS

Chitosan is a polycationic linear polysaccharide biopolymer, prepared by N-deacetylation of chitin. It is positively charged, biocompatible, nonimmunogenic, nontoxic, pH-responsive, and biodegradable, making it a choice as a component of oral drug delivery systems and attractive for biomedical and drug delivery applications (Bezerra et al., 2008). Chitosan has been extensively used in a number of areas of medicine, for example, in tissue engineering, bactericidal, fungicidal activity, bioimaging, wound-healing, and drug delivery formulations (films, hydrogel system, and particulate system) (Agnihotri et al., 2004).

Chitosan is derived from the deacetylation of Chitin. The degree of deacetylation of Chitosan (DD) from chitin is the positive charge density of chitosan. The more the deacetylation and hence the DD, the more soluble it is at acidic conditions. Higher solubility ensures better mucoadhesive properties in GI Tract for them. A high degree of deacetylation more readily allows the ionization of Chitosans, increases the viscosity, and results in an extended conformation with a more flexible chain because of the charge repulsion in the molecule (Kumirska et al., 2010). Hence, chitosan, having a reasonably high degree of deacetylation is commonly used for drug delivery research. Owing to the ability of chitosan to temporarily open tight intercellular junctions, it helps facilitate both transcellular and paracellular transport of macromolecules. They have high success in acting as delivery modes for drugs targeted at mucosal tissues in the GI tract by enhancing residence and encapsulation time within the GI tract due to their mucoadhesive property.

They are prepared by various techniques like complex coacervation, coprecipitation, polyelectrolyte complexation, solvent evaporation, emulsification solvent diffusion, microemulsion, and ionotropic gelation (Vozza et al., 2017). Among them Ionotropic gelation is the most useful method which involves interaction between the negatively charged terminal of sodium tripolyphosphate (TPP) and positively charged amino groups of Chitosan, allowing generation of hydrogel micro- and NPs and that too using mostly aqueous solutions rather than organic solvents.

7.10 REGULATION IN NANOTECHNOLOGY

Increased surface-to-volume ratio, small sizes, and reactive nature of the nanomaterials have instigated concerns about the effectiveness and safety of nanomedicine. The concern is not solely for biological applications where these materials are injected into the human body but also while manufacturing and large-scale external use. Even though it is widely believed that nanomedicine may be the driver of the global economy, concerns about potential environmental and human health, and safety needs to be mitigated simultaneously as well. We also need to anticipate and assuage adverse implications or unintended consequences.

Institute for Health and Consumer Protection (IHCP) in Europe is the major regulatory body that is diligently involved in research of nanomaterials safety, identification, and detection. The European Union (EU) has introduced a comprehensive chemical law named as REACH Directive (Registration, Evaluation, Authorization, and Restriction of Chemicals Directive) and sets an explicit responsibility on the manufacturers to divulge information to regulatory agencies regarding specified requirements.

The United States has however adopted a decentralized perspective toward the regulation of nanotechnology. The Food and Drug Administration (FDA), the Environment Protection Agency (EPA), Occupational Health and Safety Administration (OSHA), and the Consumer Product Safety (CPS) Commissioner are bestowed with the regulatory oversight authority for nanotechnology regulation. In China, the National Steering Committee for Nanoscience and Nanotechnology regulates this framework with support from the Ministry of Science and Technology, the Ministry of Education, the Chinese Academy of Sciences, and the National Natural Science Foundation of China.

India started with the launch of the Nano-Science and Technology Initiative (NSTI) in 2001 to boost research and development, infrastructure, and manpower followed by nano-mission in 2007. The country does not have any particular law dealing with toxic chemicals. However, the Environment Protection Act, of 1986 provides ample authority to the central government to draft subordinate legislation for regulating chemical substances in India. According to this, a hazardous substance is defined as ‘any substance or preparation which, by reason of its chemical or physico-chemical properties or handling, is liable to cause harm to human beings, other living creatures, plant, microorganism, property or the environment.’ In this context of hazardous wastes, the Ministry of Environment and Forests (MoEF) also has much clout to act upon. As of now, the Department of Science and Technology (DST) is the designated nodal agency for the advancement of nanotechnology research and development in India.

7.11 UPTAKE AND POTENTIAL TOXICITY CONCERNS

NPs have been shown to permeate through GI barriers better compared to micro-particles, with the former estimated to have a 15–250 times higher absorption. Ingestion, inhalation, and dermal absorption are the main physiological routes in which NPs can enter the body resulting in the potential ability to distribute themselves to a given organ over several exposure points (Smolkova et al., 2015). ADME (absorption, distribution, metabolism, and elimination) characterizes the effects of NPs exposure in a given system. The level of absorption varies dependent on the type of GI cell; for example, absorption is higher in the lumen in comparison to the oral cavity (Roblegg et al., 2012) Epithelial barriers allow for the separation of NPs from the blood and hence translocation of NPs into secondary organs can occur due to their ability to reach the connective tissue which is found beneath the epithelial layer surrounding the intestines (Smolkova et al., 2015), e.g., Titanium Oxide (TiO_2). Accumulation of TiO_2 inside the epithelial lining can give rise to detrimental outcomes such as inflammation and may be involved in inflammatory bowel disease pathogenesis. The surface properties and size of ingested NPs are influenced by several environmental factors like pH, ionic strength, protein, sugar, and lipid content that are naturally present in foods such as the mucus of the saliva (Roblegg et al., 2012). Interactions do occur between plasma proteins, platelets, and peripheral blood cells as well as with the organelles and DNA of the cells upon NP entry into the system (Smolkova et al., 2015). Inorganic particulate metals produced in a nanoparticle formulation and introduced to an environment where it solubilizes faster may produce greater toxicity compared to organic molecules, which are generally considered much safer. Increased residence time can be another property that may be responsible for greater toxicity and damage to body tissues. Moreover, particle interactions merit special consideration on a per-case basis because some NPs may be insoluble but yet biodegradable whereas others (TiO_2) can not be broken down and hence go on accumulating inside the liver and spleen. To summarize, nanoparticulates cannot be considered inherently safe or toxic. They need to be assessed with respect to their molecular structure, biological environment, degree of exposure, and host-specific considerations and consequently assigned a potential hazard threat parameter to choose wisely.

7.12 NANODELIVERY

Food Nanotechnology involves the development of new food materials, their protection, and the delivery of these bioactive compounds as well as packaging applications so as to ensure improved food safety and biosecurity.

This is more intricately involved in the delivery of vitamins, minerals, and nutraceuticals. In human bodies, vitamins, and minerals are responsible for normal physiological functioning. Nutraceuticals are however bioactive compounds that are important physiological components in food that provide physiological, behavioral, and immunological benefits apart from providing extra nutrition. Challenges like poor solubility, chemical stability, low bioavailability, taste, and flavor are important considerations for their acceptability. A combined scientific and engineering approach should be undertaken to circumvent these shortcomings. The role of nanodelivery for micronutrients can be broadly put as that which encompasses enhanced stability in the GIT, protection against oxidation, controlled, and targeted release, improved bioavailability and bioactivity, and lastly conservation of the active compound during the processes of food processing, storage, and distribution (Chen et al., 2006).

A biopharmaceutical classification system (BCS) has been commonly used by scientists to evaluate the outcome of formulation on the bioavailability of bioactive and nutrients. This denotes drugs according to their solubility and permeability characteristics. The system defines the prime factors limiting the oral bioavailability of drugs. The BCS classifies drugs into the following four major categories, depending upon their solubility and permeability virtues.

- **Class I: High Solubility, High Permeability:** Compounds that inherently have high solubility and high permeability across epithelial cell membranes. The oral bioavailability of Class I drugs is therefore limited by their solubility or permeability characteristics.
- **Class II: Low Solubility, High Permeability:** This class has compounds that show low solubility in gastrointestinal fluids but high permeability across the epithelial cell membranes. Many lipophilic drugs are classified in this category. The bioavailability of this kind of drug can be improved using methods that increase its solubility in gastrointestinal fluids, such as chemical modification (prodrugs) or lipid-based delivery systems.
- **Class III: High Solubility, Low Permeability:** Compounds having high solubility in intestinal fluids but low permeability across epithelial cell membranes. Many thoroughly hydrophilic drugs fall into this category because they are readily soluble in aqueous solutions. However, they are not easily transported across epithelium cell membranes. The bioavailability of this kind of drug may be

improved by chemical modification (prodrugs) or coadministration with material that increases cell membrane permeability.

- **Class IV: Low Solubility, Low Permeability:** This class includes compounds that have low solubility in gastrointestinal fluids. They show low permeability across epithelium cell walls as well.

But as an extension and modification of the existing BCS classification system, a new classification has been introduced known as NuBACS, which evaluates other factors limiting the oral bioavailability of food bioactive apart from traditional solubility and permeability parameters. This scheme describes and groups the factors limiting the bio-availability of nutraceuticals into 3 major classes based on their bioaccessibility, absorption, and transformation within the gut. Among these three major classes, multiple subclasses exist related to the specific physicochemical or physiological mechanisms that influence bioavailability. For designing potent delivery systems or food matrices to increase their bioavailability, knowledge of the exact and accurate mechanisms that limit the bioavailability of a particular nutraceutical is useful (McClements et al., 2015).

7.12.1 BIOACCESSIBILITY (B)*

The fraction of the active compound that is released from the food matrix or delivery system into the lumen of the GIT can be broadly defined as Bioaccessibility with regard to a particular food item. It is one of the three major constituent processes that dictate the bioavailability of a compound along with its intestinal transport and metabolism. For hydrophilic nutraceuticals, this is a fraction of the nutraceutical that is completely dissolved within the intestinal fluids (rather than present in an insoluble, complex, or crystalline form). For hydrophobic nutraceuticals, this may be a fraction of the nutraceutical solubilized inside mixed micelles in small intestinal fluids. Three phenomena regulate this entity namely: *Liberation* from the matrix, *solubilization* of compound in gastric juices, and *interactions* with other enzymes while within the GIT. A bioactive substance with high bioaccessibility (>75%) can be classified as $B^*(+)$. Conversely, a bioactive substance whose bioaccessibility is comparatively low can be classified as $B^*(-)_L$, $B^*(-)_S$, or $B(-)^*_I$ based upon whether its bioaccessibility is limited mainly by liberation, solubilization, or through interactions.

7.12.2 ABSORPTION (A*)

Once the nutrients so delivered reaches the surfaces of the GI tract, these are then absorbed. However, the rate and degree of absorption are highly dependent upon factors like the transport of compounds across the epithelium into the systemic circulation. Movement through mucus layer (ML), tight junction (TJ), permeability through bilayer (BL), active transport (AT), or efflux transport (ET) is subclasses in evaluating it. Similar to bioaccessibility it is also classified as $A^*(-)_{ML}$, $A^*(-)_{BL}$, $A^*(-)_{TJ}$, $A^*(-)_{AT}$, or $A^*(-)_{ET}$ if its absorption is restricted by transport through ML, bilayer permeability, TJs, active transporters, or efflux transporters, respectively. Nutraceuticals whose absorption is restrained by more than one physicochemical event can be classified by numerous subscripts, e.g., $A^*(-)_{BP,\ and\ TJ}$ for a nutraceutical that is restricted by poor bilayer permeability as well as low TJ transport. Nutraceuticals with comparatively high absorption (>75%) by epithelium cells can be classified as $A^*(+)$.

7.12.3 TRANSFORMATION (T*)

The bioavailability of many nutraceuticals is limited because they are converted into an inactive form within the GIT. These molecular transformations can be divided into two subclasses according to their origin; mainly Chemical degradation and metabolism. As an example, polyunsaturated lipids (like ω-3 fatty acids or carotenoids) can be oxidized due to the presence of pro-oxidants in the gut whereas proteins or peptides could also be hydrolyzed by the highly acidic and enzyme-active environment of the stomach.

Some nutraceuticals are metabolized by specific enzyme systems within the GIT, e.g., resveratrol, quercetin, and epicatechin. Many important nutraceuticals are vulnerable to phase I and phase II metabolism within the GIT. *Phase I metabolism* reactions commonly are hydrolysis, oxidation, and reduction. Multiple classes of enzymes are involved in these metabolic reactions, like reductases, esterases, CYP450 enzymes, and dehydrogenases. Phase I metabolites are often substrates for phase II enzymes, which leads to subsequent changes in their molecular, physicochemical, and biological properties. *Phase II metabolism* generally involves conjugation reactions between the parent bioactive molecules or their phase I metabolites and endogenous molecules. A few of the important kinds of enzymes involved in phase II metabolism are glutathione (GSH) *S*-transferases, glucuronyltransferases, sulfotransferases, methyl transferases, epoxide hydrolase, and acetyltransferases. Phase II metabolism *in vivo* contributes to the low bioavailability of the many bioactive phytochemicals. Based upon the commensurate bioavailability and bioactivity of the parent compound and its metabolites, the metabolism could also be either detrimental or beneficial to the overall performance of nutraceuticals.

Nutraceuticals that are not absorbed within the small intestine, on reaching the large intestine will be fermented by colonic bacteria. These fermentation reactions may change the bioavailability and bioactivity of nutraceuticals within the colon. Consequently, they enter the systemic circulation and manifest their biological activities.

A nutraceutical may therefore be classified as *T**(–)C, *T**(–)M, or *T**(–)CM if its bioaccessibility is restricted by chemical degradation, metabolism, or both, respectively. If a nutraceutical is comparatively stable to molecular transformations within the GIT (>75% remaining in a bioactive form), then it should be designated as *T**(+). Nutraceuticals that lose their bioactivity owing to chemical transformation or metabolism within the GIT are taken into account as transformation limited.

The NuBACS approach is beneficial for dividing different sorts of nutraceuticals into general categories that may require an identical food matrix design to enhance their bioavailability. For example, the bioavailability of nutraceuticals whose solubility is restricted completely by low bilayer permeability could also be improved by encapsulating them in lipid-based delivery systems. However, the bioavailability of nutraceuticals that are vulnerable to metabolism could also be improved by adding components that alter the activity of metabolic enzymes or that prevent these enzymes from interacting with the nutraceuticals (McClements et al., 2015).

The premise of classifying nutraceuticals on the basis of bioavailability is to generate enough information which can help us tide over these limitations and look for avenues to produce food items with increased bioavailability. Here, the physicochemical properties and the methodology of the nanoencapsulation technique are very important. The next part of this chapter deals with such important encapsulation techniques.

7.13 TECHNIQUES OF NANOENCAPSULATION

7.13.1 COACERVATION

Coacervation is one of the easiest techniques for the production of carbohydrate-based delivery systems. The crux of this method is electrostatic attraction between oppositely charged molecules. It involves liquid-liquid phase separation from a solution and deposition of the newly formed coacervate phase around the active ingredient. Among the two liquid phases within the colloidal system, the phase which is more concentrated is the coacervate while the other phase is the equilibrium solution. The force utilized could is induced between a charged bioactive component and an oppositely charged carbohydrate. This process is known as simple coacervation. On the other hand, if a bioactive could be trapped within a particle formed by electrostatic complexation of positively charged (e.g., chitosan) and negatively charged (e.g., pectin and alginate) biopolymers, it is known as complex coacervation (Fathi et al., 2014). Complex coacervation essentially involves the formation of immiscible phases of two or more polymers, followed by deposition and rigidization. The functional performance of the produced nanocapsules depends upon the chemical nature and the characteristics of the surface of the biopolymeric shell, rate of agitation, rate of dropwise addition of polymer solution (lower the value better the performance), and solubilities of bioactive and biopolymer in solution.

7.13.2 SPRAY DRYING

The principle of this technique depends on dispersing the active ingredient in a solution of biopolymer. The dispersion is atomized after that in a heated air chamber which rapidly extracts the solvent and creates a dried particle comprising of the active ingredient fixed in a porous wall material. Volatile and thermo-sensitive bioactives have limitations with this technique.

The efficiency of encapsulation and size of produced particles depends on varied factors: starting with the properties of the materials used, like core, and wall properties, particle size (for emulsified active ingredients); solution viscosity; spray dryers features, like atomizer type, inlet/outlet temperatures, flow rates, and humidity. While some carbohydrates may find it difficult due to the high temperatures involved in the technique, thermostable carbohydrate biopolymers such as cyclodextrins (CD) or modified materials such as hydroxypropyl cellulose are appropriate for spray drying at high temperatures.

Common spray dryers use rotary atomizers and pressure nozzles or two-fluid nozzles to form spray droplets. These dryers are unlikely to have produced submicron particles. But the recovery of particles smaller than 2 μm in traditional spray dryers is low.

However, with the introduction of the new generation of spray dryers (Nano-spray dryer B90), submicron particles can be produced relatively easier. Production of submicron particles is achieved by effective fluid breakdown combined with highly efficient particle collectors. This technology utilizes a vibrating mesh for fine droplet generation. Basically, the piezoelectric-driven actuator vibrates a slim, perforated, stainless steel membrane (spray mesh) having an assembly of precise micron-sized holes, in a small spray cap. The actuator is operated at an ultrasonic frequency, causing the membrane to vibrate upward and downward, ejecting millions of accurately sized droplets every second with very small droplet size distribution. The dried particles are finally collected by an electrostatic particle collector which is size independent in contrast to cyclones.

7.13.3 ELECTROSPINNING AND ELECTROSPRAY

This technology uses electrostatic forces to generate polymer fibers, and consists of (i) a capillary through which the liquid to be electrospun is pumped; (ii) a high voltage source (1–30 kV) with positive or negative polarity, that generates a charge in the polymer solution; and (iii) a grounded collector, that is brought into contact with the counter electrode (Fathi et al., 2014). In these processes, when a conducting polymer solution is passed through an electrified needle, two main electrostatic forces are involved. The electrostatic repulsion of like charges and the Coulombic force of the external electric field act on the polymer droplet at the needle tip. Here, the hemispherical droplet is distorted and converted into a conical shape better known as the Taylor cone. When the surface tension is tamed by electrostatic forces, a charged polymer jet is emitted from the tip of the Taylor cone. This is the actual stage where electrospinning and electrospraying processes are differentiated.

In the electrospinning process, owing to the high concentration of the polymer solution, when the jet is steered toward the collector, the extension of the jet takes place due to whipping or chaotic bending instability which leads to additional fiber thinning and solvent evaporation. This leads to the generation of solid thin fibers settled on the collector as randomly oriented nonwoven mats. But in electrospraying due to low polymer concentration, the jet unsettles due to varicose instability and forms highly charged self-dispersing droplets (Vozza et al., 2017). This droplet evaporates in its flight from the needle to the collector leaving behind solid NPs.

Varied parameters affect fiber formation and structure like applied voltage (application of higher voltage initially results in the formation of thinner and then thicker fibers), capillary–collector distance (longer the distance thinner the fiber), polymer concentration (lower the concentration thinner the fiber), solution flow rate (higher the flow rate thinner the fiber), solution properties (higher the conductivity thinner the fiber), and solvent volatility (higher the volatility thinner the fiber).

In electrospray, the electrostatic force induced by high voltage atomizes liquid into fine droplets. The solvent evaporation is done during the flight of the droplets toward the ground electrode. The most prominent and outstanding features of the electrospray technique are high encapsulation efficiency and possibility of the production in one step.

While a high voltage is exerted, the formation of whether nanofiber or nanodroplet depends on biopolymer molecular mass, biopolymer chain entanglement, and solvent evaporation rate (Fathi et al., 2014). The size of the nanoencapsulates can be managed by altering the instrumental and solution parameters such as the feed rate, electric potential, needle-to-collector distance and concentration, conductivity, and surface tension of the feed solution (Bhushani et al., 2014). The technique of electrospray is being used for the encapsulation and stabilization of food bioactives such as β-carotene, curcumin, folic acid, and lycopene, and, besides being highly useful for heat-sensitive compounds since both these methods are non-thermal and operates at normal atmospheric pressure. The technique of electrospinning has been used for the encapsulation and delivery of *bifidobacteria*, vanillin, NiO/TiO_2, lysozyme, and epigallocatechin-3-gallate (EGCG) (Vozza et al., 2017).

7.13.4 EMULSIFICATION

It is the thermodynamically stable process where two different immiscible liquids (mainly Oil and Water) are mixed and stabilized by a surfactant with droplet diameters ranging below 200 nm. The free energy of the nanoemulsion is less than that of the phase-separated components from which it is prepared, so it forms spontaneously. It is crucial to notice that such an emulsion is thermodynamically stable only under a particular set of environmental conditions (e.g., composition and temperature). If these conditions are changed, then the emulsion may not be thermodynamically stable anymore and can be converted to a different kind of system, e.g., phase separation, liquid crystalline, bicontinuous, nanoemulsion, or conventional emulsions.

NEs, similar to conventional emulsions, are often categorized according to the relative spatial organization of the oil and water phases. A system containing oil droplets dispersed within a water phase is mentioned to as an O/W nanoemulsion, vis-a-visa system that consists of water droplets dispersed in an oil phase, referred to as a water-in-oil (W/O) nanoemulsion. The liquid that creates the droplets is commonly named the dispersed phase, whereas the liquid that creates the surrounding liquid is mentioned as the continuous phase. The particles in a nanoemulsion are in the conformation of a core forged from a lipophilic material and a shell fabricated from amphiphilic surface-active material. Emulsifier can be defined as a surface-active molecule that is competent for adsorbing to droplet surfaces, thus aiding droplet disruption and securing droplets from aggregation. In high-energy approaches, the emulsifier aids droplet disruption within the homogenizer by reducing the interfacial tension, which favors the production of small droplets. In low-energy approaches, the emulsifier aids the spontaneous creation of small droplets thanks to its ability to generate very low interfacial tensions under specific environmental and solution conditions. The steadiness of a nanoemulsion to environmental stresses like heating, cooling, pH, ionic strength, or long-term storage is principally decided by the kind of emulsifier used. In the food industry, the most important types of emulsifiers are proteins, phospholipids, small-molecule surfactants, and polysaccharides.

NEs frequently have superior stability to gravitational separation, flocculation, and coalescence than conventional emulsions, but are comparatively less steady to Ostwald ripening. These are all different physicochemical processes that tend to degrade any metastable conventional or microemulsions normally (McClements et al., 2011).

NEs can be categorized as either High Energy or Low Energy approaches. High-energy approaches like ultrasonication, high-pressure homogenization, and microfluidization use mechanical forces to derange the oil and aqueous phases into steady nanoscale droplets. Here, the emulsifier in the aqueous phase helps the disruption of droplets by lowering the interfacial tension and further aids in the stabilization of the formed nanodroplets (Vozza et al., 2017).

Low-energy methods that utilize the internal chemical energy of a system to attain emulsification are sorted depending on whether or not alterations in the surfactant's spontaneous curvature are produced during the process. Nano-emulsion generation activated by rapid diffusion of surfactant and/or solvent molecules from the dispersed phase to the continuous phase without requiring a change in the spontaneous curvature of the surfactant is referred to as "self-emulsification." When changes in the surfactant spontaneous curvature are produced during the emulsification process they are called "phase inversion" methods. These are categorized as phase inversion temperature (PIT) and phase inversion composition (PIC) methods if emulsification is activated by a change in temperature or composition, respectively (Solans et al., 2012).

Examples of nanoemulsified food bioactives include β-carotene, curcumin, resveratrol, catechins, α-tocopherol, omega-3 fatty acids, and essential oils.

7.13.5 EMULSION–DIFFUSION METHOD

It is an effective technique to encapsulate both hydrophilic and lipophilic bioactives with high encapsulation efficiency, high reproducibility, narrow size distribution, less physical stress, and an easy scaling-up procedure. It starts from a two-step process that depends on a regular emulsification step followed by the elimination of part of the organic phase (diffusion step). The organic phase containing biopolymer and oil in an organic solvent and the aqueous phase are prepared separately. These are then made into an emulsion using a mechanical shearing method (emulsification step). Depending upon its polarity an encapsulant could be present in either the aqueous and/or organic phase. Water addition leads to the very fast elimination of the organic solvent contained in the oil phase and causes the separation of the biopolymer and the oil, particle size reduction, biopolymer precipitation, and finally nanocarrier formation. The organic solvent can then be evaporated under reduced pressure. This method is much milder than the direct evaporation of the organic solvent that is usually performed in microencapsulation technology. The organic solvent should have high solubility in both oil and polymer and partial solubility in water to make sure that diffusion after dilution is possible. The nanocapsule size is associated with the shear rate and temperature utilized in the emulsification process, the chemical compositions and proportions of the organic phase and emulsifier, the polymer concentration, and the oil-to-polymer ratio. The method has been used for the fabrication of carbohydrate-based delivery systems such as fish oil and eugenol-loaded cyclodextrin, β-carotene, DL-α-tocopheryl acetate, and capsicum oleoresin loaded poly-ε-caprolactone (Fathi et al., 2014).

7.13.6 EMULSION–SOLVENT EVAPORATION

After a nanoemulsion has been produced utilizing either a low-energy or high-energy approach, it is possible to alter the droplet characteristics using different methods, which are referred to as finishing techniques. Numerous such methods exist namely Solvent Evaporation, Oil exchange, emulsifier exchange, Coating, and droplet crystallization (McClements et al., 2011).

The solvent displacement and/or solvent-evaporation method is one such finishing technique where the size of the droplets in an emulsion can be decreased after an early homogenization step. It is a method for the encapsulation of bioactive molecules inside water-insoluble polymers. The principle depends upon the emulsification of non-water soluble organic solutions of lipophilic bioactive and preformed polymers into an aqueous phase containing surfactants. This procedure leads to the formation of organic solvent droplets that serve as nanocarriers for the bioactives. Following evaporation of the solvent, nanoencapsulation of the lipophilic drug occurs (Vozza et al., 2017).

In the case of hydrophilic drugs, double emulsions (water-in-oil-in-water) are produced. Here the drug is dissolved in the aqueous phase (Nicolas et al., 2013). The organic solvent is chosen based on its grade of water miscibility, safety, boiling point, and legal status. The lipid can be a liquid oil (fish, corn, sunflower, olive oil), a lipophilic nutraceutical (β-carotene, lycopene, or curcumin), a solid fat (hydrogenated palm oil), or a mixed system. But whichever lipid is selected, should be able to get itself dissolved within the organic solvent prior to homogenization. This is required since it is tough to form emulsions when the oil phase contains any crystalline material. The organic solvent can be hydrophilic, lipophilic, or amphiphilic.

In the solvent-displacement method, an amphiphilic organic solvent that is partly miscible with water is employed. To start with, the aqueous phase of the emulsion is saturated with the solvent. When this emulsion is diluted more with water, the organic solvent transfers from the oil droplets into the aqueous phase thus causing the droplets to shrink in dimension. If felt necessary, the remaining organic solvent can then be extracted by evaporation. In the solvent-evaporation method, a volatile organic solvent that is either partially miscible (amphiphilic solvent) or immiscible (lipophilic solvent) with water can be utilized (Horn et al., 2001). In this case, the organic solvent is removed from the oil droplets by evaporation, which causes the droplets to diminish in size. The final droplet size created using these two methods hinges on the initial droplet size in the original emulsion, the amount and water solubility of the organic solvent, the type, and amount of the dispersing agent, the viscosity of the organic and aqueous phases, the emulsification method adapted and the temperature of operation (McClements et al., 2011).

7.13.7 INCLUSION COMPLEXATION

The procedure of creation of a complex where one compound (the host) forms a cavity in which molecules of another compound (the guest) are located is called Inclusion complexation (Song et al., 2009). CDs are the famed pruned cone shape oligosaccharides having 6, 7, or 8 (α, β, or γ-CDs, respectively) glucose units connected by α-(1→4) glucosidic bonds. They possess a lipophilic central cavity and a hydrophilic outside surface. Hence, they are capable to form inclusion complexes with hydrophobic food bioactives ensnared in the inner cavity. The depth of the cavity for these CDs is the same (about 8 Å), but the sizes of the cavities are different for α-CD, β-CD, and γ-CD (about 6, 8, and 10 Å, respectively). The formation and stability of the cyclodextrin-active material complex strongly depend on the size and shape suitability between the CD and bioactive molecule. So, different types of CDs (α-CD, β-CD, and γ-CD) show variegated abilities to trap the same molecule. CDs are useful for entrapping poorly soluble, chemically labile, or temperature-sensitive food bioactive. They have been applied to encapsulate a variety of bioactive components, including antimicrobials, antioxidants (Singh et al., 2017a), flavors, and others like vanillin, usnic acid, and LA.

7.13.8 NANOPRECIPITATION

Nanoprecipitation is a spontaneous emulsification method in which an organic phase comprising the dissolved polymer and a bioactive compound in an organic solvent emulsifies impetuously into the aqueous external phase. The procedure involves rapid desolvation of the polymer from the organic phase and diffusion of the organic solvent in the aqueous medium. Once the polymer containing the solvent diffuses into the aqueous medium, the polymer settles as NPs and this precipitation facilitates the instantaneous entrapment of the bioactive compound into the polymeric wall material. Numerous food bioactives have been entrapped by this method, the most notable being encapsulation of curcumin into poly(lactic-*co*-glycolic acid) (PLGA) and polyethylene glycol (PEG); quercetin into Eudragit® E and polyvinyl alcohol (PVA) and α-tocopherol into polycaprolactone nanocarriers (Vozza et al., 2017).

7.13.9 SUPER CRITICAL FLUID TECHNIQUE

A SCF can be defined as any substance which at a temperature and pressure above its critical point, does not exist in distinct liquid and gas phases, but is lower than the pressure necessary to compress it into a solid. The most widely used supercritical fluid is pure Carbon dioxide (CO_2). Its supercritical region can be attained at modest temperatures and pressures (Tc = 304.2 K, Pc = 7.38 MPa); therefore, this technique is considered a promising method for encapsulating heat-sensitive food bioactive. Supercritical CO_2 also supplies an inert medium appropriate for the entrapment of easily oxidable substances such as unsaturated fatty acids. On the opposite hand, the application of this method diminishes the utilization of organic solvents and facilitates the dissociation of the SCF from the product by depressurization. The high solubility of most organic solvents in SCF permits obtaining solvent-free products as well as producing particles with controllable morphology and narrow size distribution. A few examples of the application of carbohydrate-based nanocarriers developed using the Super Critical Fluid technique comprise the encapsulation of lycopene and lutein using CDs and hydroxypropyl methylcellulose phthalate (Fathi et al., 2014). Benefits associated with Supercritical CO_2 include low cost, nontoxicity, nonflammable, easy availability, and easy recycling ability. A great advantage of SC-CO_2 is that it shows gas-liked diffusivities and liquid-like densities. Applications of supercritical CO_2 are also found in particle formation and encapsulation. Few techniques for particle formation using SCF utilize it as a *solvent* (rapid expansion of supercritical solutions-RESS, supercritical solvent impregnation) or *solute* (particles from gas saturated solutions) or *antisolvent* (SCF extraction of emulsions and supercritical antisolvent precipitation) (Vozza et al., 2017).

7.13.10 PRECIPITATION OF PRESSURIZED EMULSIONS OR HYDROGELS

The technique of precipitation from pressurized emulsions or hydrosols has also been utilized to create some water-insoluble natural compounds. This technique is based upon bringing in contiguity a hot pressurized hydrophilic solution of an active compound and a cold aqueous solution of carrier and stabilization materials. To fabricate nanocarriers the encapsulant solution is driven at an ambient temperature into the first mixer containing hot solvent and then the cold solution of carbohydrate and emulsifier is sent into the second mixer. This results in a fast and controlled enhancement of supersaturation due to the mixing of cold aqueous solution and hot solvent containing encapsulant. This process leads to a thermal shock giving rise to the formation of NPs (range of 50–100 nm). The solvent is then removed using a back pressure valve. The usual operating temperatures of this procedure are in the range of 200°C. Three components namely applied pressure, concentration of surfactant, and organic solvent/water ratio have the most perceptible consequences on product properties. The higher the pressure applied, the smaller the size of particles obtained. Increased concentration of surfactant results in augmentation of the encapsulation efficiency. Similarly, at high organic solvent/water ratios, a radical increase in the particle size is documented (Fathi et al., 2014).

7.14 FUTURE DIMENSIONS

As we step into the future, we need to imbibe newer technologies, and re-innovate the traditional outlook and processes as the emphasis on living healthy and safely gathers more clamor. With the emergence of new microbes, sicknesses, and diseases resulting from our lifestyle changes, the road to these technologies must tread very cautiously. The food industry is ready for the big jump in the days to come but important to keep in mind the major scope of potential benefits and challenges it will offer to both consumers and manufacturers alike. Decisions have to be made on the basis of individual health and desires not compromising on the safety realm as well. Along with the potential health and environmental benefits of the more efficient delivery mechanisms of nutraceuticals, the sustainability of nanomaterials in food, provided it is regulated perfectly, can hugely impact the future direction of the food industry (Ravichandran, 2010). As more and more newer products based on these technologies will hit the market, health, safety, and environmental concerns are addressed, long-term effects need to be assessed, and potential risks to be identified and rectified prior to the successful implementation of each product. Regulations should be made stringent yet fluid for innovation to take the lead and public confidence to be guaranteed so as to make these truly beneficial.

KEYWORDS

- **bioactives**
- **food**
- **industry**
- **nanodelivery**
- **nanotechnology**
- **nutrient delivery**
- **oral delivery systems**

REFERENCES

Acosta, E., (2009). Bioavailability of nanoparticles in nutrient and nutraceutical delivery. *Curr. Opin. Colloid Interf. Sci., 14*, 3–15.

Agnihotri, S. A., Mallikarjuna, N. N., & Aminabhavi, T. M., (2004). Recent advances on chitosan-based micro- and nanoparticles in drug delivery. *J. Control Release, 100*(1), 5–28.

Agrawal, U., Sharma, R., Gupta, M., & Vyas, S. P., (2014). Is nanotechnology a boon for oral drug delivery? *Drug Discovery Today, 19*(10), 1530–1546.

Almeida, A. J., & Souto, E., (2007). Solid lipid nanoparticles as a drug delivery system for peptides and proteins. *Adv. Drug Deliv. Rev., 59*(6), 478–490.

Augustin, M. A., & Hemar, Y., (2009). Nano- and micro-structured assemblies for encapsulation of food ingredients. *Chem. Soc. Rev., 38*(4), 902–912.

Benshitrit, R. C., Levi, C. S., Tal, S. L., Shimoni, E., & Lesmes, U., (2012). Development of oral food-grade delivery systems: Current knowledge and future challenges. *Food Funct., 3*(1), 10–21.

Bezerra, M. A., Santelli, R. E., Oliveira, E. P., Villar, L. S., & Escaleira, L. A., (2008). Response surface methodology (RSM) as a tool for optimization in analytical chemistry. *Talanta, 76*(5), 965–977.

Bhushani, J. A., & Anandharamakrishnan, C., (2014). Electrospinning and electrospraying techniques: Potential food-based applications. *Trends Food Sci. Technol., 38*(1), 21–33.

Chan, H. K., & Kwok, P. C. L., (2011). Production methods for nanodrug particles using the bottom-up approach. *Adv. Drug Deliv. Rev., 63*(6), 406–416.

Chen, L., Remondetto, G. E., & Subirade, M., (2006). Food protein-based materials as nutraceutical delivery systems. *Trends Food Sci. Technol., 17*(5), 272–283.

Elzoghby, A. O., Samy, W. M., & Elgindy, N. A., (2012). Protein-based nanocarriers as promising drug and gene delivery systems. *J. Control Rel., 161*, 38–49.

European Commission, (2015). *EUROPA—Food Safety—Labelling & Nutrition—Food Supplements—Introduction*. Available from: https://food.ec.europa.eu/safety/labelling-and-nutrition_en (accessed on 05 July 2023).

Fathi, M., Martin, A., & McClements, D. J., (2014). Nanoencapsulation of food ingredients using carbohydrate-based delivery systems. *Trends Food Sci. Technol., 39*(1), 18–39.

Helson, L., (2013). Curcumin (diferuloylmethane) delivery methods: A review. *Biofactors, 39*(1), 21–26.

Horn, D., & Rieger, J., (2001). Organic nanoparticles in the aqueous phase—Theory, experiment, and use. *Angew Chem. Int. Ed., 40*, 4331–4361.

Kalra, E., (2003). Nutraceutical definition and introduction. *AAPS Pharm. Sci., 5*(3), 27–28.

Keservani, R. K., & Sharma, A. K., (2014). Flavonoids: Emerging trends and potential health benefits. *Journal of Chinese Pharmaceutical Sciences, 23*(12), 815.

Keservani, R. K., Kesharwani, R. K., Sharma, A. K., Gautam, S. P., & Verma, S. K., (2016). Nutraceutical formulations and challenges. In: Debasis, B., & Nair, S., (eds.), *New Product Development in Nutraceuticals & Functional Foods: From Concept to Market* (pp. 161–177). Chapter 9, Academic Press, Elsevier.

Keservani, R. K., Kesharwani, R. K., Sharma, A. K., Vyas, N., & Chadoker, A., (2010b). Nutritional supplements: An overview. *International Journal of Current Pharmaceutical Review and Research, 1*(1), 59–75.

Keservani, R. K., Kesharwani, R. K., Vyas, N., Jain, S., Raghuvanshi, R., & Sharma, A. K., (2010a). Nutraceutical and functional food as future food: A review. *Der Pharmacia Lettre, 2*(1), 106–116.

Keservani, R. K., Sharma, A. K., & Kesharwani, R. K., (2020). *Nutraceuticals and Dietary Supplements: Applications in Health Improvement and Disease Management*. CRC Press. ISBN: 9781771888738.

Kumari, A., Yadav, S. K., & Yadav, S. C., (2010). Biodegradable polymeric nanoparticles-based drug delivery systems. *Colloids Surf. B, 75*(1), 1–18.

Kumirska, J., Czerwicka, M., Kaczynski, Z., Bychowska, A., Brzozowski, A., Thoming, J., et al., (2010). Application of spectroscopic methods for structural analysis of chitin and chitosan. *Marine Drugs, 8*(5), 1567–1636.

Lee, B. K., Yun, Y. H., & Park, K., (2015). Smart nanoparticles for drug delivery: Boundaries and opportunities. *Chem. Eng. Sci., 125*, 158–164.

Maeda, H., Wu, J., Sawa, T., Matsumura, Y., & Hori, K., (2000). Tumor vascular permeability and the EPR effect in macromolecular therapeutics: A review. *J. Control Release, 65*(1, 2), 271–284.

McClements, D. J., & Rao, J., (2011). Food-grade nanoemulsions: Formulation, fabrication, properties, performance, biological fate, and potential toxicity. *Crit. Rev. Food Sci. Nutr., 51*, 285–330.

McClements, D. J., Li, F., & Xiao, H., (2015). The nutraceutical bioavailability classification scheme: Classifying nutraceuticals according to factors limiting their oral bioavailability. *Annu Rev of Food Sci. and Technol., 6*, 299–327.

Mukherjee, S., Ray, S., & Thakur, R. S., (2009). Solid lipid nanoparticles: A modern formulation approach in drug delivery system. *Indian J. Pharm. Sci., 71*(4), 349–358.

Nicolas, J., Mura, S., Brambilla, D., Mackiewicz, N., & Couvreur, P., (2013). Design, functionalization strategies, and biomedical applications of targeted biodegradable/biocompatible polymer-based nanocarriers for drug delivery. *Chem. Soc. Rev., 42*(3), 1147–1235.

Nobs, L., Bucheggar, F., Gurny, R., & Allemann, E., (2004). Poly(lactic acid) nanoparticles labeled with biologically active neutravidin for active targeting. *Eur. J. Pharm. Biopharm, 58*(3), 483–490.

Ravichandran, R., (2010). Nanotechnology applications in food and food processing: Innovative green approaches, opportunities, and uncertainties for the global market. *Int. J. Green Nanotechnol., 1*(2), 72–96.

Roblegg, E., Frohlich, E., Meindl, C., Teubl, B., Zaversky, M., & Zimmer, A., (2012). Evaluation of a physiological in vitro system to study the transport of nanoparticles through the buccal mucosa. *Nanotoxicology, 6*(4), 399–413.

Singh, P., Kesharwani, R. K., & Keservani, R. K., (2017a). Antioxidants and vitamins: Roles in cellular function and metabolism. In: *Sustained Energy for Enhanced Human Functions and Activity* (pp. 385–407). Academic Press. ISBN:9780128054130.

Singh, P., Kesharwani, R. K., & Keservani, R. K., (2017b). Protein, carbohydrates, and fats: Energy metabolism. In: *Sustained Energy for Enhanced Human Functions and Activity* (pp. 103–115). Academic Press. ISBN:9780128054130.

Smolkova, B., El Yamani, N., Collins, A. R., Gutleb, A. C., & Dusinska, M., (2015). Nanoparticles in food. Epigenetic changes induced by nanomaterials and possible impact on health. *Food Chem. Toxicol., 77*, 64–73.

Solans, C., & Solé, I., (2012). Nano-emulsions: Formation by low-energy methods. *Curr. Opin. Colloid Interface Sci., 17*(5), 246–254.

Song, F., Zhang, L. M., Yang, C., & Yan, L., (2009). Genipin-crosslinked casein hydrogels for controlled drug delivery. *Int. J. Pharm., 373*, 41–47.

Ting, Y., Jiang, Y., Ho, C. T., & Huang, Q., (2014). Common delivery systems for enhancing in vivo bioavailability and biological efficacy of nutraceuticals. *J. Funct. Foods, 7*, 112–128.

Vozza, G., Khalid, M., Bryne, H. J., Ryan, S., & Frias, J., (2017). Nutrition-nutrient delivery. In: Grumezescu, A., (ed.), *Nutrient Delivery; Nano Technology in the Agri-food Industry* (Vol. 5, pp. 1–42). Elsevier.

CHAPTER 8

MICRONUTRIENTS AND MACRONUTRIENTS IN NUTRIGENOMICS

MEGHNA DUTTA[1] and SUDDHASATWYA CHATTERJEE[2]

[1]*Registrar, Critical Care Unit, Apollo Multispecialty Hospitals, Kolkata, West Bengal, India*

[2]*Sr. Consultant, Department of Internal Medicine and Rheumatology, Apollo Multispecialty Hospitals, Kolkata, West Bengal, India*

ABSTRACT

Picture being able to feature into medical nutrition therapeutic intervention a consumer's genetic vulnerabilities and environmental influences so that therapy could be targeted to optimize health and lessen disease. It may sound like science fiction for the immediate future, but not for the long term. Non-communicable diseases (NCDs) like diabetes, hypertension, cancers, obesity, and other cardiac ailments have become pandemics that are associated with decreasing quality of life for the masses and a soaring economic burden for their nations. NCDs are chronic diseases, the result of inappropriate lifestyle, particularly nutrition, interacting with each individual's genotype. These disease-promoting habits often unfold in childhood, and for many, prenatally. Hence primary and primordial levels of prevention will go a long way in curbing the prevalence of these chronic diseases.

Micronutrients and Macronutrients as Nutraceuticals. Prakash Chandra Gupta, PhD, Sayan Bhattacharyya, MBBS, MD, Nisha Sharma, PhD, MPharm, Rajesh K. Kesharwani, PhD, Raj K. Keservani, PhD, MPharm (Eds.)
© 2024 Apple Academic Press, Inc. Co-published with CRC Press (Taylor & Francis)

8.1 INTRODUCTION

Currently, clinical nutrition focuses on the prevention, diagnosis, and management of nutritional alterations in patients with chronic diseases or conditions primarily in health care. Clinical nutrition specialists give individual counseling and can devise meal plans for these patients. Delivering effective lifestyle rehabilitation will be the chief target of clinical nutrition in the decades ahead and will involve knowledge, skills, and tools that target the molecular, biochemical, physiologic, and social aspects of the healthy and diseased states of mind and body.

The Human Genome Project has enhanced our comprehension of how chronic disease occurs at the molecular level. At this level common variations in DNA interact with a plethora of environmental factors, including foods consumed, to influence physiologic outcomes. Prevailing healthcare practices for individuals with chronic disorders are predominantly based on management with medications. Probing into the root cause of these disorders and identification of the underlying mechanisms responsible for their development is paving the way for new approaches that raise hopes of restoring health or even preventing their development.

The goal of nutrition research is to gain insight into the mechanisms that underlie these interactions and to project these findings into clinical interventions for more effective prevention and management. Our genetic makeup determines our susceptibility to disease. However, study on chronic diseases has revealed that diet and other lifestyle choices play a vital role in ascertaining who will develop the dysfunction and the severity of the condition. Dietary patterns, level of physical activity, habits, sleep patterns, thoughts, and emotions, relationships with self and others, and one's sense of purpose in life—affect bodily function at the molecular, biochemical, and physiologic levels. These modifiable environmental factors have the potential for changing the trajectory from a poor quality of life filled with disease and disability to one of thriving and flourishing. This is enabling clinicians to tailor therapy to the client's uniqueness (Ruth DeBusk et al., 2016).

Nutritional genomics is an emerging sub-discipline of clinical nutrition that provides the tools for identifying genetic variations that hint at increasing vulnerability to developing chronic diseases and provide the knowledge of potentially modifiable lifestyle choices to promote health. The interplay among our genetic material, food habits, and other lifestyle choices, and their impact on health and disease are the focus of nutritional genomics. This can be applied as a three-pronged assessment tool that provides the ability to: (i) identify the unique genetic signature of a person; (ii) weigh the vulnerability of a person towards the state of health and disease; and (iii) project, for every variable in one's lifestyle that can be changed to influence health, which choices are most likely to improve longterm health and prevent illness throughout the lifespan (Ruth DeBusk et al., 2016).

8.2 GENE-DIET-DISEASE INTERACTION

The Human Genome Project was completed in 2003 but was just the initiation of the movement to integrate genetic principles into health care. A genome is an organism's complete set of genetic information. Many aspects of the genome are preserved during evolution, which provides useful information as to which regions of the genome are vital to life. Genomics is the study of all of a person's genes (called the genome), as well as their interactions with one another and with the person's surroundings. This study gives us the chance to weigh the size, nucleotide sequence, and organizational intricacy of the human genome against other organisms. Valuable facts pertaining to human health and disease have been obtained from model systems, such as the laboratory mouse, whose genome is similar to that of humans. The Encyclopedia of DNA Elements (ENCODE), a follow-up to the Human Genome Project, is a public research consortium that targets all functional elements in human and mouse genomes. It is a continuing international collaboration of research groups financially supported by the National Human Genome Research Institute (NHGRI). ENCODE has two levels of annotations: (i) Integrative-level annotations that integrate multiple types of experimental data and ground-level annotations; and (ii) ground-level annotations that are derived directly from the experimental data.

Whereas the Human Genome Project focused on identifying genes within the total genome, the ENCODE project investigates the non-gene sequences, which account for over 99% of the human genome. Initially thought to be "junk DNA" because the majority of this DNA does not code for proteins, these non-gene sequences are critical to regulating the expression of genes and their encoded proteins (Ruth DeBusk et al., 2016).

The Human Genome Project gave birth to many disciplines like proteomics, metabolomics, microbiomics, and bioinformatics. Proteomics centers on determining the protein that each gene encodes in an organism's genome and establishing its function. Metabolomics is involved with detecting metabolites produced in various phases of metabolism, most commonly as a result of protein function. Microbiomics emphasizes the importance of the digestive tract's microbial ecology (microbiome) and other bodily cavities. Beneficial and pathogenic microbes colonize these cavities. The huge amount of information generated by these areas has led to the rapid growth of bioinformatics, a field that sits at the crossroads of computer science, information science, biology, and medicine. Bioinformatics is an important tool for managing the colossal amounts of data produced by many "omic" disciplines. The development of sophisticated computers and software that can organize, store, and retrieve massive amounts of data has enabled researchers around the world to share and compare data on various "omic" profiles.

Dietary signatures are patterns of gene expression, protein expression, and metabolite production in response to specific nutrients or nutritional regimens. Nutrigenomics evaluates dietary signatures in specific cells, tissues, and organisms and aids with the understanding of how nutrition affects homeostasis.

Many cofactors and substrates are required for DNA replication and DNA repair. Errors during the proofreading of DNA may lead to faulty replication. Accumulation of such errors may trigger cell death by apoptosis. Genotoxic insults lead to the accumulation of errors at the base sequence or chromosome level. This is now recognized as a leading cause of developmental defects accelerated aging, and an increased risk of degenerative conditions such as infertility, immune dysfunction, cancer, and cardiovascular, and neurodegenerative disorders. Nutrigenomics and nutrigenetics are two different aspects of the same coin (Mutch et al., 2005).

DNA sequencing analysis is used to identify pathogenic organisms and is replacing rapidly the time-consuming growth assays used in the clinical laboratory to identify which microbial strains are present in a particular part of the patient's body, along with the relative concentrations of each strain. In case of the good microbes, researchers are investigating which strains promote our physiological well-being and how judicious diet and lifestyle choices can support their vitality and successful colonization in the body.

8.3 CLINICAL APPLICATIONS

Nutritional research today focuses on health promotion, performance enhancement, disease prevention, protection from toxins, and stress management.

8.3.1 PHARMACOGENOMICS

Pharmacogenomics is the use of genomics to analyze the genetic variations in the genes that control the synthesis of drug-metabolizing enzymes and use this data to scrutinize a patient's response to the drug. Genetic differences between individuals can lead to differing functions in these enzymes – this explains why certain drugs may have the intended effects in some, be ineffective in others, and be harmful to another person. Genetic polymorphisms can moderate the expression and function of enzymes and proteins concerned with drug metabolism, affecting absorption, distribution, biotransformation, and excretion over and above the drug-target interaction (Quiñones et al., 2017).

Assisting with diagnosis and choice of therapeutic interventions is another dimension of clinical application. Knowledge of the genes associated with a particular disease and the DNA sequence of those genes, their protein product, and the function of those proteins as a health booster or in promoting the disease provides the basis for diagnostic assays and optimal intervention. Genetic profiling in screening and therapy is an important tool of management for oncologists. Pathologically identical tumors can be differentiated by their genetic profiles. This is significant because different tumors respond to different therapeutic approaches. Genomic analysis has also been used by oncologists to monitor therapeutic response and for early prediction of which individuals are most likely to experience treatment failure so that they may be switched to an alternate therapy.

A genomic study can also be utilized to discover abnormal functions in people who don't have any symptoms. Assessment of genetic susceptibility facilitates early intervention before the person develops symptoms.

8.3.2 PROTEOMICS

Proteomics can be used in different aspects of human welfare such as food technology, the discovery of biomarkers, and the identification of the targets of a drug. This technology can examine and illustrate the protein content in our diet and how it varies during the manufacturing process (Amiri-Dashatan et al., 2018). Proteomic biomarkers have been discovered to be useful in a variety of conditions, including malignancy, cardiovascular disease (CVD), AIDS, and renal failure. Two-dimensional polyacrylamide gel electrophoresis (2D-PAGE) and chromatography techniques along with mass spectrometry has aided in this field of work (Stevenson et al., 2009). Proteomics can be used in the food business for detection, validation, optimization, and quality control (Cifuentes, 2012). Microbial contaminants and their toxins, detection of the origin of food allergens, quality assurance, and safety of raw materials are some of the many roles of this branch of science. A number of proteins that are important for plant growth and development have been discovered. Proteomic approaches must be used to assess genetically modified crops. It's also looked into how crops interact with other organisms to affect yield.

8.3.3 METABOLOMICS

Metabolomics synonymous with "metabonomics" and "metabolic profiling" has empowered clinicians with novel sets of diagnostic biomarkers for disease states besides quantifying treatment response to medications at an individualized level. Metabolic profiling can be carried out with the help of mass spectrometry and nuclear magnetic resonance-based methods using a variety of biological models. Targeted metabolomic profiling depends on commercial kits which offer highly standardized measurements on a classified set of known metabolites (Ashrafian et al., 2021).

8.3.4 EPIGENOMICS

Epigenomics focuses on the interplay between the genome and the information received from the environment. Epigenetic programs are essential for normal cellular development and differentiation programs. Studies of gene expression, proteins, or metabolites are essentially different from epigenetic biomarkers, which provide glimpses of functional status at a specific time point. Epigenetics provides different strata of regulatory and environmental exposure data on top of each person's unique genome. The most extensively researched epigenetic mark is the methylation of DNA cytosine residues at the carbon 5 position (5mC). This mark is made via the DNA N-methyl transferase (DNMT) enzymes. Besides 5mC which is the most well-characterized, other cytosine alterations have been discovered in recent years, such as 5-hydroxymethyl cytosine (5hmC), 5-formyl cytosine (5fC) and 5-carboxyl cytosine (5caC) (Bachman et al., 2015). Post-translational modification of histones is another example of an epigenetic mark. Mono-, di-, tri-methylation, acetylation, and phosphorylation are among the best-characterized modifications. These modifications are written and erased by certain enzyme families, such as histone acetyltransferases (HATs) and histone deacetylases (HDACs) in the case of acetylation marks, or histone methyltransferases (HMTs) and demethylases (HDMs) in the case of methylation marks (Bannister et al., 2011). RoadMap Epigenetics aims to reveal the epigenetic processes that contribute to human biology and disease. The working models generated by RoadMap Epigenomics include a core 15 chromatin state model and an expanded 18 chromatin state model with 12 active and 6 inactive states in the latter model (Kronfol et al., 2017). All of the epigenetic markings in that cell type together form the "epigenome."

8.4 GENE AND THE GENOME

A *gene* is a basic block of heredity that is located on a chromosome at a specific position. Every particular gene has nucleotides that are arranged in a linear manner. The majority of the genes code for a specific protein or part of the protein leading to a certain characteristic or function. A *genome* which is the complete set of genetic information of an organism provides the entire data an organism requires to function. The *transcriptome* is the collection of RNA that a genome expresses, whereas the complete set of proteins generated by the genome is known as *the proteome*.

Each eukaryotic chromosome is made up of DNA that is coiled and condensed around *histones*, which are nuclear proteins. Humans inherit one set of chromosomes from either biological parent. In most instances, human cells contain 46 chromosomes with 22 pairs of *autosomes* or non-sex chromosomes and 2 sex-determining chromosomes, also known as *allosomes*. X and Y chromosomes are the two allosomes in humans – females carry two X chromosomes while the males carry one X and one Y chromosome. Those cells in the body that carry two sets of chromosomes are known as *diploid cells*. Meanwhile, cells with half the number of chromosomes compared to diploid cells are known as *haploid cells*. Germline cells which later produce egg or sperm cells contain a haploid number of chromosomes.

Chromatin is a DNA and protein cluster that creates chromosomes in eukaryotic cells' nuclei. On microscopic examination, chromatin resembles a strand of beads, where the beads are the nucleosomes. A *nucleosome* is made up of a histone octamer, which is made up of a little less than two rounds of DNA wrapped around eight histone proteins. H2A, H2B, H3, and H4 each have two copies in a histone octamer. The nucleosome is the most basic component of chromatin.

Deoxyribonucleic acid (DNA) is all living organisms' genetic material. It gets its name from its structure, which consists of a sugar and a phosphate backbone with bases protruding from it. There are four kinds of bases in the DNA, namely, adenine, cytosine, guanine, and thymine. Adenine and thymine form specific base pairings, while cytosine and guanine form specific base pairings. Structurally DNA is organized into chromosomes and further wound around nucleosomes. Functionally, this genetic material is organized into genes, which leads to noticeable traits. A nitrogenous base and a five-carbon carbohydrate group, commonly a ribose or deoxyribose molecule, make up *nucleosides*. A nucleoside with one or more phosphate groups attached to it is known as *a nucleotide*. The DNA molecule is made up of two strands that create a double helix when they wind around one other. Bonds between the bases hold the two strands together. Individual nucleotides are joined together by phosphodiester bonds between their 5′- and 3′-carbons to form a polynucleotide. The two outer phosphates (the β- and the γ-phosphates) from one nucleotide are removed during the polymerization step, and the hydroxyl group linked to the 3′-carbon of the second nucleotide is replaced. As a result, the polynucleotide has a chemical orientation, which can be represented as 5′→3′ or 3′→5′. Because of the polarity of the phosphodiester link, the chemical events required to extend a DNA polymer in the 5′→3′ direction differ from those required to extend a DNA polymer in the 3′→5′ direction. Only 5′→3′ synthesis is possible with natural DNA polymerase enzymes.

There are approximately 19,000 genes in the human genome. The human DNA sequence contains ~ 3,100,000,000 base pairs (depending on the actual source of the assembled DNA sequence) which includes most of the non-heterochromatic portions of the genome and contains around 250 gaps (Little, 2005). The process of identification of genes is known as annotation and this has been achieved with the help of bioinformatics, especially by homology analyses and some de novo gene predictions. Such data is easily accessible through several large genome "browsers" (Birney et al., 2004).

One of the characteristics of eukaryotes is spliceosomal introns. *Intervening sequences*, also known as *introns*, are non-coding regions of an RNA transcript or the DNA encoding it that are spliced away before the RNA molecule is translated into a protein. *Exons* are the sections of DNA or RNA that code for proteins. Approximately more than 50% of human genetic disorders are caused by deviation from the usual splicing pattern (Chorev et al., 2012).

The *promoter region* and the *coding region* are the two primary functional elements of every gene. The promoter region determines the site and the situation of gene expression. The coding region of a gene determines the structure of its protein. The coding region is made up of a linear sequence of nucleotides that codes for a protein's amino acid sequence. The genetic code is in a triplet form and hence every 3 successive nucleotides group to code for a single amino acid. A *codon* is a three-nucleotide sequence. Some amino acids are encoded by several different triplets. Upon translation of the genetic code, we can obtain the amino acid sequence of every protein (Polyak, 2003). The *regulatory* region lies upstream of the promoter region. This site influences the ability of the polymerase to attach to the promoter, thus controlling whether transcription will occur. *Response elements* are DNA sequences located in the regulator region which serve as binding sites for regulatory proteins like transcription factors and the ligands they bind.

Gene expression, i.e., transcription is regulated at various levels. The chromosomes containing the genes to be expressed are first decondensed. The next level of transcriptional control is at the epigenetic level, which occurs at the level of the DNA. Another control mechanism is when the methyl groups have to attach to the DNA within the regulatory region of the gene. Whenever attachment of methyl groups occurs, transcription is blocked and the expression of the gene is silenced. Gene expression occurs once the methyl groups are removed. A *single nucleotide polymorphism* or *SNP* is a difference in the DNA sequence at a single point between people. A *mutation* is a permanent and irreversible change in the DNA sequence. Both SNPs and mutations are DNA variants, however, SNPs are detectable in >1% of the human population. DNA mutations, on the other hand, are detectable in <1% of the human population (Karki et al., 2015). These SNPs are the most important elements in human genetic variation and provide a biological basis for phenotypic variances.

Besides the DNA code, there are two other sources of information: the epigenetic code and the environmental factors to which the cells are exposed. Acetyl or methyl groups are covalently attached to the histone proteins associated with the DNA or directly to the DNA itself, respectively, deciding whether the DNA is available for coding. These groups can be attached or detached as per necessity and are affected by exposure to different environmental factors like toxins, cytokines, hormones, chemicals, drugs, and traditional nutrients. A majority of these environmental factors are lifestyle choices and are modifiable based on an individual's attitude toward life. Choline, methionine, folate, vitamins B12, B6, and riboflavin deficiency disrupts one-carbon metabolism, impairs DNA methylation, and raise the risk of neural tube abnormalities, cancer, and CVD.

There are three ways that traits are passed down from generation to generation: Mendelian inheritance, mitochondrial inheritance, and epigenetic inheritance. *Nutrigenetics* is concerned with how an individual's set of genetic variations affects function. Meanwhile, *nutrigenomics* is the study of how genes interact with environmental variables to cause changes in gene expression. The influence of changes in nutrient intake (food-based diets, culinary limitations, or nutritional supplements) on gene expression is referred to as nutrigenomics; it focuses on the effects of nutrition on the transcriptome, proteome, and metabolome in cells, tissues, and organisms, as well as confirming the genes that influence diet-related disease risk. It may also be useful in the understanding of how dietary intake and other deliberate lifestyle choices affect metabolic pathways and how these regulations are inhibited in the early stages of diet-related disease.

Research in the field of Nutritional Genomics is a burning subject with numerous calls for projects at the national and international levels, and relevant research networks, such as the so-called NuGO (Nutrigenomics Organization). NuGO is a European-funded Network of Excellence, the full title of which is 'The European Nutrigenomics Organization: connecting genomics, nutrition, and health research,' which conducts cutting-edge research into all "omic" areas linked to nutrition and health, as well as the ethical issues that arise from it.

8.4.1 NUTRIGENOMICS AND OBESITY

Obesity has emerged as a major public health issue in today's world. It is the outcome of a net increase in dietary energy intake and a decrease in energy expenditure, which leads to an energy imbalance and a gain in body weight. The causes of obesity can be quite complex and multifactorial. Obesity is influenced by a lot of factors such as environmental situation, behavioral responses, hormonal balance in the body, metabolic rate, and genetic predisposition of the individual (Walker et al., 2007; Sharma et al., 2010). Obesity has been linked to an increased risk of heart disease, type 2 diabetes, metabolic syndrome, and several cancers. Overweight adults with a BMI of 25 or higher accounted for 1.5 billion people worldwide, whereas obese adults with a BMI of 30 or higher accounted for 500 million (Finucane et al., 2011). A high-energy diet is a major contributor to obesity. Consumption of macronutrients in excess of the prescribed levels of protein, carbohydrate, and fat results in intake of high energy. Consumption of dietary fat and empty calorie items at an excessive level can play a major role in the development of obesity (Amato et al., 2010; Hinney et al., 2010). The total energy from solid fats and added sugars is known as empty calories (Reedy et al., 2010). On the contrary, other research has found that high-satiety foods (such as vegetables, fruits, and whole grains) are inversely connected to the global prevalence of obesity (Giskes et al., 2011; Deram et al., 2009). The diluted calorie density of the diet, poor fiber absorption, and an increase in satiety due to delayed stomach emptying of the consumed food are all probable explanations for this condition. Furthermore, indisciplined eating habits that disrupt the management of dietary energy intakes, such as missing meals, snacking, and eating away from home, were linked to obesity (Giskes et al., 2011). Low-energy diets with dietary fat limits, on the other hand, have been demonstrated to be unsustainable in obese people's long-term weight loss. It was unclear what influence fruit and vegetable consumption played in body weight as a sign of obesity (Alinia et al., 2009). Individuals responded differentially to the effect of diet composition or type on obesity, which is noteworthy to notice. Numerous studies among families, adopted kids and twin siblings suggest that genetic factors contributed to 45% to 75% of the variations in BMI among different individuals, the most often used adiposity metric (Sharma et al., 2010). Single nucleotide polymorphisms (SNPs) were identified with the completion of the Human Genome Project in 2003. Inter-individual differences in SNPs could be the reason that certain individuals are more vulnerable to common diseases.

Genome-Wide Association Studies (GWASs) have become a regularly used method for inspecting genetic variations, such as finding heritable risk factors linked with a particularly complex disease (Lewis et al., 2011; McCarthy et al., 2008). GWASs have determined the association between chromosomal loci and the concerned disease genome in an impartial manner (Chial et al., 2008). These studies have been performed for numerous diseases such as type I and type II diabetes, inflammatory bowel disease, prostate cancer, breast cancer, asthma, and coronary artery disease among others. New loci linked to BMI and obesity have also been discovered (Gresham et al., 2008). It was in the Framingham Heart Study that the first GWAS for obesity and its related traits was noted in the form of a variant 10 kb upstream of INSIG-2 (insulin-induced gene-2) (Herbert et al., 2006). Later an obesity susceptibility locus named FTO (fat mass and obesity-associated gene) was identified and has remained the first gene to be recognized as a well-known and rigorously replicated risk factor for common obesity (Hinney et al., 2007). As FTO is a locus with the maximum impact on BMI, so its locus is demarcated as a formidable susceptibility locus for obesity. Besides the *FTO* locus, near-*MC4R* loci and three other GWASs in 2009 from the GIANT (genomic investigation of anthropometric traits) fraternity, deCODE genetics, and GWAS in European populations discovered 13 new loci, near *NEGR1* (neuronal growth regulator 1), near *TMEM18* (transmembrane protein 18), in *SH2B1* (SH2B adaptor protein 1), near *GNPDA2* (glucosamine-6-phosphate deaminase 2), in *MTCH2* (mitochondrial carrier homolog 2), in *SEC16B* (SEC16 homolog B), between *ETV5* (Ets variant gene 5) and *DGKG* (diacylglycerol kinase), in *BDNF* (brain-derived neurotrophic factor), between *BCDIN3D* (BCDIN3 domain containing) and *FAIM2* (Fas apoptotic inhibitory molecule 2), in *NPC1* (Niemann-Pick disease, type C1), near *KCTD15* (potassium channel tetramerization domain containing 15), near *MAF* (v-maf musculoaponeurotic fibrosarcoma oncogene homolog), and *PTER* (phosphotriesterase related) (Loos et al., 2008; Willer et al., 2009; Thorleifsson et al., 2009; Meyre et al., 2009).

It is a well-observed and accepted fact in the absence of exposure to an obesogenic environment, genetic variants alone do not cause obesity. The impact of genetic and environmental variables on obesity may be disproportionate to the effect of specific genetic variants. As a result, obesity is a multifaceted disease influenced by both genetic and environmental variables such as nutrition and physical exercise. Genetic factors and genetic variations (SNPs) may influence an individual's susceptibility to obesity via a variety of potential mechanisms governing pathways and regulatory systems at various levels, such as energy intake and expenditure, and nutrient control and allocation between fat and lean mass tissue (Frazer et al., 2009; Hinney et al., 2010).

Obesity is linked to a higher risk of developing numerous chronic diseases such as cardiovascular disorders, type 2 diabetes, joint ailments, elevated blood pressure, and certain cancers such as esophageal carcinoma, breast carcinoma, endometrial cancer, colorectal cancer, gall bladder malignancies and other possible forms of malignancies. Interventional studies have shown inconsistent results on the quantitative and qualitative analyzes of dietary nutrients on obesity. The features of nutrigenomics have been successfully utilized to explain the association between specific dietary nutrient intake and genetic differences in obesity. There are unmistakable disparities in obesity genetic variants between races or populations. In the current world, nutrigenomics is extensively used in the study of obesity and diet-related problems. That is why, with the help of "personalized" nutritional advice and consultation according to an individual's genetic profile, the focus of nutrigenomics may move to effective obesity control (Doo et al., 2015). Further research is awaited in genetic variations on obesity, the relationship between various dietary nutrient intake, and genetic variation on obesity for the development of effective management tools of "personalized" nutrition for various races or populations.

8.5 NUTRIGENOMICS AND CARDIOVASCULAR DISEASES (CVDS)

Cardiovascular disease (CVD), the world's leading cause of morbidity and mortality, is a complicated disorder influenced heavily by environmental and hereditary variables. The link between food and the risk of CVD is well-documented. Because food is the most common way to reduce risk factors, it's critical to examine genetic implications on CVD risk. In cardiovascular genetics, there are two main approaches to finding disease-causing genes: whole-genome and candidate gene association studies (Engler, 2009). Recent nutrigenomic research has revealed that genotype has an impact on the responsiveness to dietary variables or nutrients that may reduce CVD risk.

Modifications in dietary fat and plasma lipid concentrations in relation to CVD prevention have been extensively explored due to the large differences in the influence of dietary changes on phenotypes amongst different individuals. It is common knowledge that low-fat diets are associated with a decrease in total and plasma LDL cholesterol. However, in this era of evidence-based medicine, it has been noted beyond doubt that dramatic inter-individual differences occur that are partly attributable to genetic factors. The consumption of Mediterranean products like olive oil has become popular worldwide owing to the discovery of the cardio-protective and other healthy properties of the Mediterranean diet. Studies at the molecular, clinical, and epidemiological levels have begun to provide some enlightenment about how various components of the Mediterranean diet may provide protection to the cardiovascular system and also decrease the chances of the occurrence of other ailments such as malignancy. Another likely reason is that to fully benefit from this diet, the appropriate balance of genetic, cultural, and socioeconomic factors is required. The Mediterranean diet is thought to be closer to the food consumed by our forefathers throughout the early stages of human development, and our metabolism may have evolved to perform best on this diet rather than modern diets that are higher in saturated fat and highly refined and processed foods (Ordovas et al., 2007). As a result, it's possible that alleles linked to a higher risk of disease are suppressed in the presence of a more ancestral and traditional food and lifestyle. This understanding can serve as the foundation for effective public health as well as individual illness prevention strategies.

Despite their multifactorial complexity, CVDs are the disease group in which the most progress has been made in terms of understanding their genetic risk factors, both in terms of identifying candidate genes using the traditional approach based on protein function and more recently through genome-wide association studies (GWAS). This was made possible by the previous depiction of a number of intermediate phenotypes linked to those diseases, including plasma lipid concentrations, plasma glucose, and related parameters, markers of inflammation and endothelial damage, oxidative stress, blood pressure, anthropometric measurements, and even phenotypes obtained using non-invasive imaging techniques such as measuring the intima-media thickness of arterial walls (Wilson, 2004; Lindner, 2009). Because of the relative ease of computing these intermediate phenotypes and the certainty that they are risk factors, many studies have been conducted aimed at identifying genes and genetic variants associated with each of them, allowing for a better understanding of the numerous genes involved in the final phenotypes of CVD (i.e., CHDs, cerebrovascular diseases, peripheral CVDs) (Van Horn et al., 2008).

There are many factors involved in the failure of dietary modification in reducing the risk of these diseases, among them include the inherent difficulty in changing behaviors, fashions, mass media pressure, sedentary lifestyles, and a lack of nutritional education programs, to name a few. Over the recent decades, it has been observed that quite often contradictory nutritional recommendations are made based on the findings of studies with limited external validity and a multitude of business interests (McDonald, 2004). For a long time, the benefits of a low-carbohydrate, high-fat diet against a low-fat, high-carbohydrate diet in terms of regulating body weight and accompanying cardiovascular risk factors have been a point of contention (Sacks et al., 2009). Another bone of disagreement is the best macronutrient composition in the diet to prevent or treat CVD, particularly in terms of the percentage of total fat and its various fatty acids provided, as well as the source of those fatty acids; for example, monounsaturated fatty acids (MUFA) derived from olive oil vis a vis MUFA derived from meat and other animal foods. A similar conundrum exists with the best origin and kind (omega-6 and omega-3 series) of polyunsaturated fatty acids (PUFA) in preventing or treating CVD (Jakobsen et al., 2009). Hence it has been concluded that optimal diet may differ depending on a person's attributes and the goals he or she has set for themselves.

Genetic variants have been studied to see if they can cause disease, especially SNPs, and can explain the inter-individual differences in response to dietary modification. These gene-diet tailoring techniques could also aid in deciphering the various symptoms associated with a given genotype, such as those seen in some monogenic forms of CVD. A single gene mutation may be responsible for a wide range of clinical presentations, from asymptomatic heart disease to severe CVD (Corella et al., 2005). Based on the findings of preliminary investigations, there is evidence that the APOE locus interacts with dietary saturated fat, increasing LDL-C levels and CVD risk, higher in E4 carriers than in others (E2 and E3). However, there is an extremely low level of evidence of this interaction in epidemiological studies. There is currently insufficient epidemiological evidence to justify the implementation of dietary advice on fat consumption based on the APOE genotype (Corella et al., 2009).

More study on the subject of nutrigenomics is expected to give scientific evidence for a genotype-based tailored diet to enhance health and avoid chronic disease, such as CVD. It is critical that healthcare personnel, particularly nurses assigned to cardiovascular patients, receive genetics training in order to provide competent genetic- and genomic-focused care and to assist in the integration of this new knowledge into current clinical practice, education, and research.

8.6 NUTRIGENOMICS AND DIABETES MELLITUS

Because of its polygenic and multifactorial nature, Type 2 diabetes mellitus (T2DM) is indeed a very good model for cancer, obesity, CVD, and other chronic diseases driven by nutrition and environment. As per the oral glucose tolerance test (OGTT), individuals are grouped into three classes: normal, impaired, and diabetic. Irrespective of the grouping for diabetes status, individuals may also have concurrent obesity, dyslipidemia, hypertension, insulin resistance, and/or hyperinsulinemia. The diagnosis and treatment options may be further complicated due to these physiological abnormalities that may have molecular and genetic elements that overlap. A majority of diabetic patients gradually develop complications of the disease including retinopathy, nephropathy, neuropathies, and CVD. A very small proportion of diabetics (~20%) experience control of their symptoms with dietary modifications and exercise alone (Koro et al., 2004). The patients who do not improve with these interventions or those who present with severe symptoms are treated including one or more of the various available drugs. The present-day medical management of T2DM can be a lengthy trial and error method, involving a formidable amount of time and considerable financial drain.

Geneticists have developed novel methods such as the quantitative trait locus (QTL) analysis to identify segments of chromosomes that chip into a complex trait. QTLs are statistical assessments of the frequency with which a chromosomal region is linked to a measurable phenotype, e.g., insulin levels or glucose response for T2DM. Each of the genes within QTLs may be responsible in different amounts for the expression of the phenotype. Therefore, gene variants (i.e., SNPs) may be liable for little to major contributions to the complex feature. The aggregate of these contributions from causative alleles in different QTLs causes a specific ailment or sickness (Flint et al., 2005). Each of these QTLs may be as large as ~20 million base pairs and encodes one or more genes (more specifically a variant or allele of a gene) that adds to the complex trait. However, it must be remembered that these profiles describe only the genetic risk component of the particular individual and do not account for environmental influences (Kaput et al., 2007).

A handful of these genes are believed to be influenced by the diet, as certain diets are risk factors for T2DM. This means that the vulnerability to disease in each of these individuals will also differ depending upon the pattern of nutrient intake, amount of physical activity, and other environmental factors. Some examples of many other nutrient and non-nutrient environmental factors affecting this T2DM phenotype are (i) total sleep time and sleep continuity; (ii) oxygen tension that is also affected by altitude or genetic conditions such as sickle cell disease; (iii) over the counter drugs like non-steroidal anti-inflammatory drugs (NSAIDs); (iv) water intake relative to tea and other beverages; (v) physical activity; (vi) psychological factors such as stress; (vii) exposure to allergens and pollutants; (viii) circadian rhythm and seasonal changes; and (ix) equilibrium between energy intake and expenditure. As quite a few genetic and environmental factors change over the years, during life and with aging, the net result is that health and chronic disease states are not dichotomous or discrete but rather are transitional processes (Kaput et al., 2007). Over the last few decades, the sudden and dramatic increase in the prevalence of obesity and T2DM throughout the world would suggest that a majority of individuals have a genetic susceptibility that can be influenced significantly by diet and lifestyle.

A short-term rise in plasma levels of free fatty acids (FFAs) potentiate Glucose stimulated insulin secretion (GSIS), whereas long-term exposure increases basal insulin release while impairing GSIS. The increased secretion at low plasma glucose concentrations may be a compensatory mechanism to cope with increased insulin resistance. Still, the consequential development of suppressed GSIS hinders proper regulation of plasma glucose, ultimately leading to overt T2DM. Many of the effects of FFA on pancreatic β-cell function are transcriptionally mediated. FFAs control gene expression in the following ways: (i) by increasing both the expression and activity of peroxisome proliferator-activated receptor (PPAR) transcription factors; (ii) Elevated FFA also exert effects on gene expression *via* activation of the sterol regulatory element binding protein-1c (SREBP-1c) transcription factor, which controls the expression of many lipogenic genes; (iii) Chronic exposure to elevated FFA decreases β-cell insulin content by up to 85% – these effects are mediated through changing the activation of numerous transcription factors, the most important being islet/duodenum homeobox-1 (IDX-1) and pancreas duodenum homebox-1 (PDX-1); and (iv) exposure to elevated levels of FFA can lead to β-cell apoptosis (Nino-Fong et al., 2007).

Hyperglycemia induces the expression of many genes directly involved in glucose entry into β-cells and genes involved in glucose metabolism. Phosphofructokinase, glyceraldehyde-3-phosphate dehydrogenase, L-type pyruvate kinase, and GLUT2 are all upregulated by acute elevation of plasma glucose. These changes enhance the capacity of the β-cell to import and metabolize glucose. Moreover, mRNA levels of K+ATP channels are upregulated by glucose – these channels are important in the signaling of insulin release via the metabolism-sensing mechanism. Insulin production is upregulated by glucose at both the transcriptional and translational levels. Effects on transcription may be because of glucose-induced PDX-1 and MafA activation, two crucial transcription factors regulating insulin expression. Long-term exposure to hyperglycemia contributes to the development of β-cell failure and T2DM. Thus, chronic high-glucose treatment results in impaired GSIS, lipid accumulation, and β-cell apoptosis. Thus, hyperglycemia, independent of hyperlipidemia, is sufficient to cause β-cell dysfunction (Nino-Fong et al., 2007).

We have gained quite some knowledge about the direct effects of nutrients, especially glucose and FFA, on β-cell gene expression and function but not much is known about mollifying the deleterious, chronic effects of glucolipotoxicity. Moreover, although the ability of cytokines produced in adipose tissue, including TNF-β, to induce β-cell apoptosis is known, the potential role of circulating novel adipokines such as adiponectin, visfatin, and resistin needs to be elucidated (Nino-Fong et al., 2007). To untangle the intricacy of gene-nutrient interactions underlying T2DM and its precursor, the metabolic syndrome, well-designed, and high-powered research are required.

8.7 NUTRIGENOMICS AND CANCERS

Inconsistencies in the literature, without a doubt, reflect differences in the onset and duration of exposures to specific dietary constituents, interactions with the food matrix, processing technologies, or individual genomic variation, all of which can affect absorption, metabolism, and/or the molecular target. Inter-individual differences in genetics, epigenetics, transcriptomics, proteomics, metabolomics, and microbiomics might influence the amplitude and direction of responses to bioactive food ingredients.

Epigenetic and genetic alterations that coerce the expression of oncogenes go hand-in-hand with the transition of healthy cells into cancerous ones. Nutritional deficiency states have played a crucial role in the progression and development of tumors. Cancer chemoprevention and chemotherapy have been linked to fruits, vegetables, cereals, nuts, green tea, and red wine. In both an *in vitro* and an *in vivo* model, dietary nutrients were demonstrated to significantly slow tumor growth. Dietary nutrients aid in the repair of DNA damage produced by environmental contaminants, free radicals, radiation, and pathogenic agents, to name a few. Dietary substances alter gene expression patterns, accelerating cancer growth. Natural chemicals found in food play a variety of roles, including preventing cancer cell proliferation, invasion, angiogenesis, and metastasis, as well as inhibiting stem cell growth. Polymorphisms in the genetic code play a big impact on how the body reacts (Chattopadhyay, 2020). As a result, determining the significance of natural food components as chemotherapeutic agents in cancer treatment is critical.

The large number of preclinical studies demonstrates the importance of nutrigenomics in cancer treatment. Relevant data will be transferred as therapeutic interventions for this devastating pathology and to satisfy the concept of individualized nutrition by analyzing the network of nutrient-gene linkages associated with cancer. Natural substances must also be evaluated in conjunction with conventional therapy using genomic techniques, with a focus on their ability to decrease drug resistance processes (Braicu et al., 2017).

Proof from experimental studies using a variety of models corroborates the importance of numerous dietary components that alter one or more cancer-related pathways. However, the results of epidemiologic and clinical intervention trials have been contradictory, making it difficult to make firm conclusions about which dietary changes are required to reduce cancer risk. The effect of dietary soy intake on breast cancer risk is a classic example of the variability in study conclusions. Preclinical data have demonstrated soy to have anticancer properties. According to meta-analyzes, soy isoflavone consumption is connected to a lower risk of breast cancer in Asian populations, and it may be even more protective among premenopausal women. Variability in soy and other bioactive food components and cancer prevention could be a result of individual differences in response (Nicastro et al., 2012).

Dietary factors have been discovered to have a strong relationship with methylation status. The impact of diet and multivitamin usage on the prevalence of gene promoter methylation in cells exfoliated from the aerodigestive tract was investigated in the Lovelace Smokers cohort of current and past smokers. Those who ate leafy green vegetables and folate, along with concurrent use of multivitamins on a regular basis, were found to have significant protection against methylation (Stidley et al., 2010).

By delaying the enzymatic activity of DNA methyltransferase 1, epigallocatechin-3-gallate (EGCG) from green tea can reawaken methylation-silenced genes (Fang et al., 2003). In *in-vitro* models of colorectal cancer, the Annurca polyphenol extract from the Annurca apple reversed methylation and reactivated the DNA repair mismatch gene hMLH1 (Fini et al., 2007).

HDAC alteration could be of utmost importance for changing tumor behavior (Glozak et al., 2007). Sulforaphane, which is abundant in cruciferous vegetables, is a powerful inducer of phase 2 detoxification enzymes as well as an inhibitor of HDAC. In humans, a single instance of consumption of broccoli sprouts suppressed HDAC activity for several minutes and remained for a long period, but reverted to baseline values within 24 hours. It's unclear how other food components known to change epigenetics would affect these HDAC inhibitors (Dashwood et al., 2007).

Oral, breast, cutaneous, esophageal, colorectal, prostate, pancreatic, and lung cancers are extensively prevented by dietary polyphenols. Minerals and vitamins are also involved in regulatory processes. Zinc, Selenium, and folate engage in the DNA mending process and so have anticancer characteristics.

Multivitamin supplementation stops cancer cells from becoming methylated (Nasir et al., 2020). Undeniably, understanding nutrigenomics holds much potential to reveal those individuals who will benefit the most from dietary interventions besides identifying anyone who might be put in danger as a result of excessive exposures.

8.8 CONCLUSION

Nutrigenomics holds great promise for improving personal and public health through prognostic testing of genes associated with specific classes of nutrients. The tasks facing the development of science-based gene-diet interactions, while challenging, may be conquered by implementing best practices for gene-diet-disease association studies and by fostering strategic international alliances for testing gene-nutrient associations in different ethnic populations.

The concept of personalized nutrition and its application to personalized medicine based on genetic testing requires further scientific research for ascertaining the makeup of accurate genetic tests. Nevertheless, the use of genotype-phenotype-biomarker diagnostics has started and will be of prime importance in the translation of nutrigenomics from the research lab to the consumer market. A fundamental need for converting nutrigenomic tests from academic knowledge to real-world utility will be honing nutritional genetic (nutrigenomics) practitioners capable of deducing the results of genetic tests and associating such information with specific nutrients and diets.

KEYWORDS

- **diabetes mellitus**
- **gene-diet-disease interaction**
- **metabolomics**
- **microbiomics**
- **nutrigenomics**
- **nutrition and research**
- **pharmacogenomics**

REFERENCES

Alinia, S., Hels, O., & Tetens, I., (2009). The potential association between fruit intake and body weight: A review. *Obes. Rev., 10*(6), 639–647.

Amato, R., Pinelli, M., D'Andrea, D., Miele, G., Nicodemi, M., Raiconi, G., & Cocozza, S., (2010). A novel approach to simulate gene-environment interactions in complex diseases. *BMC Bioinformatics, 11*, 8.

Amiri-Dashatan, N., Koushki, M., Abbaszadeh, H. A., Rostami-Nejad, M., & Rezaei-Tavirani, M., (2018). Proteomics applications in health: Biomarker and drug discovery and food industry. *Iran J. Pharm Res. Fall, 17*(4), 1523–1536. PMID: 30568709; PMCID: PMC6269565.

Ashrafian, H., Sounderajah, V., Glen, R., et al., (2021). Metabolomics: The stethoscope for the twenty-first century. *Med. Princ. Pract., 30*, 301–310. doi: 10.1159/000513545.

Bachman, M., Uribe-Lewis, S., Yang, X., et al., (2015). 5-Formylcytosine can be a stable DNA modification in mammals. *Nat. Chem. Biol., 11*(8), 555–557.

Bannister, A. J., & Kouzarides, T., (2011). Regulation of chromatin by histone modifications. *Cell Res., 21*(3), 381–395.

Birney, E., Andrews, T. D., Bevan, P., et al., (2004). An overview of Ensembl. *Genome Res., 14*, 925–928.

Braicu, C., Mehterov, N., Vladimirov, B., Sarafian, V., Mohammad, N. S., Atanasov, A. G., & Berindan-Neagoe, I., (2017). Nutrigenomics in cancer: Revisiting the effects of natural compounds. *Seminars in Cancer Biology, 46*, 84–106. https://doi.org/10.1016/j.semcancer.2017.06.011.

Chattopadhyay, I., (2020). Role of nutrigenetics and nutrigenomics in cancer chemoprevention. In: Kumar, M., Sharma, A., & Kumar, P., (eds.), *Pharmacotherapeutic Botanicals for Cancer Chemoprevention*. Springer, Singapore. https://doi.org/10.1007/978-981-15-5999-0_7.

Chial, H., & Craig, J., (2008). Genome-wide association studies (GWAS) and obesity. *Nature Education, 1*, 80.

Chorev, M., & Carmel, L., (2012). The function of introns. *Front. Gene, 3*, 55. doi: 10.3389/fgene.2012.00055.

Cifuentes, A., (2012). Food analysis: Present, future and foodomics. *ISRN Anal. Chem.*, 1–16.

Corella, D., & Ordovas, J. M., (2005). Single nucleotide polymorphisms that influence lipid metabolism: Interaction with dietary factors. *Annu. Rev. Nutr., 25*, 341–390.

Corella, D., & Ordovas, J. M., (2009). Nutrigenomics in cardiovascular medicine. *Circ. Cardiovasc. Genet., 2*(6), 637–651. doi: 10.1161/CIRCGENETICS.109.891366.

Dashwood, R. H., & Ho, E., (2007). Dietary histone deacetylase inhibitors: From cells to mice to man. *Semin. Cancer Biol., 17*(5), 363–369.

Deram, S., & Villares, S. M., (2009). Genetic variants influencing effectiveness of weight loss strategies. *ARQ Bras Endocrinol Metabol., 53*(2), 129–138.

Doo, M., & Kim, Y., (2015). Obesity: Interactions of genome and nutrients intake. *Prev. Nutr. Food Sci., 20*(1), 1–7. doi: 10.3746/pnf.2015.20.1.1.

Engler, M. B., (2009). Nutrigenomics in cardiovascular disease: Implications for the future. *Prog. Cardiovasc. Nurs., 24*(4), 190–195. doi: 10.1111/j.1751-7117.2009.00058.x.

Fang, M. Z., Wang, Y., Ai, N., Hou, Z., Sun, Y., Lu, H., Welsh, W., & Yang, C. S., (2003). Tea polyphenol (-)-epigallocatechin-3-gallate inhibits DNA methyltransferase and reactivates methylation-silenced genes in cancer cell lines. *Cancer Res., 63*(22), 7563–7570.

Fini, L., Selgrad, M., Fogliano, V., Graziani, G., Romano, M., Hotchkiss, E., Daoud, Y. A., et al., (2007). Annurca apple polyphenols have potent demethylating activity and can reactivate silenced tumor suppressor genes in colorectal cancer cells. *J. Nutr., 137*(12), 2622–2628.

Finucane, M. M., Stevens, G. A., Cowan, M. J., Danaei, G., Lin, J. K., Paciorek, C. J., Singh, G. M., et al., (2011). Global burden of metabolic risk factors of chronic diseases collaborating group (body mass index). National, regional, and global trends in body-mass index since 1980: Systematic analysis of health examination surveys and epidemiological studies with 960 country-years and 9·1 million participants. *Lancet, 377*(9765), 557–567.

Flint, J., Valdar, W., Shifman, S., & Mott, R., (2005). Strategies for mapping and cloning quantitative trait genes in rodents. *Nat. Rev. Genet., 6*, 271–286.

Frazer, K. A., Murray, S. S., Schork, N. J., & Topol, E. J., (2009). Human genetic variation and its contribution to complex traits. *Nat. Rev. Genet., 10*(4), 241–251.

Giskes, K., Van, L. F., Avendano-Pabon, M., & Brug, J., (2011). A systematic review of environmental factors and obesogenic dietary intakes among adults: Are we getting closer to understanding obesogenic environments? *Obes. Rev., 12*(5), e95–e106.

Glozak, M. A., & Seto, E., (2007). Histone deacetylases and cancer. *Oncogene, 26*(37), 5420–5432.

Gresham, D., Dunham, M. J., & Botstein, D., (2008). Comparing whole genomes using DNA microarrays. *Nat Rev Genet., 9*(4), 291–302.

Herbert, A., Gerry, N. P., McQueen, M. B., Heid, I. M., Pfeufer, A., Illig, T., Wichmann, H. E., et al., (2006). A common genetic variant is associated with adult and childhood obesity. *Science, 312*(5771), 279–283.

Hinney, A., Nguyen, T. T., Scherag, A., Friedel, S., Brönner, G., Müller, T. D., Grallert, H., et al., (2007). Genome wide association (GWA) study for early onset extreme obesity supports the role of fat mass and obesity associated gene (FTO) variants. *PLoS One, 2*(12), e1361.

Hinney, A., Vogel, C. I., & Hebebrand, J., (2010). From monogenic to polygenic obesity: Recent advances. *Eur. Child Adolesc. Psychiatry, 19*(3), 297–310.

Jakobsen, M. U., O'Reilly, E. J., Heitmann, B. L., Pereira, M. A., Bälter, K., Fraser, G. E., Goldbourt, U., et al., (2009). Major types of dietary fat and risk of coronary heart disease: A pooled analysis of 11 cohort studies. *Am. J. Clin. Nutr., 89*(5), 1425–1432.

Kaput, J., Noble, J., Hatipoglu, B., Kohrs, K., Dawson, K., & Bartholomew, A., (2007). Application of nutrigenomic concepts to type 2 diabetes mellitus. *Nutrition, Metabolism and Cardiovascular Dis., 17*(2), 89–103. https://doi.org/10.1016/j.numecd.2006.11.006.

Karki, R., Pandya, D., Elston, R. C., & Ferlini, C., (2015). Defining "mutation" and "polymorphism" in the era of personal genomics. *BMC Med. Genomics, 8*, 37. doi: 10.1186/s12920-015-0115-z.

Koro, C. E., Bowlin, S. J., Bourgeois, N., & Fedder, D. O., (2004). Glycemic control from 1988 to 2000 among U.S. adults diagnosed with type 2 diabetes: A preliminary report. *Diabetes Care, 27*, 17–20.

Kronfol, M. M., Dozmorov, M. G., Huang, R., Slattum, P. W., & McClay, J. L., (2017). The role of epigenomics in personalized medicine. *Expert Rev. Precis. Med. Drug Dev., 2*(1), 33–45. doi: 10.1080/23808993.2017.1284557.

Lewis, S. N., Nsoesie, E., Weeks, C., Qiao, D., & Zhang, L., (2011). Prediction of disease and phenotype associations from genome-wide association studies. *PLoS One, 6*(11), e27175.

Lindner, J. R., (2009). Molecular imaging of cardiovascular disease with contrast-enhanced ultrasonography. *Nat. Rev. Cardiol., 6*(7), 475–481.

Little, P. F., (2005). Structure and function of the human genome. *Genome Res., 15*, 1759–1766. doi: 10.1101/gr.4560905.

Loos, R. J., Lindgren, C. M., Li, S., Wheeler, E., Zhao, J. H., Prokopenko, I., Inouye, M., Freathy, R. M., et al., (2008). Common variants near MC4R are associated with fat mass, weight and risk of obesity. *Nat. Genet., 40*(6), 768–775.

McCarthy, M. I., Abecasis, G. R., Cardon, L. R., Goldstein, D. B., Little, J., Ioannidis, J. P., & Hirschhorn, J. N., (2008). Genome-wide association studies for complex traits: Consensus, uncertainty and challenges. *Nat. Rev. Genet., 9*(5), 356–369.

McDonald, B. E., (2004). The Canadian experience: Why Canada decided against an upper limit for cholesterol. *J. Am. Coll. Nutr., 23*(6 Suppl), 616S–620S.

Meyre, D., Delplanque, J., Chèvre, J. C., Lecoeur, C., Lobbens, S., Gallina, S., et al., (2009). Genome-wide association study for early-onset and morbid adult obesity identifies three new risk loci in European populations. *Nat Genet., 41*(2), 157–159.

Mutch, D., Wahli, W., & Williamson, G., (2005). Nutrigenomics and nutrigenetics: The emerging faces of nutrition. *FASEB Journal: Official Publication of the Federation of American Societies for Experimental Biology, 19*, 1602–1616. 10.1096/fj.05-3911rev.

Nasir, A., Bullo, M. M. H., Ahmed, Z., Imtiaz, A., Yaqoob, E., Jadoon, M., Ahmed, H., Afreen, A., & Yaqoob, S., (2020). Nutrigenomics: Epigenetics and cancer prevention: A comprehensive review. *Critical Reviews in Food Science and Nutrition, 60*(8), 1375–1387. doi: 10.1080/10408398.2019.1571480.

Nicastro, H. L., Trujillo, E. B., & Milner, J. A., (2012). Nutrigenomics and cancer prevention. *Curr. Nutr. Rep., 1*(1), 37–43. doi: 10.1007/s13668-011-0007-6.

Nino-Fong, R., Collins, T., & Chan, C., (2007). Nutrigenomics, beta-cell function and type 2 diabetes. *Curr. Genomics, 8*(1), 1–29. doi: 10.2174/138920207780076947.

Ordovas, J. M., Kaput, J., & Corella, D., (2007). Nutrition in the genomics era: Cardiovascular disease risk and the Mediterranean diet. *Mol. Nutr. Food Res., 51*(10), 1293–1299. doi: 10.1002/mnfr.200700041.

Polyak, K., & Meyerson, M., (2003). *Overview: Gene Structure. Holland-Frei Cancer Medicine* (6th edn.). Hamilton (ON): BC Decker. Available from: https://www.ncbi.nlm.nih.gov/books/NBK12983/ (accessed on 05 June 2023).

Quiñones, L., Roco, Á., Cayún, J. P., et al., (2017). Clinical applications of pharmacogenomics. *Rev. Med. Chil., 145*(4), 483–500. doi: 10.4067/S0034-98872017000400009.

Reedy, J., & Krebs-Smith, S. M., (2010). Dietary sources of energy, solid fats, and added sugars among children and adolescents in the United States. *J. Am. Diet Assoc., 110*(10), 1477–1484.

Ruth De, B., (2016). *Krauses' Food the Nutrition Care Process: Clinical – Nutritional Genomics* (14th, edn., pp. 64–84).

Sacks, F. M., Bray, G. A., Carey, V. J., Smith, S. R., Ryan, D. H., Anton, S. D., McManus, K., et al., (2009). Comparison of weight-loss diets with different compositions of fat, protein, and carbohydrates. *N. Engl. J. Med., 360*(9), 859–873.

Sharma, A. M., & Padwal, R., (2010). Obesity is a sign - over-eating is a symptom: An aetiological framework for the assessment and management of obesity. *Obes. Rev., 11*(5), 362–370.

Stevenson, S. E., Chu, Y., Ozias-Akins, P., & Thelen, J. J., (2009). Validation of gel-free, label-free quantitative proteomics approaches: Applications for seed allergen profiling. *J Proteomics, 72*(3), 555–566.

Stidley, C. A., Picchi, M. A., Leng, S., et al., (2010). Multivitamins, folate, and green vegetables protect against gene promoter methylation in the aerodigestive tract of smokers. *Cancer Res., 70*, 568–574.

Thorleifsson, G., Walters, G. B., Gudbjartsson, D. F., Steinthorsdottir, V., Sulem, P., Helgadottir, A., et al., (2009). Genome-wide association yields new sequence variants at seven loci that associate with measures of obesity. *Nat. Genet., 41*(1), 18–24.

Van, H. L., McCoin, M., Kris-Etherton, P. M., Burke, F., Carson, J. A., Champagne, C. M., Karmally, W., & Sikand, G., (2008). The evidence for dietary prevention and treatment of cardiovascular disease. *J. Am. Diet. Assoc., 108*(2), 287–331.

Walker, C. G., Zariwala, M. G., Holness, M. J., & Sugden, M. C., (2007). Diet, obesity and diabetes: A current update. *Clin. Sci. (Lond), 112*(2), 93–111.

Willer, C. J., Speliotes, E. K., Loos, R. J., Li, S., Lindgren, C. M., Heid, I. M., et al., (2009). Six new loci associated with body mass index highlight a neuronal influence on body weight regulation. Genetic investigation of anthropometric traits consortium. *Nat Genet., 41*(1), 25–34.

Wilson, P. W., (2004). Assessing coronary heart disease risk with traditional and novel risk factors. *Clin. Cardiol., 27*(6 Suppl 3), III7–11.

CHAPTER 9

ALTERNATIVE AND THERAPEUTIC FOODS

RASHMI JAISWAL

Applied Nutrition Consultant, Nutritionist in HealthifyMe, Kolkata, West Bengal, India

ABSTRACT

The human body's nutritional needs develop and change as we move through different life stages. To be fit and healthy, it is imperative to take into account the extra demands placed on your body by these changes. As the human lifestyle is changing drastically into more of a sedentary lifestyle, diseases are increasing rapidly. To prevent this, more importance has been given to therapeutic and alternative foods. A therapeutic nutrition edit mitigates diseases like diabetes, cardiovascular diseases (CVDs), and obesity and also provide preventive care for developing disease. Whereas, Alternative food provides a healthier alternative to food that is prepared or produced by conventional methods. Hence Alternative food as a Therapeutic diet bears high potentiality in preventing, treating, and curing many health issues and diseases.

9.1 INTRODUCTION

Food is an important part of human life, and it will not be wrong to say that the human body is made up of food. But consuming food for energy

Micronutrients and Macronutrients as Nutraceuticals. Prakash Chandra Gupta, PhD, Sayan Bhattacharyya, MBBS, MD, Nisha Sharma, PhD, MPharm, Rajesh K. Kesharwani, PhD, Raj K. Keservani, PhD, MPharm (Eds.)
© 2024 Apple Academic Press, Inc. Co-published with CRC Press (Taylor & Francis)

needs, survival, and filling stomach became the utmost priority over having a balanced diet nowadays, because of that deficiency and malnourishment are increasing, and because of it obesity (Keservani et al., 2010a), protein malnourishment, anemia, immunodeficiency, and other chronic health conditions. A therapeutic diet not only ensures optimum nourishment but extra nourishment to prevent treat and cure diseases. Whereas Alternative food gives better and healthier alternative food choices for the betterment of health and our planet.

9.2 THERAPEUTIC DIET

The human body needs a balanced diet for normal functioning in conditions like health and disease. In compromised health conditions there may be altered needs of the diet pattern, with respect to its quality as well as in quantity, a therapeutic diet ensures these altered dietary needs.

The therapeutic diet is routinely prescribed by doctors and planned by clinical dieticians in the hospitals which are difficult to get in homely settings.

Benefits of Therapeutic Diet: The prime and indispensable function of a therapeutic diet is to preserve the pre-disease health condition and tame altogether clinical or apparent and subclinical or impending nutritional deficiencies. As per precise indispensable necessities, the diet is planned based on existing scientific knowledge and technologies to conserve health along with food constituents to which individuals may be allergic and/or atopic. Additional functions of therapeutic diets are to nurture and heal the affected organs responsible for metabolism and excretion. Also, the rectification of the composition of a normal diet helps maintain homeostasis and to adjust metabolize nutrients and excrete harmful substances (www.pedagogyeducation.com, 2013).

Adaptations to Therapeutic Diet: There are various footsteps for the transformation of regular as well as normal diet to the ideal diet required for maintaining different morbidities. Firstly, the count of energy requirements based on age, gender, and other factors is derived. Non-starch polysaccharides, like roughages and fibers hence become important components along with specific nutrients considering the history of intolerance to a specific food type. Further, depending on the pathophysiology of the oropharyngeal anatomical region and esophagus, the nutrient composition and preparation method of the diet have to be altered to fit in the category of solid, semisolid, soft, semi-liquid or liquid diet needed for certain periods of illnesses. To go back to a normal diet pattern and/or normal lifestyle again, the same steps back to the regular dietary practices of the person have to be taken (www.iaps.in;www.empowerhealth.ca). Special attention is hence needed to make all therapeutic diets easily digestible and palatable with respect to various options, choices, and exchanges to modify based on the necessity of medical conditions as an addition to further medical interventions.

Common reasons why therapeutic diets may be needed:

- To maintain the nutritional status;
- To restore the deranged nutritional status of a person;
- To correct the nutritional state;
- To decrease calorie intake for weight control;
- To supplement add-on calories for weight gain;
- To balance the amounts of carbohydrates (mainly simple sugars), fat, and protein for controlling diabetes;
- To provide a greater amount of a body-building nutrient like protein;
- To decrease the amount of a particular nutrient like sodium;
- To exclude foods owing to allergies or food intolerance;
- To provide texture modifications due to any problem with chewing and/or swallowing.

Common therapeutic diets include:

1. **Nutrient Modifications:**
   - No concentrated sweets diet;
   - Diabetic diets;
   - No added salt diet;
   - Low sodium diet;
   - Low-fat diet and/or a diet low in cholesterol;
   - High fiber diet;
   - Renal diet.
2. **Texture Modification:**
   - Mechanical soft diet;
   - Puree diet.
3. **Food Allergy or Food Intolerance Modification:**
   - Food allergy;
   - Food intolerance.
4. **Tube Feedings:**
   - Liquid tube feedings instead of meals;
   - Liquid tube feedings along with meals.

5. **Additional Feedings:** In addition to food, extra nutrition may be required as:
   - Supplements – usually has as liquid nutritional shakes once, twice, or three times per day; taken either with meals or between meals.
   - Nourishments – consumed as a snack food or beverage items to be taken between meals mid-morning and/or mid-afternoon.
   - HS (hour of sleep) snack – ordered as a snack food or beverage items to be administered at the hour of sleep or before sleep.

The following list includes brief descriptions of common therapeutic diets:

1. **Clear Liquid Diet:**
   - Includes minimum residue fluids that can be seen through or transparent.
   - Examples are juices with no pulp, broth, and Jell-O.
   - Is often used as the initial step to restart oral feeding after surgery or an abdominal procedure.
   - Can also be used for fluid and electrolyte replacement in people with severe diarrhea.
   - Should not be used for an extended period of time since it does not give enough calories and nutrients.
2. **Full Liquid Diet:**
   - Includes fluids that are creamy.
   - Some examples of food allowed are ice cream, pudding, thinned hot cereal, custard, strained cream soups, and juices with the pulp.
   - Used as the second step to restarting oral feeding once clear liquids are tolerated.
   - Used for people who cannot tolerate a mechanical soft diet.
   - Should not be used for long periods.
3. **Pureed Diet:**
   - Changes the regular diet by converting it to a puree or smooth liquid consistency.
   - Indicated for those having wired jaws, extremely poor dentition in which chewing is inadequate.
   - It is often thinned down so it can pass through a straw.
   - Is for people with chewing or swallowing difficulties or with the condition of dysphasia.
   - Foods should be pureed separately.

- Avoid nuts, seeds, raw vegetables, and raw fruits.
- Is nutritionally adequate when offering all food groups.

4. **Mechanically Altered or Soft Diet:**
   - Is used when there are problems with mastication and swallowing food.
   - Changes the consistency of the regular diet to a softer texture that is easily palatable.
   - Includes finely chopped or ground meats as well as chopped or ground raw fruits and vegetables.
   - Is for people with poor oral conditions, missing teeth, no teeth, or a condition called dysphasia.
5. **High Fiber Diet:**
   - It is prescribed for the prevention or treating a number of gastrointestinal diseases (like atonic constipation), cardiovascular diseases (CVDs), diabetes, obesity, cholecystitis, cholelithiasis, and metabolic diseases.
   - Increased fiber (both soluble and insoluble) should come from a variety of sources like fruits, legumes, vegetables, whole breads, and cereals.
   - Foods included – all long-fiber vegetables such as green vegetables, amaranth, cabbage, lettuce, celery spinach, etc., have high fiber content. Special stress should be given to raw vegetable salads.

     All fruits, with skins when soft and tender, whole grain cereals and bread with bran in the fine division also have fibers. Milk, meat, fish, fowl, and eggs can be taken as desired for normal nutrition but are low in fiber.
   - Foods to avoid – highly refined and concentrated foods, fried foods, excessive amounts of coarse bran, excessive spices, and seasonings have no fiber and should be avoided.
   - Intervals of feeding – three meals daily, water, and fruit juices between meals, before breakfast, and before retiring.
6. **Very Low Residue Diet:** It is used to treat:
   - Severe diarrhea, *Ulcerative colitis* during initial stages of treatment.
   - Preceding and following operations on the colon or rectum when no movement is desired for several days.
   - Partial intestinal obstruction.

- Foods included – tender meat, fish or fowl, clear fat-free soups, fresh fruit juices, gelatin puddings, eggs: arrowroot cookies, refined cereals and bread, carbonated drinks, coffee, tea, butter, and sugar.
- Foods to avoid – coarse bread and cereals; cheese, milk, rich desserts, excessive fat, fried foods, fruits (except juices) and vegetables; tough meats, condiments, excessive sweets.
- Intervals of feeding – usually three meals daily.

7. **High-Calorie Diet:**
   - It is mainly used when it is needed to gain lost weight;
   - This diet is used in many conditions but particularly in such diseases as hyperthyroidism which has a high metabolic rate, convalescence from fevers, and under-nutrition in general.
   - Food included – a regular diet with added complex sugar, cream, butter, eggs, etc., to increase the caloric intake.
   - Foods to avoid – fried or oily foods or foods having a tendency to interfere with the appetite. Excessive quantities of bulky or heavy, low calories foods.
   - Intervals of feedings: Five to six meals daily.
8. **Low Fat/Low Cholesterol Diet:**
   - Is used to reduce fat levels and correct lipid profile and/or treat medical conditions that interfere with how the body uses fat, such as diseases of the liver, gallbladder, or pancreas.
   - Restricts fat to 50 grams or no more than 30% calories derived from fat.
   - It is low in total fat and saturated fats and contains approximately 250–300 mg of cholesterol and is rich in unsaturated fats.
9. **Renal Diet:**
   - Is for renal/kidney patients.
   - The diet plan is individualized, depending on if the person is on dialysis.
   - The diet restricts high sodium, potassium, fluid, and proteins to specified levels.
   - Lab reports are monitored closely.
10. **Low Purine Diet:**
    - It is mainly used for Gout, a condition where uric acid (UA) levels elevate in the body and UA crystal deposition starts taking place in joints and kidneys.

- Foods included: Milk, eggs, cereals and bread, potato vegetables, fruits, and moderate-butter.
- Food to avoid: Meat extracts, meats, especially organ meats, fish sardines, and alcoholic beverages. Intervals of feeding: Three meals daily.

9.3 FOOD ALLERGY MODIFICATION

- Food allergies are due to an abnormal immune response to an otherwise harmless food.
- Foods that initiate allergies should be strictly eliminated from the diet.
- Appropriate substitutions are made to ensure the meal is nutritionally adequate.
- The common food allergens are milk, egg, soy, wheat, peanuts, tree nuts, fish, and shellfish.
- A gluten-free diet means the elimination of wheat, rye, and barley. These are replaced with potato, corn, and rice products.

9.4 FOOD INTOLERANCE MODIFICATION

- The most common type of food intolerance is lactose intolerance (milk sugar) because of a decreased amount of the enzyme Lactase in the body.
- Other common types of food intolerance are adverse reactions to some products added to food to enhance taste, and color, or to protect against bacterial growth. Common symptoms involving food intolerances are vomiting, diarrhea, abdominal pain, and headaches.

1. **Acid-Ash Diet:**
    - It is mainly used to adjust the reaction of urine so that salts are held in solution. Kidney stones-calcium and magnesium phosphates, carbonates, and oxalates.
    - Foods included: Large servings of meat, fish, poultry, eggs, whole-grain cereals, bread, cake, rice, macaroni, noodles, corn, and plums.
    - Limit: Milk, fruits, vegetables.
    - Neutral foods as desired: butter sugar, plain candies, tapioca, oils, and fat, coffee, tea.

2. **Alkaline-Ash Diet:**
   - This is mainly used to adjust the reaction of the urine so that salts are held in solution. Kidney UA and cystine calculi.
   - Foods to avoid: Legumes, spinach, greens, dried fruits, salty foods such as canned fish and canned vegetables, and salted nuts.
   - Intervals of feeding: Three meals daily.
   - Foods included: Large amounts of fruits, vegetables, and milk.
   - Limited amount: Meat, 1 egg 1 serving, cereal, 1–2 slices of bread.
   - As desired: Butter, sugar, oils, and fats.
   - Foods to avoid or limit: Meat, fish, fowl, cheese, breakfast cereals, macaroni, rice, cakes, cookies.
   - Intervals of feeding: Three meals daily.
3. **Tube Feedings:**
   - Tube feedings are used mainly for people who cannot take adequate food or fluids by mouth.
   - All or a certain part of nutritional requirements are met through tube feedings.
   - Some people may consume food orally if they can swallow safely and are working to be weaned off tube feeding (www.cdss.ca.gov).

9.4.1 DIET IN SPECIFIC CONDITIONS USUALLY ENCOUNTERED IN HEALTHCARE

9.4.1.1 ADULT UNDER-NUTRITION

Undernutrition takes place in adults with chronic health conditions with and without the presence of a sufficiently healthy and balanced diet. In this situation, we need:

- Healthier food choices that are high in energy and high protein content.
- Initially light properly cooked food should be taken.
- Sufficient fluid intake should be maintained.
- Micronutrient supplementation particularly vitamins A and E, iron, and folic acid, zinc, calcium should be included.

- Include fresh pulpy and juicy fruits and green leafy vegetables.
- Meat, eggs, fish, kidney, and liver, fish liver oils (for non-vegetarians), second-generation milk products, and soybean (for vegetarians).
- Whole grains to provide adequate roughages (www.who.int).

To formulate this diet the calorie requirement of a person has to be assessed depending on age, gender, and other morphological parameters and a high-caloric diet has to be provided (www.webmd.com). A high-protein diet is usually recommended as protein deficiency is common with adult undernutrition. A protein-rich food is added to the carbohydrate-rich contents of a diet that usually triggers satisfaction and fulfillment (www.healthline.com).

9.4.1.2 LOW SODIUM AND/OR LOW SALT DIET

A low-sodium diet is mainly prescribed for generalized edema, hypertension, heart failure, and in other medical conditions. Daily intake of sodium is usually restricted to a maximum of 1,500 to 2,300 mgs in these situations with proper assessments and reassessments at different intervals (clevelandclinic.org).

9.4.1.3 LOW PROTEIN DIET

In compromised health situations for example liver and kidney diseases and with protein and other metabolic disorders consumption of protein is usually restricted to 1 to 1.5 grams/kg of body weight with evaluations at intervals (www.healthline.com).

9.4.1.4 BLAND DIET

In gastrointestinal diseases with epigastric discomfort, nausea, vomiting, acute diarrheal diseases, peptic ulcer diseases, and other conditions, a bland diet is usually suggested which is soft in texture, low in fiber, with mild seasoning. Though named a bland diet, taste, and calorie contents are not compromised (www.healthline.com).

9.4.1.5 LOW RESIDUE DIET

In postoperative cases like recent bowel surgery viz. Ileostomy, colostomy, and bowel resections, or even preparing for bowel surgeries, Crohn's or diverticular disease, and other related conditions, this low fiber diet helps. The researchers feel that this temporary diet plan intake may reduce stool amount in the large intestine, with the assumption for rest to the bowel and dietary fibers should not be more than 10 to 15 grams per day (www.webmd.com).

9.5 ALTERNATIVE FOODS IN DIABETES MELLITUS

The nutritional goal of managing diabetes is to prevent microvascular and macrovascular complications, this entails participatory learning of the patients concerning self-help lifestyle modifications, including dietary options, choices of portions, and exchanges in the planning of the diet. Dual downstream complications of dysglycemia and dyslipidemia are managed with the maintenance of overall dietary condition as per age and gender with culture-specific, religion sensitive, dietary advice. Exchange and portion list needed to respect ethno-geographical variations of availability of groups of macronutrients, having an equivalent amount of carbohydrate, protein, fat, and calories that can be easily exchanged with another in the exchange list. Healthcare providers have to orient their mindset to assemble data on the socio-economic condition of the diabetics so that they can follow the dietary advice for lifelong and can 'live with the disease' with minimum complications (Garg et al., 2018; Ghosh et al., 2018; Singh et al., 2017; Pal et al., 2010).

9.6 WILSON'S DISEASE

In Wilson's disease, dietary restriction mainly is not enough to control copper accumulation in the eyes, brain, kidneys, and liver, with local impediments in specific organs including cirrhosis of the liver. Key interventions are pharmacotherapy, to help excrete excess copper in urine and dietary zinc supplementation to reduce absorption of copper from the gut. Researchers opined that with optimum medication adherence, restriction of copper-rich food groups (except shellfish and liver) may not reduce further complications (Russel et al., 2018).

9.7 COELIAC DISEASE

In celiac disease (CD) with permanent gluten intolerance, patients must stick to a lifelong avoidance of gluten-containing foods and foods which are naturally gluten-free, but at risk of contamination. This dietary approach is all-inclusive to avoid exaggeration of symptoms and to reduce short-term and long-term complications. Firstly, excluding gluten limits food choices and exchanges with unbalanced intake of macro- and micronutrients which may lead to clinical or subclinical deficiencies. A patient-tailored dietetic protocol with strict adherence to the qualitative and quantitative composition of diets, based on rich information on gluten-containing foods is needed. Secondly, a gluten-free diet may not result in complete remission of mucosal damage caused earlier or the resolution of symptoms and long trouble-free life. Thirdly, In comorbidity of irritable bowel syndrome (IBS) with CD with comparable symptoms, a new therapeutic approach is suggested, that is avoidance of both gluten/wheat and fermentable oligosaccharides, disaccharides, monosaccharides, and polyols-short chain carbohydrates and sugar alcohol (FODMAPs) (Taus et al., 2016).

9.8 STANDARD OFFICE PROCEDURE INSTEAD OF 'LIP' SERVICES

Healthcare providers need to change themselves first to break the ice of fixed ideas and accept their limited knowledge about the art and science of diet and nutrition based on limited translational research on nutrition. That is the only way we can make sure that all healthcare professionals are to be consulted with respect to the diet plan and implementation in any case of morbidity. For this we need capacity building on diet and nutrition in undergraduate and postgraduate medical courses with the importance of therapeutic diet; special attention is needed for the family medicine specialty which is striving to ace primary care at the grassroots (Pal et al., 2014, 2016).

Further, our national policy should have been oriented to make people aware of the value of foods for comprehensive health care from womb to tomb in preventive, promotive, and curative health care services (Sharma et al., 2018).

We have to start from the likes-dislikes of individual, and family practices, religious beliefs, and other constraints by identifying of needs of the patients and extensive exchange of opinion from the community to reinforce the importance of diet and nourishment as an important component of healing. This will help implementation to overcome the hurdles in method and time of meal plan with the patient and the patient's family, encourage them to depend on themselves to eat, and provide adaptive equipment and the right atmosphere. Clinical practice guidelines have to be introduced with a comprehensive dedicated diet based on age, gender, morbidity, resource allocation, and dedicated interventions updated every day by the care-seekers to revert them with tailor-made plans and solutions (Pal, 2017).

We need to exploit our healthcare delivery system for a wider spectrum of our knowledge to the rural and urban poor people (Saxena et al., 2015).

Despite important developments in healthcare, the quality and quantity of translational research on diet, and nutrition is low on short, medium, and long-term consequences of therapeutic diets.

To sum up, we have to shift from a conventional 'ready-made diet chart approach' to a 'tailor-made approach.' (Pal & Rustagi, 2015; Pal et al., 2015).

9.9 PREOPERATIVE DIET

Patients who have lost much weight prior to surgery benefit considerably by ingesting a high protein, high-calorie diet for even a week or two prior to surgery. The diet may be of liquid or soft or even regular consistency, depending upon the nature of the pathologic condition. Parenteral nutrition or semi-synthetic fiber-free diets are sometimes used. In addition, the maintenance of metabolic equilibrium as in diabetes or other diseases must not be overlooked. Foods that provide a maximum amount of nutrients in a minimum volume are essential Small feedings at Frequent intervals are likely to be better accepted than large meals which cannot be fully consumed: For additional protein, milk beverages may be fortified with non-fat dry milk or commercial protein supplements. Fruit juices fortified with glucose or high carbohydrate food, increase carbohydrate intake and facilitate storage of glycogen. Butter incorporated into foods and light cream mixed with amounts of milk are also useful for increasing the calorie intake one should remember that the excessive use of sugars and fats may cause nausea. Food and fluids are generally allowed until midnight just preceding the day of the operation although a light breakfast may be given when the operation is scheduled for the afternoon and local anesthesia is to be used. It is essential that the stomach is kept empty prior to administering the anesthesia so as to reduce the incidence of vomiting and subsequent danger of aspiration of vomitus. When an operation is to be performed on the gastrointestinal tract

(GIT), a diet very low in residue may be given 2 to 3 days prior to the operation. In acute abdominal conditions such as appendicitis and cholecystitis, no food is allowed by mouth until nausea, vomiting, pain, and distension have passed in order to prevent the danger of peritonitis. Preoperative education includes nicotine and alcohol avoidance and nutrition support for malnourished patients. These are effective in reducing complications. Avoidance of overzealous and unnecessary preoperative starvation and preoperative carbohydrate loading is considered useful.

Reducing calorie deficit and promoting nitrogen retention and judicious use of immunonutrients, antioxidants (Keservani et al., 2015), vitamins (Singh et al., 2017), minerals (Keservani et al., 2010a) and trace elements (Keservani et al., 2010b) are to be considered in the preoperative diet.

9.10 POSTOPERATIVE DIET

In the postoperative patient, starvation leads to increased morbidity in the nutritionally depleted and stressed patient. Nutritional support in depleted patients improves clinical outcomes and reduces the length of hospital stay.

Following minor surgery, liquids are often tolerated within a few hours and rapid progression to a normal diet is made. After major surgery, oral intake may be delayed for days. Fruits like papaya and sapota are given postoperatively to initiate a bowel movement. Complete nutritional support is provided by conventional intravenous feedings, catheter jejunostomy, total parenteral nutrition (TPN), tube feedings, or semisynthetic fiber-free diets.

Medical and surgical ICU patients are insufficiently fed during their ICU stay when compared with their RDA. Feeding practices differ significantly in medical surg ICU patients. Lipid-based sedatives and dextrose-containing intravenous fluids are used more frequently in surgical populations. Until the evidence is clear regarding optimal nutrition support for severely ill ICU patients; a thorough review of all energy sources should be routinely monitored and energy delivery to be limited to 80% of the target (Srilakshmi B).

9.11 IMMUNONUTRITION

Immunonutrition refers to the addition of specific nutrients in greater than normal concentrations to modulate immune function, and thus improve clinical function. The immunonutrients comprise the following:

1. **Arginine:** This conditionally essential amino acid plays a pivotal role in maintaining body homeostasis and is prescribed when stress occurs. Arginine is a source of nitrogen in nitric oxide. It stimulates the secretion of anabolic hormones and enhances wound healing and cellular immunity. Arginine supplementation may be beneficial in patients with sepsis.
2. **Purine Nucleotides:** These are precursors for RNA and DNA. Deficiency of purine nucleotides is associated with immune suppression and supplementation improves cell-mediated immunity.
3. **Omega-3 Fatty Acids:** These have an anti-inflammatory role to play. They are especially useful in reducing the severity of sepsis. Omega-3 fatty acids are found in fish oils.
4. **Glutamine:** A nonessential amino acid present in the body. This becomes conditionally essential in the presence of severe stress and multiple organ failure. It is a major oxidative substrate and enhances the immune function essential in stressed states. It is a substrate for glutathione (GSH) synthesis. Glutamine supplementation can prevent or ameliorate the gastrointestinal mucosal atrophy seen in prolonged states of parenteral nutrition. Glutamine also helps the gastrointestinal mucosa heal more promptly after damage by either radiotherapy or chemotherapy.
5. **Whey Proteins:** It has been shown to stimulate cell-mediated and humoral immunity to improve the body's nutritional status in stressed individuals. Whey proteins have an antioxidant role by increasing tissue GSH and inhibiting the growth of several types of tumors. Commercial formulations containing immuno-nutrition are now available.

9.12 NUTRITION SUPPORT IN BURN PATIENTS

The aim is to maintain or attain nutritional status in patients will burn injury by providing nutritional support enterally and/or parenterally. Nutrition support promotes wound healing prevents infection and prevents loss of more than 10% of the patients' pre burn weight.

Metabolic changes:

- Protein lost depends upon the extent and depth of the burn. There is an increase in nitrogen excretion due to open wounds.

- The sooner a burn wound is closed the more rapidly positive nitrogen balance is reached. If an infection develops, the nitrogen losses are even greater.
- Fluids and electrolytes (K, Na, and CI) are lost through exudation. Anemia may be an additional problem. First seven days, the patient is hypercatabolic and will not retain much of the protein given.
- Six to eight weeks after the burn there is a high rate of metabolism.
- Resting energy expenditure increases by 50–125% due to increased percent release of catecholamine which is directly related to the extent of burn.
- First, adequate fluid and electrolytes are provided to maintain circulating volume and to prevent renal failure. Weight gain from the fluid is common during this stage and is generally mobilized slowly over the next two weeks. Therefore, pre-burn weight should be used in estimating kilocalories and protein needs. Serum albumin drops with post-burn even if normal on admission due to increased loss of albumin as plasma leaves from micro-vasculature to the area of damage Patient is given oral supplements. Tube feeding is resorted if the patient is unable to consume enough orally. Parenteral nutrition is the least used route as there is a high risk of catheter infection as burn wounds are likely to be colonized with microorganisms.

➢ **Energy Requirements (Major burns):**
  i. **Adults:** 25x body weight (kg) + 40x% TBSA burned (total body surface area).
  ii. **Paediatric Group:**
    - <1 year = RDA + 15x% TBSA burned;
    - 1–3 years = RDA + 25x% TBSA burned;
    - 4–15 years RDA +40x% TBSA burned.

➢ **Protein Requirement:**
  i. **Adults:**
  1.3–3.0 g/kg ideal body weight;
  Or, 20–25% of total energy.
  ii. **Children:**
    - 0–2 years-2–3 g/kg/d.
    - 3–13 years-1.5–2 g/kg/d.
    - 13–18 years 15 kg/day.

➢ **Non-Protein Energy:**
  - Fat = 15–20%
  - Carbohydrate= 50–60%

- **Micronutrients:**
  - Multivitamin = 2 x RDA;
  - Zinc = 1 x RDA.

Studies show that there is a diminished capacity to produce vitamin D among children with severe burns. Hence such children should receive vitamin D supplements. Even skin that is not burned cannot make enough vitamin D in burned children. After being burned, children lose bone mass. To prevent this, vitamin D supplements should become part of the burned children's therapy during their treatment and after they are discharged from the hospital. Indian Society for Parenteral and Enteral Nutrition (ISPEN) is an association that is involved in nutrition support services to help reduce morbidity and mortality among patients.

9.13 ALTERNATIVE FOODS

Plant-based materials like soy, wheat, peas, etc., and plant-based alternatives to animal-based food products, like 'milk' and meat, have a long history in many cultures. In recent years, their popularity has increased as consumers have begun to adopt them through personal choice. For example, vegetarians and vegans are now adopting a vegetable-based diet. Advances in science and technology imply that these plant-based foods are now more closely able to mimic animal products. Organic and vegetarian products, sometimes called alternative foods, and regarded as a healthier alternative to food that is prepared or produced by conventional methods.

9.13.1 ALTERNATIVE NUTRITION

- **What is Alternative Nutrition and Hydration?**
  If a patient is not able to swallow, there is a special treatment to help. It is called alternative nutrition (food) and hydration (fluids). It may also be called artificial nutrition. This life support treatment administers nutrition and fluid to a patient who is unable to eat or drink. This type of treatment needs to be administered by a medically trained person. The healthcare team decides how much of each one needs to be given.

- **How are Alternative Nutrition and Hydration Given?**
  Alternative nutrition and hydration are given in any one of these three ways: 1. Through a tube to the stomach or intestines (put in place by surgery) 2. Through a tube that goes through the nose to the stomach 3. Intravenously (into a vein) to the bloodstream (also called PPN or TPN) (www.cityofhope.org).

Sometimes types of diets, which differ from the nutritional habits or social beliefs of the majority of society and conventional recommendations from dietitians, usually based on the restriction of certain food groups, most often food of animal origin are also considered under Alternative Nutrition.

9.13.1.1 REASON TO ADOPT ALTERNATIVE NUTRITION

- Health reasons – the need for a change of lifestyle-obesity, gout, dyslipidemia, zoonosis;
- Moral and ethical reasons – compassion for animals;
- Ecological aspect;
- Economic reasons;
- Religion;
- Social factors – peer pressure, fashion;
- Taste preference.

9.13.1.2 TYPES OF ALTERNATIVE NUTRITION

- Vegetarianism;
- Mediterranean diet;
- Ketogenic diet;
- Atkins diet;
- Macrobiotic diet;
- Marginal types:
  - Divided food diet;
  - Diet by blood group (Adamo's diet);
  - Paleo diet (paleolithic diet);
  - Diet based on pH (alkaline diet);
  - Detox diet.
- Organic food.

9.13.2 VEGETARIANISM

- The most widespread alternative way of eating in Czechia (about 2% of the population – 2,00,000 people);
- Vegetarian – generally not consuming meat;
- Types depending on restricted consumption of food of animal origin:
    - o Semivegetarians (pulo-, pesco-) or flexitarians;
    - o Lactoovovegetarians;
    - o Lactovegetarians;
    - o Vegans;
    - o Fruitarians;
    - o Vitarians – RAW food.

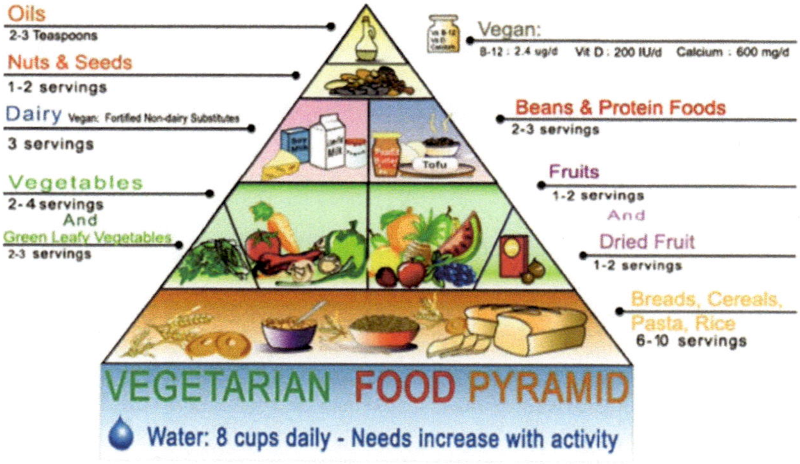

Source: 2002 Department of Nutrition, Arizona State University Art by Nick Rickert

9.13.2.1 THE BENEFITS OF VEGETARIANISM

- More frequent consumption of fruits, vegetables, cereals, sprouts, legumes, nuts, seeds, and vegetable oils;
- Lower intake of saturated fat, cholesterol, and animal proteins;
- Lower energy intake;
- Lifestyle;
- Diet based on the food of plant origin predominant consumption may reduce the risk of developing certain diseases.

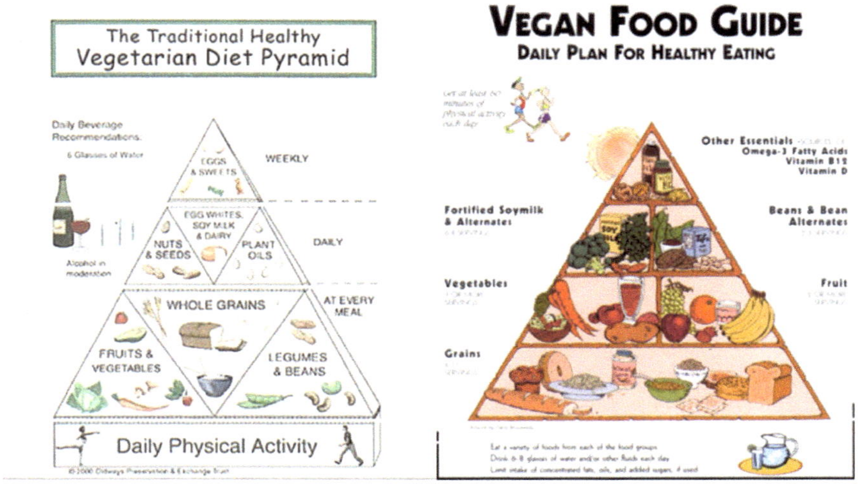

Image credit: www.oldwayspt.org and Dave Brousseau

It should be stressed, that the risk of the disease affected by dietary intake is generally low if the person has a balanced diet, regardless of whether meat is consumed or not.

9.13.2.2 THE RISKS OF VEGETARIANISM

- Incorrect food composition (poor awareness).
- Some nutrients may be missing or in reduced quantity in a plant-based diet.
- There are certain substances in the diet from plants that may reduce the absorption of certain vitamins and minerals.
- Others – inadequate intake of certain nutrients and the risks involved (www.medicalnews.com).

9.14 RULES FOR A BALANCED DIET

- One should maintain variety and balance in the diet;
- Food groups that should not be missing in a person's diet are:
  - Cereals and pseudocereals like millets;
  - Vegetables and fruits;

- o Legumes and products thereof;
- o Milk, dairy products;
- o Eggs;
- o Nuts, seeds;
- o Vegetable oils;
- o Seaweed.

A balanced meal should consist of various sources of proteins during different meals of the day.

- Cereals with legumes (bean stew with bread, leguminous soup with cereal, cholent, etc.);
- Legumes with potatoes;
- Plant protein sources with animal (cereal with dairy products – milk cereal slurry, etc.);
- Do not replace meat dishes with sweet and fried dishes.

➢ **Examples of Daily Meal Plan:**

- Breakfast – bread with egg and an apple.
- Mid-morning snack – papaya.
- Lunch – rice with rajma curry, spinach, and curd with flax seeds.
- Evening snacks – roasted peanuts.
- Dinner – wheat chapatti with brinjal curry and poached egg.

9.14.1 RAW FOOD DIET

- Vitarianism is another name for the consumption of raw food.
- Eating food without heat treatment, in the most natural form.
- Heat treatment of food at max. 42–45°C (107.6–113°F).
- According to the supporters of this type of diet, there is no loss in the nutritional value of the diet, destruction by enzymes, loss of vitamins, etc.
- The main ingredients of this diet are raw fruits and vegetables, raw nuts and seeds, legumes, various cereals, sprouts, and seaweeds, and it is predominantly supported by vegans.
- There are also some subgroups of raw food diet that allow the consumption of raw meat, eggs, and unpasteurized milk.
- Types of food processing – dipping, mixing, drying, juicing.

➢ **Example of Meals:**

- Banana oatmeal with walnuts;
- Miso soup with mushrooms;
- Cucumber soup with basil;
- Butter pumpkin noodles with sage sauce;
- Vegetable rolls;
- Guacamole with curry and vegetables;
- Chia pudink;
- Juices, cocktails, smoothies.

Image credit: vegan-raw-diet.com

9.14.1.1 THE BENEFITS AND RISKS OF RAW FOOD DIET

The benefits of raw food diet are:

- There is no reduction in nutritional value by heat treatment;
- Lower intake of fat, sodium, and cholesterol;
- Higher intake of fiber (sometimes it can be too much), potassium, and other nutrients.

The risks of raw food diet are:

- Lack of some essential nutritional substances, and energy (except for raw desserts);
- There is a risk of deficiency in the long-term of Vitamins B12, protein, iron, calcium, and omega-3 fatty acids that can be corrected by the use of dietary supplements;
- Poor digestibility of raw food;
- Lower absorption;
- Microbial infection risk (www.medicalnews.com).

9.14.2 MEDITERRANEAN DIET

- The inhabitants around the Mediterranean Sea are vaunted for their long life spans and low rates of cancer and cardiovascular ailments, the diet they follow is known as the Mediterranean diet.
- Their diet is generally low in red meat, sugar, and saturated fats. Instead, they consume a lot of fruits, vegetables, whole grains, beans, nuts, legumes, olive oil, and flavorful natural herbs and spices.
- Seafood is preferred over other types of meat.
- Moderate amounts of eggs, cheese, and yogurt are encouraged.
- Sweets and red meat are reserved for special occasions, but red wine and exercise are daily components (www.medicalnews.com).

Source: Harvard food pyramid for the Mediterranean Diet.

9.14.3 KETOGENIC DIET

- The ketogenic diet has been used for a long time for the treatment of epilepsy and is also being explored for other uses in healthcare.
- It involves reducing carbohydrate intake and upping fat intake. It sounds contrary to common sense, but it allows the body to burn fat as a fuel, rather than carbohydrates.
- Healthy fats (unsaturated fats), like those in avocados, coconuts, Brazil nuts, seeds, oily fish, and olive oil are liberally added to the diet, to maintain emphasis on fat.
- The diet causes the breakdown of fat deposits, for obtaining fue, and creates substances called ketones through a process called ketosis.
- This diet has risks of inducing ketoacidosis (formation of ketone body) in people with Type 1 Diabetes and may result in diabetic coma and death. Although most studies are 2 years or less, there is some promising research with respect to diabetes management, metabolic health, weight loss, and body composition change (www.medicalnews.com).

9.14.4 ATKINS DIET

- The Atkins diet (low carbohydrate diet), also called Atkins nutritional approach, focuses on controlling the levels of insulin in the body by a low-carbohydrate diet.
- If people eat large amounts of refined carbohydrates, their insulin levels rise and fall quickly. The rising insulin levels stimulate the body to store energy from the food that is consumed, making it less likely that the body will use stored fat as a source of energy.

Hence, people on the Atkins diet avoid carbohydrates but can eat as much protein and fat as they like.

- Although popular for some time only, the Atkins Diet has certain risks. Individuals considering adopting the Atkins Diet should take Dietitian's guidance (www.medicalnews.com).

9.14.5 MACROBIOTIC DIET

- According to macrobiotics, the food differs from the aspect of YIN and YANG;

- The yin/yang ratio gives the "value" of each food;
- Harmonic food means yin and yang are in balance;
- The most energy-balanced diet is cereals;
- Ohsawa grouped nutrition into 10° (–3 to +7).

Diet No.	Cereals	Vegetable Intake	Soup	Animal Food	Salad Fruits	Dessert	Beverages
7	100%	-	-	-	-	-	(Only when thirsty)
6	90%	10%	-	-	-	-	-
5	80%	20%	-	-	-	-	-
4	70%	20%	10%	-	-	-	-
3	60%	30%	10%	-	-	-	-
2	50%	30%	10%	10%	-	-	-
1	40%	30%	10%	20%	-	-	-
-1	30%	30%	10%	20%	10%	-	-
-2	20%	30%	10%	25%	10%	5%	-
-3	10%	30%	10%	30%	15%	5%	-

- The basic rule of macrobiotic nutrition is to consume the food slowly and to chew it very well.
- The theory of so-called biological transmutation.
- Emphasis is also put on:
  - Natural food from local sources or at least from the same climatic zone that corresponds to the season;
  - Avoid eating meat, milk, dairy products, tropical fruit, refined sugar, eggs, potatoes, and white flour;
  - Rejection of consumption of preserved, chemically treated, artificially colored, frozen, or irradiated food;
  - Denial of drug use, alcohol, and drug consumption;
  - Modesty, moderation, positive attitude.

9.14.5.1 MACROBIOTIC DIET BY MICHIO KUSHI

- Cereals, vegetables, legumes, and fermented foods make up the majority of its diet.
- The smaller ratio is represented by seaweed, seeds, nuts, temperate berries, and fish.
- Macrobiotic plate.
- Important policy – eat only in case of hunger:

- Thoroughly chew (50 times or more one bite) – eat seated, calmed, and relaxed – eat regularly 2–3 times a day, as much as needed, leaving satisfied, but not full.
- Drink liquids sparingly, only when thirsty.
- Last meal of the day should be at least 3 hours before sleeping.
- Around 50–60% Wholegrain cereals and products of wholegrain flour:
  - Daily – brown rice, barley, millet, wheat, oats, rye, corn, buckwheat;
  - Occasionally wholegrain pasta, wholegrain sliced bread, wholegrain rye bread, bulgur, couscous, and polenta.
- Around 25–30% vegetables – in each meal.
- 2/3 Cooked, fried on a small amount of vegetable oil, 1/3 raw salads, pickles (fermented):
  - Daily – cabbage, Chinese cabbage, leek, broccoli, cauliflower, carrots, onions, pumpkins, radishes, kohlrabi, cress, parsley, dandelion;
  - 2–3 times a week.
- Cucumbers, celery, mushrooms, herbs:
  - Not suitable – potatoes, spinach, red beet, peppers, tomatoes.
- Around 5%–10% Legume (not more than once a day), seaweed.
- Around 5%–10% (1–2 cups a day) soup.
- Occasionally:
  - 2–3 times per week: fish, fresh or dried seasonal fruit;
  - 1–2 cups per week: seeds (sesame, pumpkin, sunflower) and nuts (almonds, walnuts, pecans, chestnuts, peanuts, coconut);
  - Sweetening – rice syrup, barley malt;
  - Oil, preferably unrefined, corn or sesame.
- Types of drinks: clean water, some types of tea;
- Avoid:
  - Meat, animal fat, eggs, milk, and dairy products;
  - Refined sugar, honey, molasses, chocolate, simple sugars and sweetened foods, tropical fruits;
  - Artificial drinks, aromatic, and stimulating teas;
  - Technology processed grains, white flour, and products of them;
  - Hot spices, aromatic stimulating foods, artificial vinegar, hard alcohol;
- All artificially colored, preserved, and chemically treated, frozen, irradiated foods (www.clovegarden.com).

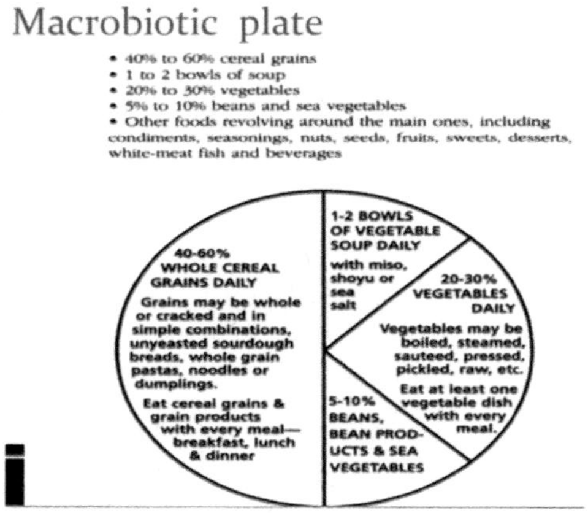

Image credit: Michio Kushi

9.14.6 PALEO DIET

- Attempt to get closer to the diet from the Palaeolithic period (Stone Age), choosing the resources from the current food.
- Basis:
  - o Lean meat, fish, seafood;
  - o Eggs;
  - o Fruit, berries;
  - o Vegetables, especially root crops;
  - o Nuts.

- Restricted:
  - o Cereals;
  - o Legumes;
  - o Potatoes;
  - o Refined oils;
  - o Sugar, salt;
  - o Milk and dairy products (wwwq.healthline.com).

9.14.7 DIVIDED FOOD DIET

Founder William Howard Hay, a physician from New York.

- Main principle is to segregate the consumption of protein-rich foods and carbohydrate-rich foods.
- Neutral foods – fats, sour milk products, fresh cheeses, curd, some vegetables, herbs, nuts, and seeds.
- Later attached policies:
  - o Do not combine protein with fruit;
  - o Fruits do not combine with vegetables;
  - o Keep drinking (2–3 liters daily);
  - o Eat at rest;
  - o Consume three meals a day with a break of 3–4 hours.

9.14.8 DIET BY BLOOD GROUP (ADAMO'S DIET)

- Book Nutrition and Blood Groups by Peter J.D. Adam.
- Individual diets by type of blood group.
- The list of foods for each blood group has been delineated.
- Three food categories are there: very beneficial, neutral, and forbidden.
- The reason for this diet pattern grouping is, according to the author, the presence of lectins in food and the human body's immune response to these substances that is affected by the blood group.

9.14.8.1 D'ADAMO RECOMMENDATION FOR EACH BLOOD TYPE

1. **Type O Blood Group:** A high-protein diet is needed, heavy on lean meat, poultry, fish, and vegetables, and light on grains, beans, and dairy. D'Adamo also recommends various supplements to help with gut troubles and other issues he says people with type O can have.

2. **Type A Blood:** A meat-free diet is required, based on fruits and vegetables, beans, and legumes, and whole grains; ideally, organic, and fresh, because D'Adamo says people with type A blood have a sensitive immune system.
3. **Type B Blood:** Avoid corn, wheat, buckwheat, lentils, tomatoes, peanuts, and sesame seeds. Chicken is also problematic, D'Adamo says. He encourages eating green vegetables, eggs, certain meats, and low-fat dairy.
4. **Type AB Blood:** Foods to emphasize include tofu, seafood, dairy, and green vegetables. He says people with type AB blood group tend to have low stomach acid levels. They need to avoid caffeine, alcohol, and smoked or cured meats (www.webmd.com).

9.14.9 DIET BY AYURVEDA

- Ayurveda is a traditional Indian medicine practiced for over 5,000 years.
- The aim is to achieve a balance between the body, soul, and senses of humans and the environment.
- The main principle is the ratio of the 3 basic types of energy Vata, Pitta, and Kapha – different for each person. They depend on:
    - o Body constitution type;
    - o The activities of humans;
    - o Choosing the right food for each person.
- Other recommendations:
    - o Prefer the plant diet as fresh as possible;
    - o Avoiding foods containing preservatives and other additives;
- o Adaptation of diet to the season, climate, and region;
- o Do not eat between meals, breaks for 3–5 hours for perfect dig (www.healthline.com).

9.14.10 DETOX DIETS

- One of the Fad diets.
- Claim to facilitate toxin elimination and weight loss promoting health and well-being.
- Recommended consumption of "alkaline foods" – such as vegetables, fruits, nuts, seeds, and herbs.

- Forbidden consumption of acidifying foods – such as meat, fish, dairy products, bakery products (products made from white flour), and refined sugar.
- Lobby of pharmaceutical firms, pharmacies, and apprentices of alternative medicine.
- Some clinical studies have shown that commercial detox diets enhance liver detoxification, BUT they have flawed methodologies and small sample sizes!
- No randomized controlled trials have been done to assess the effectiveness of commercial detox diets in humans yet (Klein et al., 2014).

9.14.11 ORGANIC FOOD

- Sometimes it's also classified under alternative nutrition.
- Consumption of foods that have been produced by methods that don't use pesticides, herbicides, or fertilizers, and which haven't been industrially processed.
- Transformed into a lifestyle.
- More friendly approach to nature and animals.

9.14.12 NUTRITIONAL AND HEALTH ASPECTS

- It depends on the degree of restriction of food of animal origin and the composition of consumed food.
- Risks of alternative nutrition:
  - Poor awareness;
  - Improper diet composition;
  - Some nutrients may be missing or in reduced quantity in a plant-based diet;
  - The presence of inhibitors of the absorption of certain nutrients;
  - Others.
- Benefits of alternative nutrition:
  - Higher consumption of fruits, vegetables, grains, sprouts, legumes, nuts, seeds, and vegetable oils.
  - Lower intake of harmful saturated fat, cholesterol, and animal protein.
  - Lower energy density.

9.14.13 SELECTED DIETARY COMPONENTS THAT ARE AT RISK

- Proteins;
- Unsaturated fatty acids;
- Iron;
- Calcium;
- Zinc;
- Iodine;
- Vitamin B12;
- Vitamin D;
- Carnitine;
- Toxic metals – cadmium;
- Advanced protein glycation end products (AGEs);
- Others – mycotoxins, nitrates.

9.14.13.1 PROTEINS

- Lower biological value of plant proteins.
- Lack of essential AAs (methionine and lysine).
- Reduced total percentage of essential AAs.
- Worse digestibility.
- To receive all essential AAs, it is necessary to consume multiple sources of vegetable protein during the day.
- The quality of plant proteins is increased when combined with one another or combined with milk, dairy products, or eggs.
- It is best to combine legumes with cereals and potatoes.

9.14.13.2 PUFA

- Vegetarian, especially vegan diet is characterized by a low intake of unsaturated long-chain fatty acids like EPA, DHA, AA, and high intakes of linoleic acid (LA).
- Vegetarians and vegans are dependent on the formation of EPA and DHA from alpha-linolenic acid.

However, this conversion is not very effective: <5–10% from EPA and 2–5% from DHA.

- The conversion also inhibits high levels of PUFA ω-6 in the diet (up to 40%), lack of energy, and protein, deficiency of pyridoxine, biotin, calcium, copper, magnesium, and zinc, and excessive ingestion of transunsaturated fatty acids.
- Iron deficiency also has a negative effect.

Increase intake of α-linolenic acid by incorporating sources with a low ω-6: ω-3 ratio such as crushed linseed oil (flaxseed), rapeseed oil, soybean oil, and walnuts.

9.14.13.3 IRON (FE)

- In plants present in non-haem form, absorption is about 3–5%, mostly depending on the overall composition of the diet.
- Absorption is negatively affected by:
  - Absorption inhibitors – phytic acid, polyphenols, oxalic acid (?), fiber;
  - Simultaneous high intake of calcium, zinc, magnesium, and potassium;
  - Plant proteins in soybean and nuts.
- Absorption is improved by:
  - Vitamin C and organic acids in fruits and vegetables;
  - Presence of AAs;
  - Soaking and germination of seeds, fermentation processes.
- The adaptation to low iron intake through increased absorption.
- Adult vegetarians have lower iron stores compared to consumers of meat, serum ferritin values are usually normal.
- Iron deficiency affects the metabolism of PUFA ω-3 with long-chain by influencing the activity of δ-6-desaturase.

9.14.13.4 CALCIUM

- Foods with a plant origin have a lower calcium availability.
  - The presence of phytates, oxalic acid, and larger content of fiber.
- Properly planned vegetarian diet can provide sufficient calcium intake, although, in the case of vegans, it is sometimes easier to use fortified foods or supplements.

9.14.13.5 ZINC

- Absorption is adversely affected by absorption inhibitors, as is the case with iron–phytic acid.
- Soya, soy products, other legumes, grains, nuts, and cheeses.
- When comparing cereals and legumes, legumes are a better source of zinc.
- Absorption is increased by seed soaking and germination, fermentation processes (disruption of zinc and phytic acid bonding), and the presence of organic acids.

9.14.13.6 IODINE

- The content of iodine in food of plant origin (*Czechia*) is less, because of its low concentration in soil.
- Vegetarians (or vegans) are hence at increased risk of Iodine deficiency.
- Increased risk of mental, psychomotor, and growth abnormalities are noted in Iodine deficiency.
- Most susceptible to iodine deficiency are newborns, infants, adolescent girls, pregnant, and lactating women.
- The risk of deficiency is increased by the so-called goitrogens, which are found in cruciferous vegetables, soybeans, and sweet potatoes.

Consumption of seaweed, mineral water containing iodine, and the use of iodinated table salt can help in managing this.

9.14.13.7 VITAMIN B12

- Plant sources of foods do not contain vitamin B12 unless they are contaminated with microorganisms that produce this vitamin.
- The main food sources are meat, fish, eggs, dairy products, yeast, and fortified foods (some soy drink substitutes, and breakfast cereals).
- Soy products like miso, tempeh, tamari, shoy, and some algae products contain only corinoids, vitamin B12 analogs, which do not normally have physiological effects comparable to those of vitamin B12.
- Vitamin B12 deficiency in adults is most likely to occur after a long-term limitation of consumption of food of animal origin rich endogenous stock up to 3 mg, unlike newborn infants only about 25 μg.
- Deficit – megaloblastic anemia, metabolic, and neurological disorders.

- Negative effects on brain development in children.
- Risk of the accumulation of homocysteine.

9.14.13.8 VITAMIN D

- Vitamin D3 or cholecalciferol is found in foods of animal origin.
- Vitamin D2 – ergocalciferol – in foods of plant origin, according to some studies absorption is less effective (?).
- Absorption is negatively affected by high phytate content in the vegetarian diet.
- Fortification of food, and supplements.
- Most at risk are individuals consuming only foods of plant origin, refusing the intake of supplements and fortified foods, particularly in the winter months.
- Children and the elderly are more prone to risks where the synthesis of vitamin D is less effective.
- Infants fed according to macrobiotic beliefs – a higher prevalence of rickets.

9.14.13.9 CARNITINE

- Essential for the metabolism of long-chain fatty acids.
- The amount of carnitine in the blood is partly dependent on food intake, about one-third of daily needs are due to endogenous synthesis.
- High carnitine content is seen in meat and is lower in dairy products. Foods of plant origin contain very little or no carnitine.
- Individuals limiting the consumption of meat, or all food of animal origin may be at risk of carnitine deficiency:
  - In terms of the insufficient amount of food;
  - Possible reduction of endogenous carnitine because of low intake of essential AAs methionine and lysine, and the possible occurrence of iron deficiency.

9.14.13.10 CADMIUM (CD)

- Toxic metal (kidney, liver) with a long biological half-life;
- Accumulation in plants;

- Higher content in cereals (grains, germ), seeds, legumes, and leafy vegetables;
- Increased levels of cadmium in vegetarians, especially vegans, in correlation with length of practice;
- Toxic effect – the presence of selenium, high saturation of the organism with iron, calcium, and zinc;
- The role in detoxification – Glutathione.

9.14.13.11 ADVANCED PROTEIN GLYCATION END PRODUCTS (AGES)

- The non-enzymatic reaction of reducing sugars with a free amino group.
- Formed during heat treatment of food.
- Negatively affect the functional properties of proteins, lipids, and DNA.
- They play an important role in the process of atherosclerosis, complications of diabetes, aging, and chronic renal insufficiency.
- For vegetarians measured higher values – cause higher intake of foods with a predominance of fructose compared to glucose which is more reactive due to a higher proportion of acyclic forms.
- AGEs may pose a certain risk, but the protective effect of eating enough fruit and vegetables is dominant from a nutritional point of view.

9.14.13.12 HEALTH ASPECTS

- Nutrition along with health style plays an important role in protecting health and preventing chronic degenerative diseases and metabolic disorders.
- Mortality – for vegetarians, lower mortality rates from ICHS.
- Cardiovascular disease.
- Blood pressure values 5–10 mmHg, occurrence of hypertension.
- Weight, BMI 1–2 kg/m^2 lower.
- Lower total cholesterol, LDL, and TAG, TNMK, higher HDL.
- Type 2 diabetes mellitus.
- Cancer – colorectal, prostate, and breast cancer (?).
- another disease – prevention of constipation, diverticulosis, and developing gallstones.

9.14.13.13 RISK POPULATION GROUPS

- Children;
- Adolescents;
- Pregnant and lactating women;
- Athletes;
- Seniors.

9.15 CONCLUSION

Thus there are many alternative foods and therapeutic foods that must be chosen keeping in mind the benefits of consuming these types of foods and the background health status of the individual. These foods are either macronutrients or micronutrients and have tremendous nutraceutical potential.

KEYWORDS

- **advanced protein glycation end products**
- **alternative**
- **diet and diseases**
- **dietary components**
- **organic food**
- **paleo diet**
- **therapeutic**

REFERENCES

17 Best Foods for Dieters. Retrieved from: https://www.webmd.com/diet/obesity/features/17-best-foods-for-dieters#4 (accessed on 05 June 2023).

A Complete Guide to a Low-Protein Diet. Retrieved from: https://www.healthline.com/nutrition/low-protein-diet (accessed on 05 June 2023).

A High-Protein Diet Plan to Lose Weight and Improve Health. Retrieved from: https://www.healthline.com/nutrition/high-protein-diet-plan (accessed on 05 June 2023).

Alternative Nutrition and Hydration – City of Hope. Retrieved from: https://www.cityofhope.org (accessed on 05 June 2023).

Atkins Diet: What Is It, and Should I Try It? Retrieved from: https://www.medicalnewstoday.com/articles/7379 (accessed on 05 June 2023).

Bland Diet: What to Eat and What to Avoid. Retrieved from: https://www.healthline.com/health/food-nutrition/bland-diet (accessed on 05 June 2023).

Fad Food Diets vs Custom Therapeutic Diets. Retrieved from: https://www.empowerhealth.ca/fad-food-vs-therapeutic-diet/ (accessed on 05 June 2023).

Garg, R., Chawla, S. P. S., Pal, R., & Ghosh, A., (2018). Impact of health education on knowledge, attitude, practices and glycemic control in type 2 diabetes mellitus. *J. Fam. Med. Primary Care.* [In Press].

Ghosh, A., & Pal, R., (2018). *Glycemic Control in Hypothyroidism and Type 2 Diabetes Mellitus.* Published in the proceedings of WONCA 15th World Rural Health Conference WRHC 2018, New Delhi.

Keservani, R. K., Kesharwani, R. K., & Sharma, A. K., (2015). Pulmonary and respiratory health: Antioxidants and nutraceuticals. In: Debasis, B., Harry, G. P., & Anand, S., (eds.), *Nutraceutical and Functional Foods in Human Life and Disease Prevention* (pp. 279–295). Chapter 17, Boca Raton, Fl. CRC Press, Taylor, and Francis. ISBN: 9781482237214.

Keservani, R. K., Kesharwani, R. K., Sharma, A. K., Vyas, N., & Chadoker, A., (2010b). Nutritional supplements: An overview. *International Journal of Current Pharmaceutical Review and Research, 1*(1), 59–75.

Keservani, R. K., Kesharwani, R. K., Vyas, N., Jain, S., Raghuvanshi, R., & Sharma, A. K., (2010a). Nutraceutical and functional food as future food: A review. *Der Pharmacia Lettre, 2*(1), 106–116.

Klein, A. V., & Kiat, H., (2014). Detox diets for toxin elimination and weight management: A critical review of the evidence. *Nutritional Sciences.* https://doi.org/10.1111/jhn.12286.

Low Copper Diet for Wilson's Disease. Retrieved from: https://www.gicare.com/gi-health-resources/copper-restriction/ (accessed on 05 June 2023).

Macrobiotics. Retrieved from: http://www.clovegarden.com/diet/macro.html (accessed on 05 June 2023).

Our Guide to the Mediterranean Diet. Retrieved from; https://www.medicalnewstoday.com/articles/324221 (accessed on 05 June 2023).

Pal, R., & Rustagi, N., (2015). Food safety: Who's who. *J. Med. Nutr. Nutraceut., 4*, 27–31.

Pal, R., (2017). Everything is learning. *Educ. Health (Abingdon), 29*(3), 271, 272.

Pal, R., Kumar, R., Pal, S., Vidyasagar, Mukherjee, B., & Sarbapalli, D., (2016). Medical education: The hot seat. *J. Family Med. Prim. Care, 5*, 20–23.

Pal, R., Kumar, R., Vidyasagar, Rustagi, N., Mukherjee, B., & Sarbapalli, D., (2014). The future direction of family medicine training in India. *J. Family Med. Prim. Care, 3*(4), 295–299.

Pal, R., Mohanta, P. K., Rustagi, N., Sarbapalli, D., Ghosh, A., & Sarker, G., (2015). Knowledge is power if you hone it. *Int. J. Green Pharm., 9*(2), 86–89.

Pal, R., Pal, S., Barua, & Ghosh, M. K., (2010). Health education intervention on diabetes in Sikkim. *Indian J. Endocr. Metab., 14*(1), 3–7.

Russel, K., Gillanders, L. K., & Plank, L. D., (2018). Dietary copper restriction in Wilson's disease. *Eur. J. Clin. Nutri., 2*, 326–331.

Saxena, V., Kumar, P., Kumari, R., Nath, B., & Pal, R., (2015). Availability of village health and nutrition day services in Uttarakhand, India. *J. Fam. Med Primary Care, 4*, 251–256.

Sharma, S., Singh, M., Pal, R., Ranjan, R., Pal, S., & Ghosh, A., (2018). National health policy 2017: Can it lead to the achievement of sustainable development goals? *Al. Ameen J. Med. Sci., 11*(1), 4-11.

Should You Try a Low-Residue Diet? Retrieved from: https://www.webmd.com/ibd-crohns-disease/crohns-disease/low-residue-diet-foods (accessed on 05 June 2023).
Singh, M., Pal, R., Ranjan, R., Sarker, G., Bharati, D. R., & Pal, S., (2017). Diabetes and dementia: Myth and reality. *J. Krishna Inst. Med. Sci. Uni., 6*(4), 12–17.
Singh, P., Kesharwani, R. K., & Keservani, R. K., (2017). Antioxidants and vitamins: Roles in cellular function and metabolism. In: *Sustained Energy for Enhanced Human Functions and Activity* (pp. 385–407). Academic Press. ISBN:9780128054130.
Sodium-Controlled Diet. Retrieved from: https://my.clevelandclinic.org/health/articles/15426-sodium-controlled-diet (accessed on 05 June 2023).
Taus, M., Mignini, E. V., Fumelli, D., Busni, D., Nicolai, G., Carletti, C., et al., (2016). Celiac disease: Gluten-free diet and what else? *Open J. Gastroenterol., 6*, 157–170.
The Blood Type Diet. Retrieved from: https://www.webmd.com/diet/a-z/blood-type-diet (accessed on 05 June 2023).
The Importance of Therapeutic Diets. Retrieved from: https://pedagogyeducation.com/PharMerica/Courses/Therapeutic-Diets-An-Introduction (accessed on 05 June 2023).
The Paleo Diet—A Beginner's Guide Plus Meal Plan. Retrieved from https://www.healthline.com/nutrition/paleo-diet-meal-plan-and-menu (accessed on 05 June 2023).
The Raw Food Diet: Should I Try It? Retrieved from: https://www.medicalnewstoday.com/articles/7381 (accessed on 05 June 2023).
Therapeutic Diets. Retrieved from: https://en.wikipedia.org/wiki/Therapeutic_food (accessed on 05 June 2023).
What Is the Ayurvedic Diet? Benefits, Downsides, and More. Retrieved from https://www.healthline.com/nutrition/ayurvedic-diet (accessed on 05 June 2023).
What to Know About Ketosis. Retrieved from: https://www.medicalnewstoday.com/articles/180858 (accessed on 05 June 2023).
What to Know About the Vegetarian Diet. Retrieved from: https://www.medicalnewstoday.com/articles/8749 (accessed on 05 June 2023).
What to Know About Vegan Diets. Retrieved from: https://www.medicalnewstoday.com/articles/149636 (accessed on 05 June 2023).
WHO/Nutrition for Older Persons. Retrieved from: http://www.who.int/nutrition/topics/ageing/en/index1.html (accessed on 05 June 2023).
www.cdss.ca.gov (accessed on 05 June 2023).

CHAPTER 10

MICRONUTRIENTS AND MACRONUTRIENTS IN FOOD ENGINEERING

H. HARIPRIYA[1] and PRAGYA KUMAR[2]

[1]*Junior Resident, Department of Community and Family Medicine, AIIMS, Phulwari Sharif, Patna, Bihar, India*

[2]*Additional Professor, Department of Community and Family Medicine, AIIMS, Phuwari Sharif, Patna, Bihar, India*

ABSTRACT

Food engineering is an upcoming branch of biotechnology as well as food technology. It renders the food fit for sale and commercialization, without hindering its quality or properties much. Presently it has many significant and novel applications. Many innovations are also being tried in this realm of food science. All these things are important for future studies and experiments. In this chapter, the authors have tried to collate and present all scientific information on this interesting aspect of biomedical research.

10.1 INTRODUCTION

The term food engineering originated from a Greek word and it is a combination of εν(ι) – eni: in or into, and the verb γείνομαι (geinomae): to create, to en-gender, to bring into being (Boom & Janssen, 2014). Food

Micronutrients and Macronutrients as Nutraceuticals. Prakash Chandra Gupta, PhD, Sayan Bhattacharyya, MBBS, MD, Nisha Sharma, PhD, MPharm, Rajesh K. Kesharwani, PhD, Raj K. Keservani, PhD, MPharm (Eds.)
© 2024 Apple Academic Press, Inc. Co-published with CRC Press (Taylor & Francis)

engineering is a multidisciplinary field including applied physical sciences, microbiology, chemistry, and engineering for food and related industries. It's a scientific and academic field of profession which applies the principles of engineering for manufacturing food and for operations like production, processing, handling, storage, packing, control, and distribution of food products (Droixhe, 2021). Engineering is defined by the European Academy of Food Engineering (EAFE) as "the study, modeling, and design of ingredients and foods at all scales using technological innovations and engineering principles in the development, manufacturing, use, and understanding of existing and emerging food processes, food packaging, and food materials from food production to digestion and satiation, enabling the development and design, production, and availability of sustainable food products." (Saguy et al., 2018).

Food engineering is a rapidly evolving discipline in the current societal scenario. Supplying food to the exploding population is one of the oldest business activities which is being done from the past till now. The major necessity is the mass production and transportation of food supplies requiring the support of a diversified good team of engineers, scientists, and marketing specialists. Food engineering is an ever-evolving field. Traditionally, food preservation and stabilization were the primary concerns, but current trends emphasize diversity, health, taste, and sustainable production. This chapter provides a historical overview of food engineering over time, as well as a discussion of the major fields within the discipline, such as thermal and nonthermal stabilization processes, separation, and isolation processes, chemical, and biochemical conversions, and mixing and structuring processes. A separate focus is paid to food engineering education. Finally, there is a discussion of potential advances and challenges. The critical issue faced by society is the production of enough amount of safe and palatable food. Hence the food engineers are appointed not just by the processing companies or the suppliers of packaging but also by the government and consulting companies. The activities of food engineering which are comprised these days are not new only the techniques have just developed from the millennium. For example, the baking of bread has begun as early as 30,000 years back. All these are done with the purpose of making safe food for consumption (Boom & Janssen, 2014).

Food processing involves the understanding of various fields of engineering. Food is too much complex and composed of a variety of physical and chemical materials which have to be learned on both macro- and microscale which involves disciplines like biochemistry, microbiology, and food chemistry at the micro-level whereas thermodynamics, rheology, and transport phenomenon at the macro-level. It requires scientists and engineers with knowledge in various aspects of food manufacturing like civil, mechanical, or chemical engineering. These fields of engineering will be helpful in devising a new procedure in food handling. Most of the food processing problems can be resolved if we enable the principles of the traditional field of engineering including the preservation and safety of food with the different new disciplines of management (Barbosa-Canovas & Juliano, 2009). A good food engineer will be well-versed in the knowledge of biology and chemical sciences applied to the food industry which includes sanitation, public health, environment control, food shelf life extension, and biological process engineering. The emerging food processes and technologies include methods like microwaves, pulse electricity radiofrequency, etc. Each topic needs food engineering calculations to strengthen food engineering and technology. The principles of food engineering are to deal with the main objectives like basic elements of food processing, transformation, conservation, food equipment, and material science like properties of food and food equipment (Kostaropoulos, 2012).

10.2 HISTORY OF FOOD ENGINEERING

Food engineering was introduced into academics as a discipline in the year 1950. In older times the trade and development of commercial systems happened because of the huge harvest and production of food grains. Specific areas were particular for the production of certain food items. Preservation of this food became equally important and the introduction of new techniques related to it also led to the development of civilization. Proper storage of the harvested food for a longer duration in central locations was the major concern of that time, which required favorable conditions for storage. Thus it became a relevant topic in the field of food processing and later in food engineering. Food preservation by definition is the process of storing the food for a longer time in the edible form. Thus all the techniques in food engineering were adopted from ancient times and now those techniques have advanced over the years (Boom & Janssen, 2014; Barbosa-Canovas & Juliano, 2009). Around the 20th centuries gave light on protecting food from the attack of microorganisms and preserving all the nutrients of the food items, making them edible as such when they are out of the package. This remained the main focus till the middle of the 20s. It focused more on the post-harvest processes which were included under food processing later after the inculcation of technology of computer science and other branches of engineering, microbiology, molecular biology, etc., over decades have made it latest. The improvement in agriculture globally by the 1960s, its huge production, and industrialization demanded a good quality of food sensorial and nutritional wise (Boom & Janssen, 2014).

The contribution made by Nicholas Appert is remarkable in the field of food engineering where he demonstrated the thermal food processing technique (Heldman & Lund, 2011). During the late 20s, the population has grown enormously and people were becoming well affluent financially leading to the usage of chemical and mechanical products like fertilizers and pesticides in the field of agriculture and also the usage of refrigerators facilitated the production of more frozen and chilled products. Thus, there was a production of diverse food items during that period (Boom & Janssen, 2014; Heldman & Lund, 2011). The production of instant and chilled food products represented the growth of food engineering at that time. The availability of ready-made products like soup, whole meals, mashed potato, etc., made life more comfortable and consumable to all. All these were made possible by Von Linde who was behind the invention and development of mechanical refrigeration. The significant contribution made by Birdseye is the basis modern frozen food industry. The contribution by Louis Pasteur led to the foundation of the reasons for food spoilage by the presence of microorganisms (Heldman & Lund, 2011).

Beyond 2000s the people were more affluent enough to live in comfort in the world which created an enormous change in the variety of food they consumed. They consumed more energy-rich food. The economic growth made people lessen their daily physical activities leading to public health problems like obesity and other related issues. The increased concern about nutrition-related health issues led to the development of this new field of engineering, which was beneficial for the consumer's health (Boom & Janssen, 2014).

10.3 EVOLUTION OF FOOD ENGINEERING AND RESEARCH

Although food engineering is the latest in the field of engineering, it had previously emerged as a discipline in academics in the 1950s itself. Various cultural practices and experiences from ancient times were taken up in the initial process of development which mostly concentrated on the food processing equipment and post-harvest techniques. During the earlier times, there were many concerns like taboos and myths related to food processing techniques and the consumption of food items which later got resolved by the evolution of Food processing and engineering practices (Barbosa-Canovas & Juliano, 2009). At first food engineering was initiated by a few engineers who had knowledge and experience in the field of the food industry. Later the engineers thought of implementing these known ideas and information into a complex item, i.e., food. The evolution of food engineering was remarkable for the past 50 years. There were several renowned scientists and engineers whose contributions were crucial for the progress in this field of science related to food. Their findings regarding food engineering topics were pen downed in various textbooks from those days to recent times. These laid the foundation for formulating the guidelines for many major topics of food engineering and were accepted by the famous institutes of food science and technology (Heldman & Lund, 2011).

Much research were conducted in the field of food engineering and technology for the past 50 years. Even though there was a wide range of research in different aspects of food engineering and technology, the core focus of the researchers was to integrate the reaction kinetics and the transport phenomenon in order to achieve the process design. The research process model is illustrated in Figure 10.1.

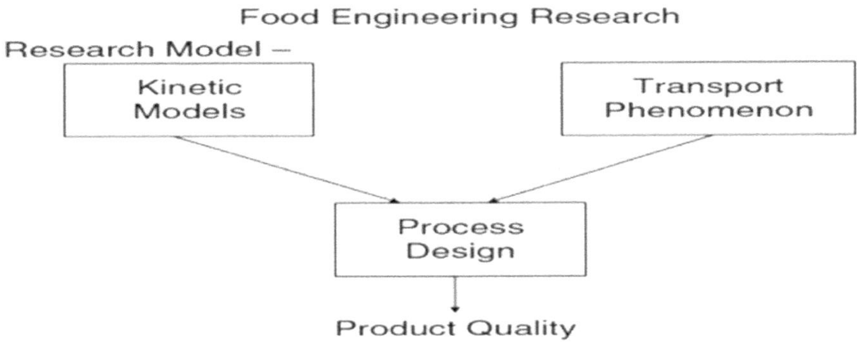

FIGURE 10.1 Flow chart of models of food engineering research.

The concept of the research model explains the three salient topics kinetic parameters, transport phenomenon, and process design. Kinetic model research explored the availability of kinetic parameters and enhanced the usage of appropriate models for food products. The kinetic models are still evolving. The research in the transport phenomenon has enhanced the understanding of heat and mass transfer in food products with a special focus on the unique properties of the food products and the mathematical models required to describe the phenomenon in a food system. It is the process design of the research which clubs the kinetic models and the appropriate transport phenomenon models and predicts the parameter to quantify the quality of the food product. Ideally, the concept is for the optimization of the process by maximizing the attributes of the product quality. Food engineering is still evolving as a discipline, and as a result, it is evolving in different ways in different regions of the world. In the past 5075 years, much of the documented evolution in food engineering research and education has occurred. Many institutions throughout the world have launched food engineering educational programs. The beginnings of these programs may be traced back to the 1950s, and the curricula are still evolving today. Some of the programs are for food engineering degrees, while others demonstrate the growing importance of food engineering in undergraduate and graduate programs leading to food science degrees. Food engineering research has been carried out at universities and in the food sector. At first, the focus was on researching specific food commodities like fruits, vegetables, dairy products, meats, and other raw food materials, but more recently, the focus has shifted to engineering principles applied to the processes required to convert raw food materials into safe food products for consumers. With a greater emphasis on new product development in the food business and a change in research to characteristics affecting ingredients and formulation, the evolution is continuing. As a result, a study has shifted to focus on the physical qualities of the compositional components of foods and food ingredients. More fundamental fields of food engineering study, such as molecular biology and nanoscale science, have become more popular in recent years. During the 20th century, the global increase in the population had increased by a factor of 6 and the production of food also exploded by a factor of 7. The consumers were purchasing more meat and other luxury food items. In recent years food engineers and food manufacturers are striving more for efficient processes. Process integration and process intensification are carried out simultaneously in the process of production of food items (Heldman & Lund, 2011). Food bio-refining is defined as a method where the raw food materials not only aim to make the basic ingredient but also for making all the side streams of high-quality food products. This is also called streamlined valorization. For example, different components of proteins are isolated after refining the whey protein (Keservani et al., 2015; Jarouliya & Keservani,

2019a; Singh et al., 2015), which is useful energy. The leftover products can be used in the production of bio-energy (Boom & Janssen, 2014).

10.4 CONTRIBUTIONS OF FOOD ENGINEERING TO RESEARCH

The process design model has many positive contributions to the manufacturing of good products. It facilitated qualitatively by predicting the process outcomes and evaluating the parameters without consuming experimental trials. Process design has a major impact on the wholesomeness and safety of food products. During the time of Nicholas Appert, it took 14 years for the process of food canning which in contrast takes only a few hours these days due to the development of process designs. All these techniques target maintaining the quality of the food products by ensuring the safety from the attack of microorganisms and thereby providing all the original attributes of the food. It is done by supplying adequate high temperature which is essential to retain the quality of food. Food engineering research ensures that quality food reaches consumers at a very modest and affordable price. All the improvements in the quality of food have come by improving the efficiency of manufacturing products. These are because of the specific improvements made at the different levels of process designs. A very good example of this is food packaging and aseptic processing (Heldman & Lund, 2011).

The concept of aseptic processing and packaging includes key components shown in Figure 10.2. Thus the availability of food products at low and affordable costs allowed the food products to be transported to distant places without compromising on their quality (Heldman & Lund, 2011).

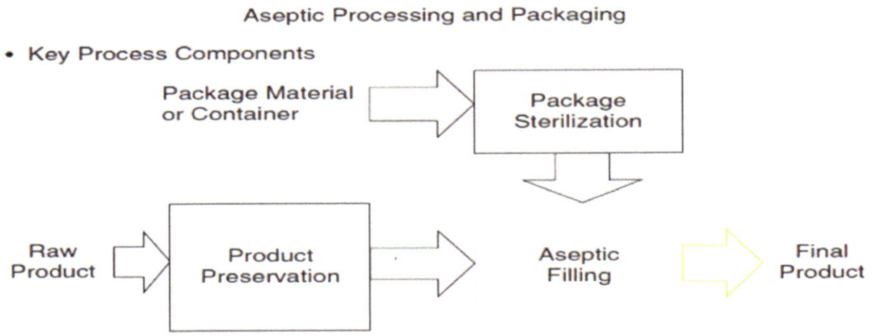

FIGURE 10.2 Aseptic processing and packaging.

Production of convenient food items through the process design technique like dry freezing was another major achievement in the field of food engineering. It helped by reducing the preparation time of the food products, availability of products with extended shelf-life, and maintaining all the quality attributes of the food making life easier and more convenient. The dry freezing technique is not a new one, its application was known since the 1930s in the production of freeze-dried coffee (Kesharwani et al., 2016). It gained attention during the 1960s during the production of low-mass food items for space flights.

Several researches are made in this field for ensuring the high quality of food by manufacturing new and unique ingredients through the processes of process design. New technologies are being used to improve product flavors by preserving volatile flavors, with a focus on encapsulation methods. Process design has created a variety of new and enhanced product textures based on new physical characteristics and information. The concept of water activity has offered a unique foundation for the creation of products and processes. Many new connections between food reaction rates and water activity have aided in the development of meals with longer shelf lives and distinct quality characteristics. These early developments paved the way for more contemporary glass transition approaches.

Many new and unique items are being developed as a result of the process design. Opportunities for new and diverse products emerge as new process technologies are developed and used in food manufacturing. As new packaging and packaging technologies are developed, similar opportunities emerge. In many circumstances, these innovative process technologies are critical for product quality improvement.

One of the best examples of a processing technology that has a great impact on food product development is extrusion. Extrusion originated in the polymer industry, and it has now been adapted for a wide range of applications in the food industry. Pressure, temperature, and time are combined in this process to create products with distinct traits and properties. The opportunities for product qualities are practically limitless when you consider the large range of each parameter and the combinations of parameters for each one. Bruin & Jongenin (2001) underline the prospects for future advancements in food extrusion, but knowledge of these principles has improved in recent years. The extrusion process involves changing the product materials from a glass to a rubber condition, then raising the temperature before expanding and drying back to a glass state. The final cooling stage gives the product the same composition and temperature as before, but it has entirely new features and characteristics. Having a thorough understanding of the process and the result opens up a world of possibilities for new and inventive goods (Heldman & Lund, 2011). The recent contribution of food engineering is to understand the breakdown of food matrices and the release and bioavailability of ingredients using food engineering concepts in the modeling and simulation of the processes taking place in the gastrointestinal tract (GIT) (Boom & Janssen, 2014).

Today Food engineering is a well-established branch of food sciences that is spread worldwide. It has made tremendous contributions and it is nowadays taught in a number of Universities there are various committees, bodies, boards, scientific journals, etc., dealing with its research.

10.5 HEALTH AND FOOD ENGINEERING

Beyond 2000 years people started consuming more energy-rich food. Everyone was affluent enough to purchase the commodities that were new in the market. It was observed that the trend of engaging in some or other physical activities was also declining. This was responsible for the emergence of health-related issues during that period which was a major public health concern. There was a rising incidence of obesity, diabetes mellitus, and cardiovascular diseases (CVDs) in some regions whereas other regions of the world were suffering from poverty, malnutrition, and famine. All these concerns related to health led to the new idea of including specific ingredients that were claimed to be beneficial for the consumer's health and well-being known as functional food. The functional foods included that are fortified with vitamins (Singh et al., 2017a) or calcium, or with probiotic ingredients (living microorganisms that are claimed to positively complement the natural ecology in the gut), prebiotics ingredients (the indigestible fiber that is claimed to positively affect the ecological balance in the human gut), unsaturated fats and oils (e.g., those containing ω-3 and ω-6 fatty acids), ingredients helping in reducing weight (e.g., fiber, but also encapsulated emulsions) or reducing cholesterols levels (phytosterols, e.g., in margarine) (Keservani & Sharma, 2014; Boom & Janssen, 2014; Heldman & Lund, 2011).

Food engineering has severed a significant role, first by producing the ingredients which deliver health benefits. In many cases, this needs separation and isolation methods to purify the components from their raw materials. Examples are the isolation of lactoferrin, lactoperoxidase, and several fatty acids from milk, the isolation of fiber from milk and other streams, and the isolation of, for example, fish oil. In addition, fermentation provides new strains of probiotic microorganisms (mostly lactic acid bacteria (LAB) and bifidobacteria), whereas enzymatic conversion of lactose yields prebiotic oligosaccharides. The techniques like stabilization of the ingredients (e.g., by freeze drying), encapsulation (e.g., for fish oil), and structuring (e.g., incorporation of phytosterols in margarine) are required to add taste and quality to the food products (Keservani et al., 2010a, b).

Dining at restaurants is common these days where most people spend their leisure time with their family and friends with good and variety of foods. It also had the positive effect that consumers worldwide have more time and resources to spend on high-quality food. This led to the emergence of molecular gastronomy and gastronomic engineering which is the latest branch of food engineering that gives an enormous scope to create innovative textures with various ingredients. This branch of food engineering explicitly aims at the creation of new sensorial experiences in a restaurant setting (i.e., not foods that are stored over time, but products that are created and immediately served to the consumer). Food engineering gives a better understanding of the process of the food matrix taking place during the preparation of the food. The field of gastronomy engineering doesn't aim at the production of food at a larger scale, but it focuses on the production of good quality food products at an individual level which brings it closer to the homes of the individual. Thus the field of food engineering increased the palatability of food products (Boom & Janssen, 2014).

After the trend of health and gastronomy, the production of food mainly focussed on its sustainability. Many efficient processes like process integration, process intensification, and processes that use less water and energy for production like drying are introduced in the field of food engineering in recent years to maintain the sustainability of food products. Process integration is reusing heat and water on several spots in the production of the process; process intensification means more than one step of the process is simultaneously carried out in small and the same equipment. The constraint of sustainability is an emerging challenge for food processing and distribution. The fundamental concepts of material and energy balances will become common tools in evaluating operations at all scales, from raw material production to product delivery to the consumer and beyond. New information transfers from process design to commercial manufacture transition will remain a significant challenge. These transitions must be achieved in a more efficient manner (Heldman & Lund, 2011).

10.6 FOOD ENGINEERING IN CURRENT FIELDS AND ITS SCOPE

Food engineering's current goal is to produce food products from ingredients that will be edible and can be preserved for a longer duration. This production of food products from the raw materials follows a sequence of steps collectively making it a process. For example, the production of beer, fructose, flour, beverages, etc., by processing the plants. There is a large-scale production of such items making use of the branches of food engineering in a wide range. The production of food ingredients, processing, and assembly of the end products includes a sequence of steps as mentioned below:

- Thermal stabilization/preservation process;
- Non-thermal stabilization process;
- Cooling and freezing;
- Dehydration process;
- Separation and isolation process;
- Chemical and biochemical conversions;
- Mixing and structuring of the process (Boom & Janssen, 2014).

The beginning of the 21st century was accompanied by the rise of micro- and nanotechnology – the technology of systems that have a dimension between 0.1 and 100 mm (Microtechnology) and between 1 and 100 nm (nanotechnology). Microfluidics will also have an impact on the production of foods. Micro-sieves are membranes made photolithographically from silicon wafers and can have an active layer as thin as 0.1 mm, have a porosity of up to 80%, and have a completely mono disperse pore size down to approximately 20 nm. Metallic micro-sieves, made through electrodeposition, are now available, which will have an impact on membrane separations (Boom & Janssen, 2014). Food engineering has a very bright future with evolving areas of research in nanoscale science and its application in the food industry. It adds another scale of consideration to the existing scales in molecular scales of the supply chain scale. As its applications are still in the evolving stage, the outcome of these will be humungous and potential in the field of food engineering. The application of the latest nano-sensors technology has increased the shelf life of food products as it detects the early deterioration of food products and helps in the immediate intervention to prevent this deterioration. The creation of innovative and improved packaging materials for food products has been aided by knowledge of packaging materials at the nanoscale. This technology also improved the understanding of the textures of the properties of the food items at a nanoscale level which improves the quality of the food (Heldman & Lund, 2011).

10.7 FUTURE CHALLENGES FOR FOOD ENGINEERING

Although food engineering is a mature field, there are major societal challenges that will transform the field in the next decade's global concerns about health, sustainability, and the coproduction of food and non-food products. It should offer suitable operations for the production of a very wide range of products and ingredients, there are large societal challenges that need to be addressed by the field. Nowadays the people with overweight are over-numbered more than the people suffering from malnutrition. Obesity is the largest pandemic in history and its incidence is increasing fast everywhere in the world. This has led to major diseases like cardiovascular disease, diabetes mellitus, hypertension, cancers, and a variety of symptoms of metabolic syndromes. Understanding how the production of new foods by new processes may help to mitigate this problem is becoming more and more urgent (Keservani et al., 2016a; Keservani & Sharma, 2016b).

The emergence of the bio-based economy is a major challenge for food engineering. As the development of fossil resources is a major force in our global economy, one has known that these fossil resources are finite. The un-bounded rate at which these resources were made available has led to an explosion of the global population and has altered the face of the planet. Now, one faces a transition back from a fossil resources-based economy, toward a bio-based economy, in which one has to use bio-based materials and energy, also for non-food purposes. But because our population has grown by at least an order of magnitude over this period, one needs to become much better in the use of these biological resources. Already one has witnessed tension over the 'food or fuel' issue; this may get more serious if one does not develop an economy in which one can produce both foods and materials and fuels from the same resources. There are a number of scientific and technological developments that may help us in approaching these challenges with some vigor.

10.8 CONCLUSION

As seen by the origins of various preservation procedures such as drying, canning, and refrigeration, many engineering impacts on the food sector have a long history. Food engineering research has recently added a quantitative dimension to the creation of both products and processes. As a result, the future of food engineering research will be focused on knowledge transfer and attempts to speed up the transition of novel processes from the laboratory bench to the production scale. Future research should focus on achieving the consumer's expectations for safe and convenient foods that will support a healthy lifestyle for long periods of time. The field can respond by absorbing the new possibilities that arise from new scientific and technological possibilities new simulation instruments based on the unprecedented computational power available, and new possibilities from the fields of micro-and nanotechnology. Food engineering education will also progress by employing more digital facilities, shifting the emphasis from understanding the detailed mechanisms, toward synthesizing process systems, and predicting the taste and health impact of products even before the products have been synthesized.

KEYWORDS

- **bio-based materials**
- **computational power**
- **food and principle**
- **food engineering**
- **healthy lifestyle**
- **palatable**
- **safe**

REFERENCES

Barbosa-Canovas, G. V., & Juliano, P., (2009). Food engineering. *Food Eng., 1*, 1–108.

Boom, R. M., & Janssen, A. E. M., (2014). Food engineering. In: Van, A. N. K., (ed.). *Encyclopedia of Agriculture and Food Systems [Internet]* (pp. 154–166). Oxford: Academic Press.

Droixhe, D., (2021). Food engineering high impact list of articles PPts journals. *Research Article: Journal of Nutrition & Food Sciences,* 826.

Heldman, D. R., & Lund, D. B., (2011). The beginning, current, and future of food engineering: A perspective. In: Aguilera, J. M., Simpson, R., Welti-Chanes, J., Bermudez-Aguirre, D., & Barbosa-Canovas, G., (eds.), *Food Engineering Interfaces (Food Engineering Series)* (pp. 3–18). New York, NY: Springer.

Jarouliya, U., & Keservani, R. K., (2019a). Protein function as a cell surface and nuclear receptor in human diseases. In: Sanjay, B., & Debarshi, K. M., (eds.), *Medicinal Chemistry with Pharmaceutical Product Development* (pp. 1–32). Chapter 1, Apple Academic Press, CRC Press, Taylor & Francis Group. ISBN: 9781771887106.

Keservani, R. K., & Sharma, A. K., (2014). Flavonoids: Emerging trends and potential health benefits. *Journal of Chinese Pharmaceutical Sciences, 23*(12), 815.

Keservani, R. K., Kesharwani, R. K., Sharma, A. K., Vyas, N., & Chadoker, A., (2010b). Nutritional supplements: An overview. *International Journal of Current Pharmaceutical Review and Research, 1*(1), 59–75.

Keservani, R. K., Kesharwani, R. K., Vyas, N., Jain, S., Raghuvanshi, R., & Sharma, A. K., (2010a). Nutraceutical and functional food as future food: A review. *Der Pharmacia Lettre, 2*(1), 106–116.

Keservani, R. K., Sharma, A. K., & Jarouliya, U., (2015). Protein and peptide in drug targeting and its therapeutic approach. *Ars Pharmaceutica (Internet), 56*(3), 165–177.

Keservani, R. K., Sharma, A. K., & Kesharwani, R. K., (2016a). Nutraceutical and functional foods for cardiovascular health. In: *Food Process Engineering* (pp. 257–278, 291–312). Apple Academic Press, CRC Press. ISBN: 9781771884020.

Keservani, R. K., Sharma, A. K., & Kesharwani, R. K., (2016b). The medicinal effect of nutraceutical fruits for cognition and brain health. *Scientifica,* 10. Article ID 3109254. doi: 10.1155/2016/3109254.

Kesharwani, R. K., Singh, P., & Keservani, R. K., (2016). Green coffee bean extract and chlorogenic acids: Chemistry and novel antioxidant benefits. In: Bagchi, D., Moriyama, H., & Swaroop, A., (eds.), *Green Coffee Bean Extract in Human Health* (pp. 1–17). Chapter 1, Boca Raton, Taylor and Francis, USA. ISBN: 9781498716376.

Kostaropoulos, A., (2012). *Food Engineering Within Sciences of Food* (p. 111).

Saguy, I. S., Roos, Y. H., & Cohen, E., (2018). Food engineering and food science and technology: Forward-looking journey to future new horizons. *Innov. Food Sci. Emerg. Technol., 47,* 326–334.

Singh, P., Kesharwani, R. K., & Keservani, R. K., (2017a). Antioxidants and vitamins: Roles in cellular function and metabolism. In: *Sustained Energy for Enhanced Human Functions and Activity* (pp. 385–407). Academic Press. ISBN: 9780128054130.

Singh, P., Kesharwani, R. K., & Keservani, R. K., (2017b). Protein, carbohydrates, and fats: Energy metabolism. In: *Sustained Energy for Enhanced Human Functions and Activity* (pp. 103–115). Academic Press. ISBN: 9780128054130.

CHAPTER 11

FOOD PROCESSING OF MICRONUTRIENTS AND MACRONUTRIENTS

K. G. NAVEEN,[1] SHAMSHAD AHMED,[2] and PRAGYA KUMAR[3]

[1]*Junior Resident, Department of Community and Family Medicine, AIIMS, Patna, Bihar, India*

[2]*Assistant Professor, Department of Community and Family Medicine, AIIMS, Patna, Bihar, India*

[3]*Additional Professor, Department of Community and Family Medicine, AIIMS, Patna, Bihar, India*

ABSTRACT

Many food processing technologies are available which make the foods last longer. However, some of these methods may rob the food of its vital nutrients. In this chapter, the authors have made a humble effort to enlist the available food processing techniques, while delineating the effects of these techniques on the food quality.

11.1 INTRODUCTION

The present mechanical world revolves around food. Food is literally in everyone's mind. This perception is imparted at least in social media. However, the healthier-looking and picture-perfect food advertised and projected on

Micronutrients and Macronutrients as Nutraceuticals. Prakash Chandra Gupta, PhD, Sayan Bhattacharyya, MBBS, MD, Nisha Sharma, PhD, MPharm, Rajesh K. Kesharwani, PhD, Raj K. Keservani, PhD, MPharm (Eds.)
© 2024 Apple Academic Press, Inc. Co-published with CRC Press (Taylor & Francis)

social media is far off from the foods people usually get on their plates daily. Few foods can be consumed raw like vegetables and fruits. But almost all foods except very few need to undergo a cumbersome process to make them more digestible, and palatable, and to enhance texture, flavor, or color to meet consumer expectations (Knorr & Watzke, 2019). Food processing may be defined as any procedure that changes or modifies food from its original state, such as adding additives (might be salt, sugar, etc.) or preservatives, or spray drying, Now, what is processed food mean? The US government has defined processed food "as any kind of food other than a raw agricultural commodity." It may include a wide variety of foodstuffs like candy, dries or candied fruits, soda, etc., which would have undergone any sort of processing at any point in time (Poti et al., 2015).

Food processing helps us to ensure safe, abundant, and readily accessible food materials. Recently food processing acquired negative connotations among consumers and experts regarding nutritional and health imbalances. Some researchers believe that the consumption of excessive processed food might lead to poor dietary quality and obesity. So, it becomes necessary to understand food processing, in brief, its vital role in meeting customer requirements, and its impact on health and nutrition.

11.2 HISTORY

Food processing existed since pre-historic times. Hunted food was eaten raw only until the fire was discovered. Then, the very first and basic technique of processing food, cooking came to play. It improves palatability, digestibility, and safety which were the major reason for processing food in those days. The consumption of processed food increases whenever it becomes difficult to consume food raw or to sustain farming and agricultural activities due to many unforeseen reasons like seasonal changes, etc. Army and navy as they have to travel huge distances for long durations, were fully reliant on processed food (Knorr & Watzke, 2019; APEC, 2020; https://ncert.nic.in/textbook/pdf/lehe105).

Over time, along with transportation and industrialization, the needs of the consumer have also increased and become more diverse. Now the demand for foods that are safe, healthy, organic, and at the same time with adequate shelf life has increased. Along with the evolution of the needs of mankind, techniques for processing food also evolved. As years passed, still more cumbersome and complex processing techniques like baking bread, cheese production, pickling vegetables using vinegar, salting or smoking the meat, and more entered the game then and thereby fulfilling the customer needs.

The advent of canned foods (Nicolas Appert) and pasteurization (Louis Pasteur) in the 18th and 19th centuries resulted in a breakthrough revolution in the food processing industry after which processing food on very large scales was introduced. *La MaisonAppert* (The House of Appert) became the very first factory to bottle food items in the whole world.

In the 20th century, a term called 'Ready-to-Eat Meals' was introduced. Consumers always look for high-quality foods that concurrently retain their form, nutritive properties, taste, and consistency, along with being simple to store and transport and simultaneously having a long shelf life. Factors deciding a consumer's food choice like cost, place of origin, cultural preferences, ease of access, way of lifestyle, health status, specific allergies, etc., also play a role in processing the food. Novel techniques, equipment, and different types of new constituents/additives all added to the food processing history of the 20th century. Further, packing and storage became so easier with the arrival of new techniques such as evaporation, the addition of new effective preservatives, and drying processes (both sun and freeze-drying).

11.3 SIGNIFICANCE

India, being a developing country progressed from an agro-deficit to an agro-surplus state which ultimately increased the demand for processing and storage of produced food particles. Indian food industry ranks 5th in terms of size with a slashing 6% GDP and is known as a significant manufacturer of processed foods:

1. **Fortification:** As most people's diet in India is their staple food (cereal, etc.), they are almost always inadequate for a few nutritive elements. To deal with this, fortification came into play which means the addition of inadequate or missing nutritive elements to the foodstuff so that the daily minimum requirements are met. In October 2016, FSSAI brought into use the Food Safety and Standards Regulations – 2016 (Fortification of Foods) for fortifying staples like wheat, flour, and rice (with Iron, Vitamin B12, and Folic Acid), Milk and Edible Oil (with Vitamins A and D) and Double Fortified Salt (with Iodine and Iron) to reduce the high burden of micronutrient

deficiency in India. The '+F' logo has been notified to help identify such fortified foods.

2. **Benefits of Fortification:** The cost-benefit ratio of food fortification is so favorable. According to the Copenhagen Consensus, every rupee spent on fortification generates a 9-rupee financial return. The equipment as well as the vitamin and mineral premix necessitate this initial investment, but the total cost of fortification is quite minimal. Nutrients are incorporated into staple ingredients in view that they may be broadly consumed. Thus, that is a very economical and effective technique to ensure that the daily nutritive requirements of a massive segment of the population are met all at once.
   - The addition of micronutrients to staple foodstuffs does not pose any health threat to people. The amount introduced is small and is based on the Recommended Daily Allowances (RDA). Fortification is regulated by an appropriate independent body to ensure secure consumption.
   - It does not alter the natural characteristics of the food itself like taste, texture, odor, color, etc. (https://ffrc.fssai.gov.in/aboutus?about=ff).
3. **Consumer Demands:** As already mentioned above, consumer perceptions about food and their demands dynamically change. Foods that are free of chemicals, pesticides, and preservatives while still having a long shelf life and retaining their natural taste and look are in high demand. All of this has raised the necessity of food processing and technology and food technologists are in high demand (https://ncert.nic.in/textbook/pdf/lehe105).
4. **Preservation:** Foods are amenable to physical, chemical, and biological deterioration. Food deterioration leads to spoilage, off-flavors, loss of textures, discoloration, and poor nutritional value, reducing the esthetic appearance and rendering it unfit/unsafe for consumption. Factors responsible for food deterioration include pests, inappropriate climatic conditions, inadequate storage, excessive moisture or heat, and so on. Food might get contaminated by microorganisms (bacteria, fungi, and molds) or chemicals such as pesticides. Food can also go bad when it is broken down by natural enzymes (a specific class of protein molecules that act as biological catalysts to speed up chemical reactions) (Keservani et al., 2010a). Some components of food spoil shortly after harvest or slaughter,

changing the quality of the food. Therefore, food processing and preservation are necessary to keep it in a safe and inedible form. Food preservation methods have prevailed since pre-historic times. The most seasoned techniques were sun drying, controlled maturation, salting/pickling, candy-coating, cooking, smoking, heating, and utilizing flavors of spices as additives. Albeit these demonstrated old procedures are being utilized even presently, with the appearance of the hi-tech transformation and modernization, new techniques have been created. Food processing encompasses and unifies the general characteristics of different classes of food and the principles of food science, chemistry, food microbiology, nutrition, sensory analysis, and statistics, including good practices in manufacturing following regulations.

11.4 TYPES OF PROCESSED FOOD

In 2009, the NOVA classification, a popular technique for classifying processed foods, was introduced. It divides foods into four categories based on how processed they are (Coetzee et al., 2020):

1. **Unprocessed or Minimally Processed Foods:** Foods that are minimally processed constitute the natural parts of plants and animals that can be consumed without any processing. Minimally processed foods have been slightly modified for the primary purpose of preservation, but this does not significantly change the nutritional value of the foods. Cleaning and removing inedible or undesired portions, as well as grinding, cooling, pasteurization, fermentation, freezing, and vacuum packaging, are just a few examples.
2. **Processed Culinary Ingredients:** Food ingredients are obtained from minimally processed foods by grinding, pressing, etc. They are generally not consumed on their own but are used in the preparation of minimally processed foods. Some examples are the oils of plants, seeds, and nuts or the flour and pasta made with whole-grain products. Also commonly referred to as "highly processed foods." These are items from the preceding group that uses artificial colors, flavors, and preservatives to extend shelf life, preserve texture, and improve palatability beyond the addition of salt, sugars, or fats. Multiple processing steps using multiple ingredients include ultra-processed foods. It is thought that these foods are specifically

designed to increase cravings so that people would possibly overeat and buy more. They are usually ready to eat with minimal additional preparation.

3. **Processed Foods:** Foods in either of the previous groups have salt, sugar, or fat added to them. Some canned fruits and vegetables, cheeses, freshly baked bread, and canned fish are examples of such processed food. These foods usually consist of at least two to three ingredients and can be consumed without any additional preparation.
4. **Ultra-Processed Foods:** Also commonly referred to as "highly processed foods" is a portion of food in the former group that includes not only salt, sweeteners, or fat, but also artificial colors, flavors, and preservatives that promote storage stability, maintain texture and improve palatability. Many of the processing steps with multiple ingredients ultimately lead to ultra-processed foods. These foods are believed to be specifically designed to increase the appetite for food, to make people overeat and buy more. They are usually ready to eat with minimal additional preparation.

This categorization, however, has since been criticized as being too broad. Consumers have been left guessing because NOVA has not supplied complete lists of individual foods in each category. Given that UPFDs are defined by food macronutrient content, it's difficult to see how this classification could aid researchers in investigating dietary links to diseases that aren't strongly linked to overall energy intakes, such as neural tube defects, cognitive growth, and decline, age-related macular degeneration (ARMD), sarcopenia, stunting, goiter, anemia, and food allergies and intolerances. Iodine, iron, and a number of micronutrients are all linked to these disorders, as are folic acid, long-chain n–3 PUFAs, protein, soluble fiber, and sugar intake frequency. In epidemiologic and genetic studies, the NOVA categorization is ineffective (Devi, 2015).

11.5 TYPES OF FOOD PROCESSING TECHNIQUES

1. **Primary Food Processing:** Food processing at its most basic level converts agricultural items such as raw wheat grains or cattle into something that can be consumed. Ingredients obtained through archaic techniques such as drying and grinding grains, shelling nuts, and cutting animals into meat fall into this category. Boning and cutting meat, freezing, and smoking fish and meat, extracting, and

filtering oils, preserving food by irradiation and translucent eggs, and homogenizing and pasteurizing milk are all part of this process.

2. **Secondary Food Processing:** It is the daily process of making food from ready-to-eat ingredients. Making bread is an example of secondary food processing, whether done at home, in a small bakery, or a huge facility. Traditional kinds of secondary food processing include fish fermentation, wine, beer, and other alcoholic beverages. Sausages are a common type of secondary food processing of meat obtained by chopping and grinding meat that has already undergone primary treatment. Most of these methods known to mankind are commonly referred to as cooking methods.
3. **Tertiary Food Processing:** It is the production for a profit of what is commonly known as processed food. These are ready-to-eat or reheated meals, such as TV and hot aero plane dishes.

11.6 ADVANTAGES OF THE FOOD PROCESSING INDUSTRY

Significant benefits for the various stakeholders involved in the food industry are:

i. **Farmer:** Higher yields, better farm utilization, less risk.
ii. **Consumer:** More variety, lower prices, new products.
iii. **Companies:** New business opportunities, increased demand economy.
iv. **Government:** Creation of jobs, reduction of migration from the countryside.

11.7 EFFECTS OF PROCESSING AND STORAGE OF FOOD ON NUTRIENTS

Certain methods of processing, such as pasteurization, cooking, and drying, can kill or inhibit the growth of harmful bacteria. Emulsifiers, for example, keep the texture of the food together by preventing solid and liquid peanut butter from separating. Other processing functions include delaying food spoilage, preserving desired sensory qualities of food (taste, texture, odor, appearance, and consistency), and providing greater convenience when preparing a complete meal. However, there are some drawbacks to food processing. Many nutrients may be destroyed or removed depending on the level of processing. Removing the outer layers of fruits, vegetables, and whole grains can extract plant nutrients and fiber. Some vitamins and minerals can be destroyed when food is heated or dried. Although food producers can recover some of the lost nutrients, it is difficult.

Almost every cooking process depletes the nutrients in food. The greatest loss of nutrients occurs during processes in which food is exposed to heat, light, and/or oxygen. Fluids added during cooking can also leach nutrients from food. Cooking potatoes, for example, can transfer the majority of the vitamin B and C in the potato into boiling water. If you consume fluids (that is, if the potatoes and water combine to make potato soup), you will still benefit from these nutrients, but not if you throw the fluid away. Similarly, when frying, baking, or frying in oil and then draining the drops, similar losses occur. Some vitamins have a higher level of stability than others. However, the majority of them are highly unstable and are lost or degraded during processing:

1. **Unstable Vitamins:** These are as follows:
   i. Folate;
   ii. Thiamine;
   iii. Vitamin C.
2. **Stable Vitamins:** Vitamins that are relatively stable are:
   i. Niacin – vitamin B3;
   ii. Vitamin K;
   iii. Vitamin D;
   iv. Biotin – vitamin B7;
   v. Pantothenic acid – vitamin B5.

11.7.1 EFFECT OF FOOD PROCESSING ON CARBOHYDRATES

Food processing has a great influence on the content and quality of the carbohydrate in the ingredient. Soluble carbohydrates such as sucrose dissolve upon cooking. Some polysaccharides are hydrolyzed. This can alter the rate and extent to which starch is digested and the properties of the dietary fiber (Keservani et al., 2020).

11.7.1.1 EFFECT ON STARCH

Heating food to prepare it and then cooling it down before consumption has a significant effect on the starch content of the food. They can be understood in terms of two important phenomena. These are given below:

1. **Gelatinization:** The crystalline structure of starch granules is irreversibly lost when heated in the presence of water, a process known as gelatinization. Because the starch granules absorb water and transform into a jelly-like substance, this occurs. Amylopectin forms a gel during this process, and the amylase dissolves. More soluble starch constituents enter the solution to form a paste when heated in excess of water. The starch granules do not completely dissolve during food processing, but their partial gelatinization is sufficient to quickly digest a significant portion of the starch. Because little gelatinization occurs during steaming, the majority of the slow-digestible starch is preserved.
2. **Retrogradation:** It is the process of recombining starch granules while cooling gelatinized starch or starch paste. Because linear amylose molecules rebind faster than highly branched amylopectin molecules, the relative proportions of amylopectin and amylose in the starch are completely dependent on this. Heating starchy foods also has an effect on this process. Because of the degree of processing and retrogradation, the digestibility of starch in the small intestine will be greatly reduced. The retrogradation of starch results in starch bread, and the rate of retrogradation depends on the temperature.

11.7.1.2 EFFECT ON DIETARY FIBER

Cereal grains are typically ground to produce refined flour, which is then processed for use in cooking. The grinding process removes the grain's outer fibrous layers, lowering the total fiber content. Wheat, rye, and corn flours are high in cellulose and hemicellulose. During the grinding process, oats and barley also lose some fiber. Furthermore, heat treatment can have an impact on the physical structure and functional properties of dietary fiber. Pectic substances thicken the juice while also softening the vegetables. The deterioration of fat and oil properties and their prevention Food processes such as heating and frying cause fat polymerization, which alters the molecular weight, color, viscosity, and refractive index of the fat or oil used. The enzymes, atmospheric oxygen, and high temperatures are the causes of these changes. Rancidity refers to the degradation of the properties of fats and oils. Those high in PUFA (linolenic acid) may lose their aroma, which gives them flavor in some cases. This is known as taste inversion. This is a serious economic issue for the food industry because it causes different tastes and tastes in cooking oils and fatty foods, making these foods less acceptable. Food spoilage is largely caused by lipid oxidation. The nutritional value of foods can be reduced by oxidative reactions, and some oxidation products are potentially toxic. On the other hand, a certain degree of lipid oxidation is sometimes desirable under certain conditions, such as in aged cheeses and some fried foods.

11.7.2 HEALTH RISKS OF HEAVILY PROCESSED FOODS

Highly processed foods tend to taste good and are often inexpensive. However, they generally have some ingredients that can be harmful if consumed in excess, like saturated fat, added sugar, and salt. These foods also contain less fiber and vitamins than all foods (LHSFNA, 2021; Eating, 2018).

11.7.2.1 INCREASED CANCER RISK

A study involving over 1,00,000 people over 5 years found that every 10% increase in the consumption of ultra-processed foods is associated with a 12% higher risk of cancer.

1. **Excess Sugar, Sodium, and Fat:** Foods that have been heavily processed often have unhealthy levels of added sugar, sodium, and fat. These ingredients make our food taste better, but too much of them can cause serious health problems such as obesity, heart disease, high blood pressure, and diabetes.
   High-glycemic white flour products with added sweeteners cause an increase in blood glucose levels without providing fiber, nutrients, or phytochemicals. These baked goods are also high in acrylamides and advanced glycation end (AGE) products, which contribute to an increase in glycoproteins in our tissues. The resulting rise in blood glucose levels causes abnormally high levels of insulin, which promotes angiogenesis, fuels fat cell growth, increases cellular replication, and promotes tumor growth. Most Americans consume an excessive amount of animal protein (such as chicken meat, which is incorrectly thought to be the healthier meat), which creates a synergistic "sandwich" of insulin and IGF-1, which may accelerate brain aging, interfere with cellular detoxification and repair, and promote cancer.

Highly processed foods are frequently high in unhealthy and inexpensive fats. For example, they primarily use refined seeds or vegetable oils that are simple to use, inexpensive, and long-lasting. Artificial trans fats are created by adding hydrogen to liquid vegetable oils, which makes them more difficult to produce. Trans fats contribute to increased inflammation in the body. They also raise levels of low-density lipoproteins, or "bad" cholesterol, while decreasing levels of high-density lipoproteins, or "good" cholesterol.
The consumption of trans-fat is associated with an increased risk of heart disease, stroke, and type 2 diabetes. For example, according to a recent study, a 2% increase in energy intake of trans fats is associated with a 23% increased risk of CVD (Islam et al., 2019; FDA, 2021).

2. **Nutritional Value is Lacking:** Heavy processing robs many foods of their essential nutrients. Hence many foods today are fortified with fiber, vitamins, and minerals (Processed, 2020).
   Processes affecting the nutrient content of foods when growing, harvesting, storing, and preparing food, various processes can take place that can affect the nutritional value of the food. The processes in which food is exposed to heat, light, or oxygen cause the greatest loss of nutrients.
   i. **Fertilizers:** Most of the plants are grown using oil fertilizers. The widespread use of nitrogen fertilizers reduces the vitamin C content in many fruit and vegetable crops. It seems that for the nutritional value of the plant, it does not matter whether the fertilizer is organic or not.
   ii. **Milling:** Cereals, such as wheat, can be crushed to remove fibrous scales. The husk contains most of the plant's dietary fiber, B vitamins, photochemical, and some minerals. This is why foods like white bread are less healthy than whole grains, even though they have been artificially fortified with certain nutrients that have been lost after grinding. Everything removed, especially phytochemicals, cannot be returned. The "fiber" that is added back to some foods often comes in the form of resistant starch, and it may not be as beneficial as the removed fiber.
   iii. **Blanching:** Food is usually heated very quickly with steam or water before canning or freezing. Water-soluble vitamins, including vitamin C and B complex, are sensitive and break down easily during blanching.

iv. **Canning:** Food is heated in a jar to kill any harmful microorganisms and extend the shelf life. Some types of microorganisms require harsh cooking, which can affect the taste and texture of food, making it less attractive. Preservatives are usually not needed or used in canned foods. Water-soluble vitamins are especially sensitive to high temperatures. Many people opine that canned foods are not as nutritious as their fresh counterparts, but this is not always the case as fresh food often spoils faster than canned food.

v. **Freezing:** The nutritional value of the food is preserved after freezing. Any loss of nutrients is related to the pre-freezing and cooking after the thawing of frozen food.

vi. **Pasteurization:** It involves heating liquid foods, such as milk and fruit juices, to a certain temperature to kill microorganisms. The nutritional value of milk usually does not change after Pasteurization. Vitamin C, however, can be lost with pasteurizing fruit juices.

vii. **High-Pressure Processing:** This alternative method of preservation puts food under pressure, with or without heat, to kill microorganisms. This method has been used in products such as fruit juices. As no heating is needed, this process has less of an impact on the vitamin content, flavor, and color of the food.

viii. **Dehydrating:** Drying foods like fruit can lower the amounts of vitamin C they contain, but it can also concentrate other nutrients, especially fiber, in plant-based foods. Food dehydration also renders the food more nutritious, which can contribute to weight gain. If dehydrated food is reconstituted and then cooked in water, further nutrients are leached from the food and are lost in the cooking water.

3. **Calorie Dense and Addictive:** It is very easy to overeat unhealthy foods and We eat more calories than we realize. For example, a popular branded cookie has about 50 calories (Green, 2021), while a whole cup of green beans has only 44 calories (Rauber, 2021). Processed foods like this are also designed to trigger the dopamine "feel-good" center of our brains, prompting us to seek more of them in the future, leading to overeating. There are significant links between ultra-processed food consumption and numerous indices of obesity, according to the literature. Although the link between ultra-processed food consumption and overall obesity isn't statistically

significant, a 10% increase in ultra-processed food consumption was linked to an increased risk of abdominal obesity (Warner, 2013).

4. **Quick to Digest:** Foods that have been processed are easier to digest as compared to unprocessed whole foods. This implies that our bodies burn less energy by digesting them. It is estimated that we burn half of the calories by digesting processed foods as compared to unprocessed foods. This fact, combined with the overall caloric density of processed foods, can facilitate weight gain. In other spectrums, processed foods might be very difficult to digest too due to their varied range of ingredients and additives.
5. **Full of Artificial Ingredients:** Approximately 5,000 compounds are introduced to our meals from outside sources (How, 2021). Most of them have never been tested outside of the company that uses them. Additives that alter color, texture, flavor, and aroma, as well as preservatives and sweeteners, fall into this category.
   Worse, the majority of these food additives have not been subjected to any safety testing, and only a minority have been evaluated in the form in which they are consumed, i.e., in combination with other compounds. Some of these ingredients are harmful. The following are some examples (More et al., 2012):
   - Diacetyl and 2,3-pentanedione (PD), which are both used to give microwave popcorn a buttery aroma, have been linked to brain functioning, Alzheimer's disease (AD), and lung toxicity (Race, 2009).
   - Butylated hydroxyanisole (BHA) and butylated hydroxyl toluene (BHT) are preservatives that can disrupt your neurological system, resulting in disruptive behaviors and even cancer. Tertiary butylhydroquinone (TBHQ), another preservative, is equally dangerous. According to the FDA, just five grams can kill you (Race, 2009).

According to the research on UPFDs, they are hyper-palatable, encourage food addiction, and have a lesser capacity to produce satiety. "Ultra-processed items are meant to satiate food demands, they are often hyper-palatable and habit building, and sometimes even quasi-addictive," according to a recent report by the Pan American Health Organization. The term "hyper palatability" refers to sensory combinations of UPFDs that override the body's and brain's natural satiation processes to encourage overconsumption.

11.8 CONCLUSION

If one is deciding whether or to not include highly processed food in the diet, it is going to be useful to judge its nutritional value and long-term health effects Ultra-processed foods with an unevenly high calorie-to-nutrient ratio are likewise deemed unhealthy. According to research, a high intake of sugar-sweetened beverages is linked to an increased risk of obesity, diabetes, and CVD. However, some processed foods, such as oil or meal, that contain beneficial nutrients have been associated with lower risks of those chronic diseases. So, the sole thanks to avoid the deleterious effects of processed food is to own a wise knowledge of the ingredients and therefore the classification.

KEYWORDS

- **additives**
- **food addiction**
- **food processing**
- **long-term health**
- **nutritional value**
- **preservatives**
- **quality**

REFERENCES

About Us: Food Fortification Resource Centre [Internet], (2021). Available from: 6. Boston 677 Huntington Avenue, Ma 02115 +1495-1000. *Processed Foods and Health* [Internet]. The Nutrition Source–2019 [cited 2021 Aug 13]. Available from: https://www.hsph.harvard.edu/nutritionsource/processed-foods/; https://ffrc.fssai.gov.in/aboutus?about=ff, (accessed on 05 June 2023).

APEC USA, (2020). *The History of Food Processing: How We Got to What We Eat [Internet].* Available from: https://www.apecusa.com/blog/the-history-of-food-processing-how-we-got-to-what-we-eat/ (accessed on 05 June 2023).

Coetzee, L. M., Cassim, N., & Glencross, D. K., (2020). Weekly laboratory turn-around time identifies poor performance masked by aggregated reporting. *Afr. J. Lab. Med. [Internet], 9*(1). Available from: https://www.ncbi.nlm.nih.gov/pmc/articles/PMC7756605/ (accessed on 05 June 2023).

Devi, R., (2015). Food processing and impact on nutrition. *Sch. J. Agric. Vet. Sci., 2*(4A), 304–311.

Eating Highly Processed Foods May Raise Cancer Risk [Internet], (2018). Harvard Health. Available from: https://www.health.harvard.edu/cancer/eating-highly-processed-foods-may-raise-cancer-risk (accessed on 05 June 2023).

FDA Takes Aim at Heart Disease, Cuts Trans Fats – LHSFNA [Internet]. Available from: https://www.lhsfna.org/index.cfm/lifelines/august-2015/fda-takes-aim-at-heart-disease-cuts-trans-fats/ (accessed on 05 June 2023).

Green Bean Nutrition Facts and Health Benefits, Facebook, Twitter, Linkedin. [Internet]. Verywell Fit. Available from: https://www.verywellfit.com/green-beans-nutrition-facts-calories-carbs-and-health-benefits-4169523 (accessed on 05 June 2023).

How Processed Foods Wreak Havoc on Your Health [Internet]. Available from: https://www.organicconsumers.org/essays/how-processed-foods-wreak-havoc-your-health (accessed on 05 June 2023).

https://ncert.nic.in/textbook/pdf/lehe105.pdf [Internet] (accessed on 05 June 2023).

Islam, M. A., Amin, M. N., Siddiqui, S. A., Hossain, M. P., Sultana, F., & Kabir, M. R., (2019). Trans fatty acids and lipid profile: A serious risk factor for cardiovascular disease, cancer, and diabetes. *Diabetes Metab. Syndr., 13*(2), 1643–1647.

Keservani, R. K., Kesharwani, R. K., Vyas, N., Jain, S., Raghuvanshi, R., & Sharma, A. K., (2010a). Nutraceutical and functional food as future food: A review. *Der Pharmacia Lettre, 2*(1), 106–116.

Keservani, R. K., Sharma, A. K., & Kesharwani, R. K., (2020). *Nutraceuticals and Dietary Supplements: Applications in Health Improvement and Disease Management.* CRC Press. ISBN: 9781771888738.

Knorr, D., & Watzke, H., (2019). Food processing at a crossroads. *Front Nutr., 6*, 85.

LHSFNA, *Dietary Habits That Make a Difference – [Internet].* Available from: https://www.lhsfna.org/index.cfm/lifelines/august-2013/dietary-habits-that-make-a-difference/ (accessed on 05 June 2023).

More, S. S., et al., (2012). The butter flavorant, diacetyl, exacerbates β-amyloid cytotoxicity. *Chem. Res. Toxicol., 25*(10). Available from: https://pubmed.ncbi.nlm.nih.gov/22731744/ (accessed on 05 June 2023).

Nutritious Cookies, Facebook, Twitter, Linkedin. The Most (and Least) [Internet]. Verywell Fit. Available from: https://www.verywellfit.com/the-best-cookies-for-weight-loss-3495635 (accessed on 05 June 2023).

Poti, J. M., Mendez, M. A., Ng, S. W., & Popkin, B. M., (2015). Is the degree of food processing and convenience linked with the nutritional quality of foods purchased by US households? *Am. J. Clin. Nutr., 101*(6), 1251–1262.

Processed Foods: Health Risks and What to Avoid [Internet]. (2020). Available from: https://www.medicalnewstoday.com/articles/318630 (accessed on 05 June 2023).

Race, S., (2009). *Antioxidants: The Truth About BHA, BHT, TBHQ, and Other Antioxidants Used as Food Additives.* Tigmors Books, United Kingdom. http://www.foodcanmakeyouill.co.uk/uploads/1/2/7/4/12746572/antioxidants.pdf (accessed on 05 June 2023).

Rauber, F., (2021). *Ultra-processed Food Consumption and Risk of Obesity: A Prospective Cohort Study of UK Biobank,* 2169–2180. https://link.springer.com/article/10.1007/s00394-020-02367-1 (accessed on 05 June 2023).

Warner, M., (2013). *Our Unsafe Food Supply Is Killing Us.* The Daily Beast [Internet]. Available from: https://www.thedailybeast.com/articles/2013/03/01/our-unsafe-food-supply-is-killing-us (accessed on 05 June 2023).

CHAPTER 12

NUTRITION, DIET, AND HEALTH: ROLE OF MACRONUTRIENTS, MICRONUTRIENTS, AND NUTRACEUTICALS

MADHUMITA DOBE

Director/Professor (Public Health), All India Institute of Hygiene and Public Health (AIIH&PH), Kolkata, West Bengal, India

ABSTRACT

Food is known to provide energy and building blocks for the growth and survival of living beings. Physiological and biochemical processes convert complex foodstuffs into particular nutrients that can then be used in metabolism. Nutraceuticals, micronutrients, and macronutrients are all essential for the sound health of the consumer.

12.1 INTRODUCTION

Nutrition is defined to be the biochemical as well as physiological process by means by which a given organism can use up food. It includes consumption and digestion after consumption inside the alimentary tract. Nutrition is a fundamental requisite, right from the first stage of development of the fetus, throughout life. It is needed for optimum physical growth, survival,

Micronutrients and Macronutrients as Nutraceuticals. Prakash Chandra Gupta, PhD, Sayan Bhattacharyya, MBBS, MD, Nisha Sharma, PhD, MPharm, Rajesh K. Kesharwani, PhD, Raj K. Keservani, PhD, MPharm (Eds.)
© 2024 Apple Academic Press, Inc. Co-published with CRC Press (Taylor & Francis)

development of the psyche, performance, and productivity (globalnutrition.org, 2021).

'Diet' refers to the kinds of food that we eat, in any kind, shape, or appearance. Human beings need a defined quota of calories and nutrients in order to maintain good health. A balanced diet generally supplies all the requisite nutrients in concordance with the recommended daily calorie intake.

Our diet has a lot of nutrients. Nutrients are molecules required by our body to carry out its primary functions. Nutrients are grouped into two main classes: Macronutrients (needed in large quantities) and micronutrients (needed in minute quantities). Macronutrients are those nutrients that our body needs in large quotas. They are carbohydrates, protein, and fats. They supply the body energy, or calories when metabolized (WHO EMRO, 2021).

Micronutrients are the nutrients (like vitamins and minerals) that we require in smaller amounts (WHO, 2021). They act as coenzymes, co-catalysts, and buffers in metabolism. Because our body does not produce these nutrients, they must be got from the diet, making them imperative components of the diet. Inadequate consumption of micronutrients can lead to poor health.

Nutraceutical is a term derived from adding the 2 words: "nutrition" with "pharmaceutical." Stephen DeFelicehas defined nutraceutical to be, "a type of food (or a component of food) which can confer some health benefits, like preventing and/or treating of a disease" (Keservani et al., 2010a, b, 2020; Das et al., 2012). Food-based nutraceuticals can be classified as:

1. **Fiber of Diet:** It is the plant matter which is not hydrolyzed by the enzymes secreted within the alimentary tract. However, they are digested by the microbes present in our gut. Dietary fibers are commonly non-starchy polysaccharides (NSP), like celluloses, lignin, hemicelluloses, gums, pectin, resistant dextrin, and resistant starches. Foods that are high in dietary fiber (Keservani et al., 2020) include oats, fruits, barley, and beans. They have been discussed in detail later on in this chapter.
2. **Probiotics:** They are food supplements that are indeed live microbes. They are administered in order to improve the microbial balance of the intestine. This has also been discussed in detail in later chapters.
3. **Prebiotics:** They are dietary ingredients that can selectively change the constitution or metabolism of the gut microbiome (discussed in detail in later chapters).
4. **Polyunsaturated Fatty Acids (PUFAs):** (discussed in detail later in this chapter).

5. **Antioxidant Vitamins:** Vitamins E and C (also called ascorbic acid) and carotenoid compounds are known to be potent antioxidant vitamins. These types of vitamins can act both singly and in synergy with each other for preventing oxidative reactions which can culminate in various degenerative diseases. These diseases include cancer, cardiovascular diseases (CVDs), and cataracts. They are found in high amounts in many different fruits and vegetables.
6. **Polyphenols:** There are about 8,000 different varieties of polyphenols. These are produced by various plants. The most common dietary polyphenols are flavonoids and phenolic acids (Keservani & Sharma, 2014). Polyphenols in diet possess powerful anti-carcinogenic and anti-atherogenic activities. The paradoxical phenomenon of relatively lesser occurrence of cardiovascular ailments among the French people, despite their consumption of diets relatively loaded in saturated fats, is due to the effect of their high consumption of red wine. The latter has antioxidant and anti-inflammatory properties, due to the presence of the polyphenol resveratrol (found in the black skin of grapes) and also proanthocyanidin, which strongly inhibits the production of Endothelin-1. The latter is a vasoactive peptide pivotal for developing coronary arteriosclerosis. Tea (also termed *Camellia sinensis)* is also a luxuriant source of polyphenols, like catechin compounds. Tea also contains the flavonols quercetin and myricetin. Black tea and green tea possess anticancer activities.
7. **Spices:** Dietary spices have antioxidative, antimutagenic, anti-inflammatory, and immunomodulatory effects.

12.2 MACRONUTRIENTS

Nutrients that are required in bigger amounts are called macronutrients. There are three broad classes of macronutrients – carbohydrates, lipids, and proteins. They are metabolically processed to derive cellular energy and thence utilized to perform work. Energy requirements are the dietary energy intake required in order to maintain the balance of energy in a healthy individual of a given age, gender, height, weight, and degree of bodily activity. The unit for measuring food-derived energy is the calorie. In children and pregnant or lactating women, energy requirements encompass the extra demands that accompany the development of tissues or milk production. Body weight indicates energy sufficiency or inadequacy. Consuming too much or too less of energy over time can result in body weight changes (globalnutrition.org, 2021). However, body weight is not a trustworthy indicator of either macronutrient or micronutrient sufficiency.

12.2.1 CARBOHYDRATES

Carbohydrates are prime sources of energy in our diet and contribute about half of the total calories consumed. Carbohydrates are made up of hydrogen, carbon, and oxygen. Important dietary carbohydrates can be grouped as:

1. **Monosaccharides:** Common dietary monosaccharides are glucose, galactose, and fructose. Of these, the most noteworthy is obviously glucose. "Blood sugar" thus generally refers to glucose only. The body has got highly adapted physiologic ways to maintain optimum blood glucose levels because the brain is highly reliant upon a regular glucose supply. Fructose (which is also termed laevulose and fruit sugar) is the sweetest among all monosaccharides. Most fruits have 1% to 7% fructose. Fructose makes up about 40% of honey. As fruits ripen, their enzymes break down the sucrose to liberate glucose and fructose, which gives a sweeter taste. Dietary galactose is generated from lactose (or milk sugar) by hydrolysis in the digestive process. Some infants have an inherent inability to metabolize galactose, an entity called galactosemia. Recent epidemiological studies have proven that high-fructose diets (along with some other factors) may be responsible for type 2 diabetes and syndrome X.
2. **Disaccharides:** The three commonest disaccharides seen in our diet are lactose, sucrose, and maltose. Sucrose (or cane sugar) is formed by the linking of glucose with fructose. It is found naturally in many foods. It is also used as an adjunct in so many commercially processed items. Honey is prepared by the action of sucrase and amylase enzymes on glucose and fructose found in the sucrose of nectar. Lactose, or milk sugar is synthesized by glucose and galactose. It is produced in the apocrine mammary glands of animals that lactate. Human and cow's milk, contain 7.5% and 4.5% of lactose, respectively. Maltose (malt sugar) is seldom seen naturally in foods but is formed by the hydrolysis of starch polymers during the process of digestion. Maltose can also be used in the form of an additive agent in a range of food products.

3. **Oligosaccharides:** These are readily water soluble and are sweet in taste. Larger carbohydrates that are not digestible, are branded as dietary fiber. Fiber is a slow-releasing carbohydrate. It cannot be metabolized in the human body; hence it passes unchanged through the gut, and remains undigested, if not the bacteria which are found in the gut can catabolize it. Dietary guidelines have hence recommended the consumption of carbohydrates like fibrous fruits, veggies, and whole grains. Dietary fibers have beneficial effects on the GI tract and help in lowering the risk of certain diseases. For example, whole wheat has cellulose which increases the water-holding capacity of the gut, increases stool volume, and decreases the transit time of the gut. Apples contain pectin which binds lipids, and bile acids increases the excretion of lipids, and reduces cholesterol levels in serum.
4. **Polysaccharides:** These are carbohydrates having more than 10 monosaccharide units, and are stored by plants as starch. Plants make two different types of starch: amylose as well as amylopectin. Amylopectin, commonly found in food, constitutes the greater portion of the starch in grains and starch-rich tubers. Raw starch (raw potato and grains) is not so well digested. Moist cooking makes these granules swell and then gelatinizes the starch. It also softens and breaks the cell wall, and makes the starch much more digestible by the action of enzymes like pancreatic amylase.

So, 1 gram weight of carbohydrates gives us four kilocalories of energy. Besides supplying energy and acting as building blocks for larger macromolecules, carbohydrates are required for the optimum functions of the nervous system, heart, and also kidneys.

12.2.1.1 GLUCOSE ABSORPTION AND THE GLYCEMIC INDEX (GI)

Dietary carbohydrates are simplified into glucose, fructose, and galactose by the alpha-amylase and digestive enzymes of the brush border present in the upper GI tract. This is modified by: (i) the susceptibility of starch to enzymatic action; (ii) the actions of the digestive enzymes present at the mucosal brush border; and (iii) the presence of other dietary factors like fat which may slow the stomach emptying. Viscous dietary fibers such as pectin and gums can dilute the enzyme concentration. So, a diet abundant in fruits, vegetables, legumes, nuts, and also minimally processed grains helps retard the speed of absorption of glucose.

Once digested, glucose is absorbed via the intestinal epithelial cells and then transferred through the portal veins to the liver. The liver then removes about 50% of this absorbed glucose for oxidation and storage as glycogen. Galactose and fructose are taken up by the hepatic cells and then incorporated into the glucose metabolism pathways. Glucose then leaves the liver and enters systemic circulation for uptake by peripheral tissues. This process is insulin-dependent.

Therefore, the major factors regulating glucose concentration in blood, after a meal are: (i) the quantity and digestibility of the ingested carbohydrate; (ii) the absorption and degree of uptake by liver cells; and (iii) insulin secretion and the proneness of peripheral tissues to the insulin action. In 1981, Jenkins defined the 'glycemic index' (GI) to categorize different dietary carbohydrates, based on their ability to elevate blood glucose levels, comparing them with a reference food. The GI of any given diet has a foreseeable effect on blood glucose levels and is useful in the dietary management of diabetes and hyperlipidemia.

12.2.2 FATS/LIPIDS

Lipids comprise hydrogen, carbon, and oxygen atoms. They are not water-soluble. Lipids can be found in butter, meats, oils, nuts, dairy products, many seeds, and in most foods that are processed. The three major groups of dietary lipids are phospholipids, triglycerides, and sterols. The major role played by lipids is to conserve energy. Dietary fat is stored in the adipose (fat) cells of the body. Lipids also serve as components of the cell membranes. They also surround and protect the visceral organs. They help in thermoregulation and control several other body functions. Lipids yield 9 KCal of energy per gram of energy. Dietary fat is also required for the digestion, absorption, and transportation of lipid-soluble vitamins and plant-derived chemicals like carotenoids and lycopene.

12.2.2.1 FATTY ACIDS

Another important group of nutrients is the fatty acids. Saturated fats are all tightly packed. The number of double bonds present between the fatty acids is nil here. Most saturated fats are usually solidified at ambient temperature. Sources of saturated fats are fat-laden pieces of meat, dairy products like cream, butter, whole milk and cheese, palm, and coconut oils. Unsaturated fats have a loosely arranged structure. They are mostly liquid at room temperature. There are two main groups of unsaturated fats: monounsaturated fatty acids (MUFA) and polyunsaturated fatty acids (PUFA). Foods abundant in monounsaturated fats include Avocado, olive oil, peanut oil, most nuts, and edible seeds.

12.2.2.1.1 Polyunsaturated Fatty Acids (PUFAs)

Polyunsaturated fatty acids (PUFAs) help in muscle movement and blood clotting. Polyunsaturated fats have two classes of fatty acids, namely omega-3 and omega-6. Linoleic acid (LA) is a vital omega-6 fatty acid, and α-linolenic acid (also called ALA) is an omega-3 fatty acid that is important for health. These are also called essential fatty acids since they cannot be produced by the human body.

Both dietary intake and endogenous metabolisms determine the level of essential fatty acids that are found in the body. Long-chain-bearing omega-3 fatty acids, namely eicosapentaenoic acid (EPA) as well as docosahexaenoic acid (DHA), can also be generated from ALA. However, since this conversion efficiency is low, foods that are rich in EPA and DHA should be consumed. Omega-6, as well as omega-3 fatty acids, are vital structural blocks of the cell membranes. They act as predecessors of lipid mediators and also supply energy. Long-chain bearing omega-3 PUFA produce anti-inflammatory effects and supplementing them can decrease mortality in patients having coronary heart disease (CHD) as well as heart failure with no preserved ventricular function (National Institute of Health, 2016). Studies have found that fish intake is associated with lower risks of decline in cognition and Alzheimer's dementia. DHA supplementation during pregnancy lowers the chances of early premature birth (before 34 weeks of pregnancy) and also very low birth weight (<1.5 kg) (National Institute of Health, 2016). DHA is also required for the visual and neurological growth of the child.

The omega-6/omega-3 ratio in our diet can affect neurotransmission and brain functioning. The ideal omega-6 fatty acid:omega-3 fatty acid ratio has been calculated to be 2: 1 to 3:1. However, the usual dietary intake is much lower. Humans should consume more omega-3 fatty acids from vegetable and marine sources like sardines, tuna, trout, salmon, mackerel, and herring, as well as from crab, oysters, and shrimp. ALA can be obtained from ground flax and flaxseed oil and soybeans. Soybean oil and sunflower oil are good sources of omega-6 fatty acids.

12.2.2.1.2 *Trans-Fatty Acids*

In naturally found unsaturated fatty acids, there are 2 carbon atoms in the double bond. Of these, each binds a hydrogen atom which is present on the ipsilateral aspect of the bond. This is called the cis-isomer form. Hydrogen can be added both in the natural cis position (with two hydrogens on the same side of the double bond) and in the trans position also (with one hydrogen present on the opposing sides of the double bond). Hydrogenation of the unsaturated fatty acids can generate stable, solid fats like margarine.

The major sources of trans-fatty acids in the diet are chemically hydrogenated margarine, commercial frying fat, baked goods, and salted snacks containing these fats. Butter and animal fats also contain trans-fatty acids which are derived from fermentation by bacteria present in the gut of cows and sheep. Studies show that a higher intake of trans-fatty acids is associated with a higher risk of CHD, cancer, and chronic diseases (like type 2 diabetes and allergies).

12.2.2.2 RECOMMENDED LEVELS OF FAT INTAKE

Fats and oils should be consumed in moderation, varying the types of oils. The Heart Association in America (AHA) advocates eating saturated fat at an amount of more than 6% of one's daily calorie needs. However, certain high-fat diets, like the Mediterranean diet, are beneficial for health. Energy demand, genetics, and lifestyle together decide these needs.

12.2.3 PROTEINS

The term 'protein' is deduced from the Greek term which implies "of primary importance." Proteins are macromolecules that consist of chains of individual subunits called amino acids. These Amino acids contain carbon, oxygen, hydrogen, and nitrogen. Dietary proteins are obtained from animal-origin foods such as meats, seafood, dairy products, and a range of different plant-based foods, such as soy, beans, and nuts. Proteins generate 4 kilocalories of energy per gram. They help in developing muscle tissue, bones, and skin, and play a role in carrying out most of the chemical reactions which occur in the body. Over 1,00,000 different proteins are seen inside the human body.

12.2.3.1 ESSENTIAL AMINO ACIDS

Our body requires 20 different amino acids in order to thrive and function properly. Though all of these are essential for health, only nine are defined as essential. They are valine, histidine, leucine, isoleucine, lysine, methionine, phenylalanine, tryptophan, and threonine. Essential amino acids cannot be synthesized by the human body; they must be obtained by diet. The most well-recognized sources for getting the essential amino acids in food are the animal proteins sourced from meat, eggs, and poultry (Healthline, 2018).

The essential amino acids can perform many important jobs:

1. **Phenylalanine:** It is the predecessor of the neurotransmitter amino acids tyrosine, epinephrine, dopamine, and norepinephrine. It serves a very important function in the structure as well as functions of proteins and enzymes. It also helps in synthesizing other amino acids.
2. **Valine:** It helps in muscle growth and regeneration. It is also involved in the production of energy.
3. **Threonine:** It is the major constituent of structural proteins in our skin and also connective tissue, like elastin and collagen. It also plays important roles in the metabolism of fat, and immunity.
4. **Tryptophan:** It is needed to maintain the proper balance of nitrogen. It is a precursor of serotonin, which is a neurotransmitter able to regulate hunger, sleep, and mood.
5. **Methionine:** It has a very important role in the control of metabolism as well as detoxification. It is also required for the optimum growth of tissues and absorption of the trace elements Zinc and Selenium.
6. **Leucine:** It is crucial for protein synthesis and muscle repair. It controls blood glucose levels, stimulates wound healing, and also produces growth hormones.
7. **Isoleucine:** It is needed for muscle metabolism. It is vital for carrying out various immune functions. It also facilitates hemoglobin production and energy regulation.
8. **Lysine:** It plays important roles in protein production, hormone and enzyme production, and the absorption of Calcium. It is needed for the production of energy, immune function, and producing collagen and elastin.
9. **Histidine:** It is required to form histamine, a neurotransmitter essential for maintaining digestion, sexual functions, immune functions, and the sleep-wake cycle. It is required for maintaining the protective myelin sheath covering the nerve cells.

Foods having all these 9 molecules, or essential amino acids can be called complete proteins (Lopez et al., 2021). Sources of complete protein are seafood, meat, poultry, eggs, and also dairy produce. Quinoa, soy, and buckwheat are plant-derived sources of such complete proteins. Plant-based proteins like beans and nuts are termed incomplete sources. This is because they lack a few of the essential amino acids. Diets based on single plant food do not have enough amino acids for protein synthesis. If another plant protein containing an excess of these limiting amino acids is mixed, the protein combination can be complemented. One can guarantee intake of all these essential amino acids, by taking a diverse diet having a number of plant protein sources like beans, seeds, nuts, whole grains, and vegetables. Protein complementarity is very important for populations with limited dietary diversity.

12.2.3.2 DIETARY PROTEIN QUALITY

The ability to produce all the required proteins for the body is dependent upon the availability of all the necessary amino acids. So, the quality of a given protein present in a diet depends on the amino acid content of that diet, and also its bioavailability.

Protein quality can be evaluated by measurement of the protein quantity that is actually used by the body (or net protein utilization (NPU)). This can be calculated by measuring Nitrogen in the diet and then converting it to the quantity of protein on the basis of the formula (nitrogen [grams] x 6.25= protein [in grams]). NPU varies from 40 to 94. Generally, proteins from animal products score higher, and proteins from vegetables have a lower NPU.

Digestibility is a major factor that affects protein quality. Meat preparations involve marination and moist heat to make meat tender through denaturation. These methods release muscle proteins and make them available to the enzymes entrusted with the digestion, of the gut. Vegetable proteins are usually less well-digested as compared to animal-source proteins. This is because they are confined within tough complex carbohydrate cell walls.

For a long time, macronutrients have remained the main focus in 'Malnutrition,' which refers to the lack, excesses, or imbalances in one's intake of energy and/or macronutrients. Malnutrition is usually manifested as:

i. **Undernutrition:** It is characterized by stunting (or low height as compared to age), wasting (defined as low weight as compared to height), and being underweight (or low weight with respect to the age of the child).

ii. **Overweight and Obesity:** The definitions of overweight and obesity include abnormal or excessive fat accumulation that could harm one's health (Jarouliya & Keservani, 2019).

12.2.3.3 PROTEIN ENERGY MALNUTRITION (PEM)

Protein-energy malnutrition (PEM) is defined as an imbalance seen between the supply of protein in food and energy, and the body's demand for these to make sure good growth and function (Jarouliya & Keservani, 2019a). It was, until recently a major public health problem in our country. It is manifested either as marasmus or as kwashiorkor (wikidoc.org, 2021). Kwashiorkor, in the "Ga" dialect of Accra, which is the capital of Ghana in Africa, means 'the disease of the deposed baby.' So, this refers literally to the sickness of an elder child when a younger sibling is born. Kwashiorkor was first described in the year 1932, by Dr. Cicely Williams, when she was working with African children on the African Gold Coast. Kwashiokor is characterized by edema (particularly ascites) resulting from severe restrictions in oral protein intake. The term marasmus, on the contrary, is a derivative of the word 'marasmos,' a Greek word meaning withering or wasting. In Marasmus, there is characteristic wasting of body tissue, especially muscles and subcutaneous fatty tissue. It results from drastic restrictions on energy intake. However, marasmus and kwashiorkor both may be present at the same time (marasmic kwashiorkor) masking the malnutrition due to the presence of edema.

Anthropometry **is** the science of measuring weight and height. It is the most commonly used method to assess malnutrition in a particular set of people. The 3 most commonly utilized anthropometric indices are: weight-for-age (WFA); height/length-for-age (HFA); weight-for-height/length (WFH).

12.2.3.4 TYPES OF UNDERNUTRITION

Three indices, namely – WFA, HFA, and WFH can be used to identify and define underweight or less body weight, stunting, and wasting, respectively. All of these indices can be calculated and written as standard deviation units (Z-scores) from the median of the reference population. Depending on this, they are further grouped as moderate or severe.

1. **Underweight:** Based on WFA, can result from either chronic or acute malnutrition, or both of these. An underweight child typically shows a WFA Z score that is at least 2 standard deviations (–2SD) less than the median according to the World Health Organization (WHO) Child Growth Standards (WHO, 2021).
2. **Stunting:** It actually indicates linear growth retardation. It can occur due to factors like long-term insufficient nutrient intake, frequently occurring infections, persistent improper feeding practices, and poverty. A stunted child shows a height-for-age Z score which is at least –2SD lesser than the median, as per the standards of WHO Child Growth (WHO, 2021).
3. **Wasting:** It indicates a failure to receive optimum nutrition in the recent past. Causes of wasting may include insufficient food intake, 9 feeding practices, different diseases, infection, or a combination of all these factors. A wasted child shows a weight-for-height Z score that is a minimum of two standard deviations (–2SD) below the median value as per the WHO Child Growth Standards (WHO, 2021).
4. **Severe Acute Malnutrition (SAM):** It means a very low WFH value (Z-score, that is below –3 SD of the median WHO child growth standards). It is assessed by a circumference of the mid-upper arm less than 115 mm, or by finding nutritional edema. SAM increases significantly the odds of death in under-five children. Lack of exclusive breastfeeding, late familiarity with complementary feeds, use of diluted feeds with lesser amounts of nutrients, repeated gut, bronchial, and lung infections (Keservani et al., 2015c), poverty, and ignorance are some important factors leading to severe acute malnutrition (SAM).
5. **Moderate Acute Level Malnutrition:** A child having 70–80% of the median value for weight-for-height (or Z score of <–3SD to <–2SD), or a Mid Upper Arm Circumference of 115–125 centimeters, without any edema, is defined as having moderate acute malnutrition (MAM). The child should possess a good appetite and should be alert and clinically good. Children with MAM can be managed in outpatient clinics with provisions for supplementary feeding.
6. **Micronutrient Malnutrition:** A number of initiatives have been undertaken to address the problems of macronutrient malnutrition, but growing evidence worldwide has shown that billions of people are also having micronutrient deficiencies. This is called "hidden

hunger," since here the victims do not feel hungry. They subsist on starchy diets lacking diversity. Such diets provide the requisite calories but deprive them of optimum nutrition. Micronutrient malnutrition is the lack of nutrients that are needed by our body in minute amounts, but are imperative for carrying out various physiological functions of the body. The World Health Report of 2000 has identified Iron, Iodine, Zinc, and vitamin A deficiencies as serious concerns in all countries of the world. Micronutrient malnutrition can lead to a plethora of clinical manifestations, physiological derangements, diminished resistance to infections as well as metabolic disorders. It leads to delayed or impaired physical and psychomotor development (Kathleen et al., 2008). A significant chunk of the burden of micronutrient deficiencies is due to poor dietary diversity of diet, lesser bioavailability of the nutrients in diets, and nutrient losses during illnesses.

7. **Vitamins:** These are water-soluble or fat-soluble organic compounds (Singh et al., 2017a). The water-soluble vitamins are vitamin C as also the B-complex group of vitamins (like thiamine, Niacin, riboflavin, pantothenic acid, biotin, Pyridoxine or Vitamin B6, folate, and cyanocobalamin). The fat-soluble vitamins are Vitamins A (retinol), E (tocopherol), D (cholecalciferol), and K (2-methyl-1,4-naphthoquinone). The body needs Vitamins to execute many regulatory and metabolic functions in the body, such as the formation of red blood cells (RBC), synthesis of bones, maintaining normal vision, nervous system functions, as well as immunity. Vitamins facilitate catalysis in chemical reactions occurring in the body and act as coenzymes to help to release energy from carbohydrates, fat, and protein (Singh et al., 2017b).
8. **Minerals:** Essential minerals can be grouped into major minerals (another name is macrominerals), and also trace minerals (also called microminerals). Both these categories of minerals seem to be equally vital for health, but microminerals are required in smaller amounts. A balanced diet contains all the essential minerals, like Calcium, Potassium, Phosphorus, Sodium, chloride, Magnesium, Zinc, Iron, Iodine, copper, Chromium, fluoride, Manganese, Molybdenum, and Selenium.

The first 50 years of the 20th century witnessed the discovery, isolation, and manufacturing of essential micronutrients and describing their role in various deficiency diseases. By the mid-20th century, all major vitamins had been discovered, described, and produced. This led to the discovery of so many dietary strategies meant to tackle conditions such as Beriberi (deficiency of vitamin B1), Scurvy (due to lack of Vitamin C), Pellagra (due to lack of vitamin B3), pernicious anemia (lack of vitamin B12), Rickets (inadequate Vitamin D), and several other deficiency disorders.

With an evolving focus on deficiency disorders of single nutrients, fortification of some chosen staple foods with micronutrients was undertaken. For instance, adding Iodine in salt, Niacin (vitamin B3), and iron in wheat flour and bread is being done. These approaches have been very effective in lowering the prevalence of common deficiency diseases like goiter (iodine deficiency), xerophthalmia (deficiency of Vitamin A), rickets (low vitamin D), and anemia (due to lack of Iron) (Mozaffarian et al., 2018). Since then, local staple foods have been supplemented with Calcium, phosphorus, iron, and specific vitamins (like A, B, C, D) (India Newborn Action Plan, NHM, 2021).

Negative Consequences of Micronutrient Deficiencies:

- Vitamin A deficiency affects eye health, resulting in xerosis and night blindness.
- Deficiency of Iron during the fetal developmental period and also early childhood can change greatly a child's growth and psychological development.
- Iodine deficiency in the gestational period and the first 12 months of life leads to permanent physical, mental, and cognitive damage to the child.
- Zinc deficiency hampers immunity and contributes to stunting in childhood period.
- Lack of folic acid and Cyanocobalamin (vitamin B12) in the early pregnancy period can lead to neural-tube defects occurring in the developing fetus.
- Lack of Calcium and Vitamin D can lead to rickets as well as retarded growth in children. It also leads to osteoporosis and osteopenia or weak bones in adults. Vitamin D deficiency is accompanied by an elevated risk of common cancers, autoimmune diseases, and hypertension.

Source: WHO & FAO (2004).

12.3 IMPORTANT ISSUES RELATED TO MICRONUTRIENT DEFICIENCIES IN INDIA

About 50% of the world's micronutrient-lacking persons stay in our country (USAID OMNI, 2005). India's most recent National Health Survey (NFHS-5, 2019–2020) results show that 57.2% women of aged 15 to 49 years and 68.9% of children aged between 6 months and 5 years are anemic. Deficiencies during pregnancy and childhood have severe consequences like more mortality, morbidity, and physical as well as psychological disorders (NFHS-5, 2021). India has a very high reported prevalence of stunting and wasting in childhood in the world. These are found in approximately 33% of all children in India (FAO, 2013).

More than 80% of the adolescent population in India has many micronutrient deficiencies (UNICEF 2019 report 'Adolescents, Diets, and Nutrition: Growing Well in the Changing World') (www.unicef.org, 2021). Such deficiencies at a young age can lead to severe effects like increased mortality, and morbidity along with physical and mental defects. A higher percentage of girls suffer from a range of micronutrient deficiencies as compared to boys of similar age, during adolescence. This adversely affects their health when they bear children later. Their children are more prone to become undernourished, leading to a vicious cycle of malnutrition occurring through generations (AIIH&PH annual report, 2016).

12.3.1 INTERVENTIONS FOR DELIVERING MICRONUTRIENTS

Interventions for the delivery of micronutrients can be categorized into 5 broad classes: nutrition-sensitive approaches, breastfeeding, diversification, and modification of diet, strengthening food with micronutrients, and providing mineral and vitamin fortifications to the targeted populations (USAID, 2021; WHO, 2021).

12.3.1.1 NUTRITION-SENSITIVE APPROACHES

Nutrition-sensitive approaches are needed to address and ameliorate the underlying systematic causes behind malnutrition, like food security, enough caregiving at maternal, household, and community strata, and timing and spacing of pregnancies. Interventions also address the promotion of safe and hygienic environments.

Agriculture interventions is a key Nutrition-sensitive approach, and it operates via three main pathways:

1. **Food Production** – increasing availability and also reducing the price of foods that are loaded with nutrients, thereby improving the diversity of diet and nutrient intake.
2. **Income** – increasing the expenditure on food.
3. **Women's Empowerment** – increasing the women's decision-making power regarding using their income for nutrition, feeding, and caregiving practices and efficient female energy expenditure.
4. **Biofortification (increasing the micronutrient concentration of common crops and their bioavailability)** – increasing the micronutrient intake by increasing the acceptability of bio-fortified crops.

12.3.1.2 DIETARY DIVERSIFICATION AND MODIFICATION

In most parts of India, diets mainly consist of starches, with low consumption of eggs, meats, dairy, fruits, and fresh vegetables. Improving household diets hence also means an increase in these categories of foods in the diet:

1. **Diversification of Diet:** It includes all interventions which increase the variety of the foods eaten at the household level. An example is the increasing consumption of animal-origin foods, especially in complementary feeds for infants. Incorporating animal-source foods, in fresh or in dried form, can increase the bioavailability of Retinol (vitamin A), iron, zinc, and Vitamins B2 as well as Vitamin B12 in the staple diet.
2. **Dietary Modification:** It means alterations in food preparation, processing, and consumption at the individual, commercial, or household levels. It improves the bioavailability of the micronutrients present in food. For instance, coffee and tea have polyphenols, which inhibit iron absorption. Hence, their consumption of food while having meals should be discouraged. Food can be processed at the household level, in order to improve bioavailability by means of germination, fermentation, and soaking. This diminishes the phytate level which can interfere with the absorption of iron and Zinc.

   Understanding food practices and access to foods, and also the study of the social and cultural barriers and other factors that can affect diet at the household and individual levels can be helpful in making changes in household dietary practices via need-based, context-specific nutrition education, counseling, and social marketing campaigns.

12.3.1.3 BREASTFEEDING

Human milk is the ideal, single source of nutrition for children less than 6 months of age. It continues to be a good complement to their diet until the second year of life. The content of some micronutrients present in breast milk depends also on the mother's intake. The breast milk concentration of some water-soluble vitamins like thiamine (B1), riboflavin (B2), pyridoxine (B6), niacin (B3), Ascorbic acid, and vitamin B12 also depends on maternal intake. More consumption of retinol and cholecalciferol by women leads to increased levels of these vitamins in breast milk. Consumption of foods rich in Iodine and Selenium can enhance the amount of these elements in breast milk. The content of Calcium as seen in breast milk is enough and comes from the mother's osteal tissue.

Improving the overall nutrition of all adolescent females who are passing through reproductive age can ensure that women maintain adequate micronutrient reserves in pregnancy and lactation period.

12.3.1.4 Food Fortification

Food fortification means adding a few essential nutrients to an industry-made food to prevent or correct a proven deficiency of those nutrient substances in the population. There are 3 main types of food fortification: (i) mass fortification or adding micronutrients and minerals to commonly consumed products, like cereal staples, milk, rice, oil, sugar, and salt; (ii) targeted fortification which involves the addition of some micronutrients to some foods that a particular population or age group eats; and (iii) commercially-driven fortification, in which food manufacturers can supplement their products with different micronutrients.

12.3.1.4.1 Mass Fortification (for Edible Products with Wide Consumption)

Mass fortification is considered when micronutrient inadequacy is widely spread in the population. Staple foods, like wheat flour and rice, vegetable oil, sugar, milk, and salt, are commonly fortified. The universal iodization of Salt program is a classical example of a successful large-scale fortification program.

Mass fortification thrives on the normal eating practices of the population. However, mass fortification does not involve promoting consumption of the fortified foods. Also, this method only partially helps benefit those people whose micronutrient needs are more than the needs of the general population, for instance, young children or pregnant and lactating women. For these categories, it is useful to utilize targeted fortification or supplementation to complete mass fortification (National Institute of Health, 2016).

12.3.1.4.2 Targeted Fortification (or Adding Micronutrients to Food for Specific Populations)

In targeted fortification, special types of foods are designed to meet the micronutrient requirements of specific populations or age groups, like.: Ready-to-Use Therapeutic Foods (also termed as RUTF) and ready-to-use supplementary foods (RUSF). These products are quite rich in both energy and protein. They supply the fatty acids and micronutrients, which are essential, to meet the needs of malnourished persons. RUTFs are used to cure simple cases of severe acute malnutrition (SAM). RUSFs on the other hand, are used for the management of MAM.

12.3.1.4.3 Commercial Fortification

When food-manufacturing people add a few micronutrients to a particular food product for commerce or brand advantage, it is called commercial fortification. Examples of commercially fortified products are fortified breakfast cereals. These products reach a smaller portion of the population than those reached by fortified staple foods. Also, they often do not reach the most susceptible groups in the population.

Government regulation is always needed to make sure that this fortification does not cause undue high consumption of micronutrients. This is a possibility when individuals consume a variety of fortified products.

12.3.1.5 SUPPLEMENTATION

A supplement is a food product having a dietary ingredient used to add more nutritional value to a diet (National Institute of Health, internet). Supplements include tablets, capsules, gelcaps, soft gels, liquids, or powders (FDA, 2015). Supplements may be consumed alone or with other foods. In public health, there are two main strategies for using micronutrient supplements:

first, traditional supplementation programs, and second, supplementation for point-of-use "fortification."

1. **Traditional Supplementation Programs:** Supplementation is targeted at the vulnerable groups whose micronutrient needs are not adequately met from the food sources alone. Pregnant women are targeted for supplementation, because their micronutrient needs, especially iron and folate requirements increase during pregnancy. Preventing anemia in adolescent female subjects of reproductive age to ensure that their iron needs are met if and when these women conceive, is achieved through intermittent (once a week) IFA supplementation for pubertal girls and menstruating women living in settings with the prevalence of anemia of 20% or higher. WHO recommends intermittent iron supplementation of preschool (aged 24–59 months) and school-age (aged 5–12 years) children if anemia prevalence is higher than 20%. Daily supplementation of iron is advised for children aged 6–23 months, in those places where anemia in children of about one year of age is above 40% in prevalence, or where the diet does not include foods strengthened with iron. Folic acid supplements are recommended before pregnancy, to prevent the development of neural-tube defects occurring in the fetus. Since the neural tube of the fetus develops in the first 14–21 days of gestation, waiting till a woman confirms she is pregnant can be too late. Calcium supplementation during pregnancy is recommended to lower the risks of eclampsia. Vitamin A supplementation is provided to children aged 6 months, until 5 years of age.

Supplementation Interventions advocated by the honorable Ministry of Health and Family Welfare (MoHFW), Government of India:

- ➢ **Iron folic Acid (IFA) Supplementation.**
- **Children:** For 0–5 years of age, 20 mg of iron and 100 micrograms (mcg) folate per ml of liquid formulation, and also removing gut worms for 100 days apt to age.
  In 6–10 years 30 mg of iron + 250 mcg folate for each child per day for a period of 100 days in the full year is advocated.
- **Adolescents 10–19 Years:** A weekly dose of 100 mg iron with 500 mcg folate +2 times-a-year deworming is beneficial.

- **Pregnant as well as Lactating Women:** For this group, 100 mg of Fe and 500 mcg folate per day, for 100 days in pregnancy, and then the same dosage for a span of 100 days in post-partum times, is postulated.
  Long-lasting insecticide-treated nets (LLINs)/insecticide-treated bed nets (ITBNs) should also be provided for pregnant women to avoid malaria-associated anemia.

MOHFW, GOI has developed the National Iron+ Initiative to bring together these extant programs (IFA supplementation for pregnant as well as lactating women and also children in the age stratum of 6–60 months) and also introduce new age groups. The National Iron+ Initiative targets the following people for supplementation/preventive programs:

- Bi-weekly iron supplementation for preschool children aged 6 months to up to 5 years;
- Weekly supplementation for children during 1st to 5th grade in government and government-aided schools;
- Weekly supplementation for the out-of-school children (5–10 years) at the various Anganwadi Centers;
- Weekly supplementation for adolescents (10–19 years);
- Weekly supplementation for women of reproductive age;
- Supplementation for pregnant + lactating women.

2. **Point-of-Use "Fortification":** Multiple single-use micronutrient powder (MNP) packets, rich in vitamins and minerals, can be used as an adjunct to any semi-solid type of food as supplements. They are targeted for children of age 6–23 months. Since these supplements are added to foods for imminent consumption, it is also termed "point-of-use fortification" or "home fortification." Depending on the identified micronutrient deficiencies of the target group, other micronutrients may also be added to the MNP.

12.4 RECOMMENDED DIETARY INTAKES OF NUTRIENTS

The spotlight on single nutrients and their deficiencies coincided with the second world war. At that time, there was widespread fear of deficiency diseases owing to shortages of food. The first rules for framing recommended dietary allowance (RDAs) were promulgated in 1941, when the

League of Nations, the British Medical Association, and The US Government separately employed scientists to yield data on new minimum dietary requirements to be prepared for war.

During the 1980s, more evidence was generated on the overall importance of micro-nutrients for immune function, physical work capacity, cognitive development, and learning prowess in children. Accordingly, the WHO and FAO organized a jointly held expert meeting in Bangkok, Thailand (September 1998) to review data on specific nutrient needs and also started preparing new recommendations for daily intake of nutrients for pregnant as well as lactating women, infants, children, and adults.

Using datasets like the National Family-Health Survey 4 (or NFHS-4, 2015–2016), and other databases like the National Nutrition Monitoring Bureau (or NNMB, 2015–2016), the WHO (WHO, 2006–2007) and Indian Academy of Paediatrics (or IAP, 2015), acceptable standard body weight figures were assessed through a person's lifetime. Reference height, then, was considered as the 95th centile for an adult male and female. With a normal BMI ranging from 18.5 to 22.9 kg/m^2, a reference/standard body weight was found. Body weights of 65 kg and 55 kg for males and females, respectively have been fixed for a normal BMI. In the year 2020, RDA has also been formulated by NIN-ICMR.

12.5 NUTRITION ACROSS ONE'S LIFE CYCLE

Good nutrition is very important during specific periods of growth, development, and aging. The impact of proper nutrition in pregnancy is pivotal for the health of the infant and mother. We know that the establishment of good dietary habits during childhood reduces the chances of inappropriate eating behavior and allays the occurrence of degenerative diseases later on in life. During early adulthood, some changes take place, that lead to the occurrence of many diseases related to aging. Many of these changes can either be hastened or slowed by modifying the type of a person's diet, gut health, and immune functions. With an increasing life span now, the elderly population is on the rise. Hence, there is a need to add to the limited nutrition data currently available for this category of individuals. Although it is known that energy needs to go down with aging, much more needs to be known about the requirements for specific nutrients. Identification of these unique differences in nutritional needs in the various stages of life is hence becoming more important and challenging.

12.5.1 THE FIRST ONE THOUSAND DAYS OF LIFE

The first one thousand days of life or the period between conception and the baby's second birthday (270 days of pregnancy+365 days in infancy+365 days in the second year of life), is a unique period. At this time, the foundation-stone for good health and development through one's lifespan is established. Right nutrition during this 1,000-day window period can improve child survival and their potential to grow, learn, and thereafter earn. Evidence shows that it is more effective and prudent to support the development of the brain by preventing nutritional gaps during this period than to go for replacement therapies after the deficit has occurred. This requires:

1. **Improved Nutrition Meant for Adolescent Females and Also Young Women, Prior to, During, and After Gestation:** The mother's diet should be diverse with at least one type of food from each group. It should include:
   - i. **Cereals:** With no white bread, biscuits, and other foods made from refined flour (Maida); seasonal fruits and vegetables to be consumed liberally with at least 2 bowls of green leafy vegetables, 1 bowl of starchy vegetables like potato, beet, carrot, and 1 bowl of other vegetables, like okra, brinjal, tomatoes, and beans. Colorful fruits and veggies are usually more nutritious.

> **Flavor Bridge:**
> Scientists at Monel Chemical Senses Center, Philadelphia, USA have discovered that babies prefer the foods consumed by their mothers during pregnancy and breastfeeding. When pregnant women have a variety of foods with diverse essential nutrients, their children adopt this habit later. They do not turn into picky eaters and stay healthy.

   - ii. **Protein:** Non-vegetarians can choose fish, full-boiled eggs, and cooked meat. Vegetarians can choose any of the pulses like moong, masoor, tuvar, or rajma, at least 2 bowls a day.
   - iii. **Fat:** Its intake should be more from vegetable sources such as mustard oil and rice bran oil. Mixing mustard oil with oil derived from rice bran, or mustard oil with groundnut oil, is advised for the optimum supply of n–3 and n–6 fatty acids in the diet. One can also eat flax seeds (or Alsi) to obtain a sufficient quota of n–3 fatty acids in the diet.

iv. **Water:** Pregnant ladies should drink at least 10–12 glasses of water every day.

v. **Iodine:** Even mild iodine deficiency during pregnancy and breast-feeding affects the brain of the unborn child and growing infants, reducing the intelligence quotient (IQ). To meet the daily requirements, pregnant women should drink 2 glassfuls of milk, 2 bowls of curd, or 2 full-boiled eggs with 1 bowl of spinach daily. Total daily salt intake should not be more than 2 tsp. of iodized salt every day. There can be a loss of iodine during storing and cooking. Salt should not be kept in hot and moist conditions, and should always be stored in airtight glass jars.

vi. **Iron:** A marked increase in the maternal blood supply during pregnancy greatly enhances the demand for iron. So, a pregnant woman must consume an additional 700 to 800 mg of iron; 500 mg is needed for hematopoiesis and 250 to 300 mg for the fetal and placental tissues. Maternal and fetal demands for Iron are the most after the 20th week of pregnancy. Pregnant women in India rarely have sufficient iron reserves to cover the demands of pregnancy. Poor iron consumption leads to poor hemoglobin production, resulting in compromised delivery of oxygen to the uterus, placenta, and developing fetus. Maternal anemia usually means a hemoglobin level lower than 11 gm/dl of blood. An anemic woman tolerates hemorrhage during delivery poorly.

Under the National Iron+ Initiative undertaken by The MOHFW, Government of India, iron supplementation for pregnant women is done with 100 mg elemental Fe (Iron) and 500 micrograms of folic acid daily, for 100 days in pregnancy. Then they are administered the same drugs in the same dosage for a period of 100 days during the post-delivery period. For good absorption, the iron supplement should be consumed between meals and not with milk, tea, or coffee, since they interfere with absorption. Vitamin C or Ascorbic acid enhances Iron absorption.

vii. **Peri-Conceptional Folic Acid Supplementation:** About 9% of neonatal deaths in India occur due to congenital anomalies (India-INAP). Defects of the neural tube (called NTDs) are reported as the commonest major congenital CNS anomaly worldwide (0.5–14/1,000 live births). Adequate Folic Acid or folate consumption before and during early pregnancy facilitates the prevention of NTD. Women need to start a diet rich in folic acid, and also folate supplementation

(400 micrograms), once they have planned for pregnancy and should continue for 3 months following confirmation of pregnancy.

viii. **Hypertensive Disorders:** This threatens the lives of mothers during pregnancy as well as lactation. Calcium prevents preeclampsia. It also facilitates the enhancement of maternal bone mineralization, thus increasing breast milk content of Calcium which in turn helps in the bone development of neonates. Recently, 'National Guidelines for Calcium Supplement during Pregnancy and Lactation for prevention of pregnancy-induced High BP' has advocated routine intake of Calcium (500 mg) + Vit D3 (250 IU) containing tablets [I tab twice daily] after 3.5 months of gestation to 1.5 to 3 months post-partum.

2. **Screening and Assessment of All Women for Pre-Conceptional Nutritional Status:** Maternal nutritional status plays a significant role in favorable outcomes of pregnancy. Body-mass index (BMI) is taken as an accepted measurement of nutritional status. Normal maternal BMI usually ranges between 18.5 kg/m^2 and 24.9 kg/m^2. An underweight mother has more risk of pre-term delivery. On the other hand, obesity can be associated with many complications like miscarriage, fetal congenital anomalies, thrombo-embolism, gestational diabetes, preeclamptic states, abnormal labor, postpartum bleeding, wound infections, stillbirth, and neonatal death.
3. **Exclusive Breastfeeding in the First Semester of Life:** As a baby's first vaccine, breastfeeding is deemed the first and best protection against illnesses, disease, and death.
4. **Provision of Nutritious, Safe, and Appropriate Food to Complement Breastmilk, as the Baby Ages:** Timely, enough, and appropriate complementary feeding is to be practiced. When breast milk *perse* does not suffice in meeting the nutritional needs of infants, other foods and liquid diets are also needed, along with breast milk. The transition from only breastfeeding to other family foods occurring over a period from 180 days to two years of age is termed complementary feeding. This is a critical period during which nutrient deficiencies and illnesses contribute to high rates of undernutrition. Complementary feeds should guarantee diversity of diet to meet the child's nutrient needs adequately. Minimum dietary diversity (MDD) implies the consumption of 4 or more food categories from the 7 established food groups so as to meet the recommended need for daily energy and nutrients. This also assures that the child has a high chance of having at least one animal-origin food and 1 fruit or

vegetable that day, along with a staple food (such as grains, roots, or tuber). In India, only about 33% of all children of age 6–23 months receive the minimum meal frequency (MMF). Also, less than one-fourth (22%) of children meet the minimum diversity of diet, with large discrepancies taking place across and within regions (NHM.gov.in, 2021). MDD with a minimum frequency of complementary feeds and up to minimum adequate diet. Less than one-tenth of all children receive a minimum requisite or adequate diet in India.

5. Availability of safe water, improved hygiene and sanitation norms, and regular assessment for tracking growth and development.

12.6 GERIATRIC NUTRITION

With increasing life expectancy, the number of people aged 60 years and above is on the rise. The composition of the body changes with increasing age affecting the nutritional needs of the elderly. Elderly people require adequate amounts of protein, carbohydrates, fats, vitamins, minerals, and dietary fiber in their diet. Nutritional anemia is a very common problem among the elderly in India and ranges from 17%–89%. For healthy aging, the elderly need more Iron, calcium, zinc, vitamin A, and various antioxidants to prevent age-related degenerative diseases. NNMB (2000) has reported inadequate intake of vitamin A, riboflavin, niacin, vitamin B12, and folates. Cholecalciferol deficiency is also quite commonplace in the geriatric population (20-90%), who have low calcium intake.

12.7 EMERGING NUTRITION AND DIET-RELATED DISEASES

Industrialization, the development of more cities, the development of the economy, and more globalization have had significant effects on the nutritional levels of people, with negative consequences in terms of inappropriate patterns of diet and the resultant increase in nutritional and diet-related disorders. With increasing incomes and more spending on foods, Indians have also been swaying away from traditional diets to more processed, fast foods high in calories and lacking in micro-nutrients. This is particularly evident in India's young population. The country is now facing a three-pronged attack of undernourishment, micronutrient deficiencies, as well as obesity (Global Nutrition Report, 2021).

In India, only about 9.6% of children between 6 and 23 months of age have been found to have minimum acceptable diets. This implies that children, irrespective of their economic status, can suffer from malnutrition during the most critical period of their mental and physical development. Other important nutritional problems include low rates of early starting and exclusive breastfeeding, poor complementary feeding indicators, folate deficiency in women of childbearing age and adolescent females, and zinc deficiency contributing to growth retardation, diarrhea, immune deficiency, and skin lesions.

Consuming a healthy diet throughout life helps to prevent malnutrition and non-communicable diseases (NCDs). However, more production of food matters that are processed, speedy city-building, and altered lifestyles have shifted our dietary patterns. People now eat foods that are loaded with fats, energy, and free sugars as well as high in salt/sodium, and do not eat sufficient quota of fruits, green vegetables, or other dietary fiber in the shape of whole grains (WHO, 2021).

The upgoing numbers of overweight and obese children and adolescents also imply the evolution of a cohort who are at higher risk to develop serious ailments like hypertension, type 2 diabetes, bronchial asthma, many other respiratory ailments, various sleep disorders, and liver problems. They may also suffer from various psychological effects like low self-esteem, depression, and social isolation along with premature death as well as disability in adulthood (Kathleen et al., 2008). In May 2014, the WHO established CECO, or Commission for Ending Childhood Obesity. In 2016, this Commission suggested a few recommendations to successfully manage obesity in childhood and adolescent period, from different perspectives, all around the world.

12.8 A DIVERSIFIED, SAFE, AND HEALTHY DIET IS NEEDED IN ALL PERIODS OF LIFE

In November 2014, the WHO and the Food and Agriculture Organization, a body of the United Nations (also called FAO), jointly held the 2nd International Conference on Nutrition (also called ICN2). Here they postulated the Declaration of Rome on Nutrition and postulated the Framework needed for the Action, which advises some policies and tactics so as to promote a diverse, secure, and healthy diet in all stages of life.

The exact formula of a diverse, healthy, and balanced diet can vary, depending on a person's characteristics (like age, sex, lifestyle, and pattern of physical activity), culture, foods available locally, and also extant dietary customs.

Basic Principles and Tenets of a Healthy Diet

- For Adults and Children: A good diet should have the following:
  - Fruits, veggies, legumes (like beans and lentils), nuts, and some whole grains (like unprocessed maize, oats, millet, brown rice, and wheat).
  - A minimum of 400 grams (or five servings) of fruit and veggies per day, other than potatoes, sweet potatoes, and other starch-laden roots, is recommended.
  - Less than one-tenth of the total intake of energy should take place from free sugars. Ideally, it should be lower than 5% of the total energy intake. Free sugars are the sugars that are generally added to food or also drinks by the maker, cook, or consumer, and also the sugars that are otherwise present in syrup, honey, fruit juices, and various fruit juice concentrates. Consuming free sugars enhances the risk of caries of tooth (or tooth decay). Too much of calories from foods and drinks that are loaded with free sugars can contribute to excess weight and obesity. Recent studies show that free sugars can affect blood pressure and serum fat levels so a reduction in free sugars intake reduces the risk factors for cardiovascular diseases (Keservani et al., 2016a). In 2010, the Health Assembly endorsed a slew of recommendations for the marketing of foods and non-alcoholic drinks which are meant for children.
  - Decreasing the amount of total fat intake to about 30% of the total calorie intake may help control weight gain in adult persons. The risk of incurring NCDs can be minimized by reducing the saturated fat to lower than one-tenth of our total daily calorie intake. One should lower trans-fats to less than 1% of the total energy intake, and replace saturated fats as well as the trans-fats in food with unsaturated fats – preferably with polyunsaturated fats. Unsaturated fats (as seen in fish, nuts, avocado, sunflower, soybean, canola, and olive oils) are to be preferred to saturated fats (found in fatty meat, and coconut oil, butter, palm oil, cheese, cream, ghee or clarified butter, and

lard) and *trans*-fats, including industrially-produced *trans*-fats (in baked or even fried food matters, pre-packed snacks, food items like frozen pizza, cookies, pies, wafers, biscuits, a few cooking oils, and spreads) and ruminant *trans*-fats (seen in meat and dairy foods obtained from ruminants, such as cows, goats, sheep, and camels). The WHO has developed a roadmap for all countries (or REPLACE action package) which provides a somewhat planned approach for eliminating all industrially-produced trans-fat from all the supply chains, aiming to achieve the target of worldwide elimination of obesity and NCD by the year 2023.

- Less than 5 gm of salt (or equivalent to almost one teaspoon) per day. Salt should always be iodized. Most people eat excess ofSodium by salt (an average of 9–12 gm of salt every day) and not so much potassium (lower than 3.5 g). High sodium and concomitant low potassium intake may also lead to hypertension, which increases the risks of cardiac disorders and stroke. Most of the salt in food comes from eating processed foods (or ready-to-eat foods, processed meat products like bacon, salami, ham, cheese, and all salty snacks). Salt is also added by us in foods at the time of cooking (like in stock cubes, soy sauce, or fish sauce) or while consuming (like table salt). In 2013, the World Health Assembly agreed to 9 global voluntary targets set with the intention to prevent and control various NCDs. These targets aim for a 30% relative lowering in the intake of salt by 2025.

➤ Additional Focus for All Children: In the initial 2 years of the child's life, proper nutrition facilitates healthy growth, improves cognitive development, and also reduces the chances of being overweight/having obesity and acquiring NCDs later on in life. A healthy diet meant for infants as well as children is somewhat the same as that for adults, along with:

- Exclusive breastfeeding during the first 180 days of life.
- Continued breastfeeding until 2 years of age and beyond.

Complementing breastmilk after 6 months of age, with a number of adequate, safe, and nutrient-dense foods. Salt and sugar should not be added to complementary food.

12.8.1 PRACTICAL SUGGESTIONS FOR MAINTAINING A HEALTHY DIET

1. **Fruit and Vegetables:** Consuming at least 400 g, or five portions, of different varieties of fresh seasonal fruits and veggies per day, lowers the risk of NCDs (noncommunicable diseases). It also helps ensure an adequate daily supply of fiber in the diet. A healthy diet should also have fruits (Keservani et al., 2016b) and vegetables as snack items.
2. **Fats:** Lowering Fat intake, particularly saturated fat and industrially-made *trans*-fat by:
   - Steaming or boiling and not frying during cooking;
   - Discarding butter, lard, and ghee and adopting oils rich in poly-unsaturated fats;
   - Eating less fat in food;
     Restricting the consumption of fried and baked foods, and pre-packed snacks and foods (such as cakes, biscuits, and cookies) that have industrially produced *trans*-fats.
3. **Salt, Sodium, and Potassium:** Lowering salt intake by:
   - Limiting the amounts of salt and Sodium-rich condiments (like sauce) while cooking and preparing foods;
   - Not eating salt or high-sodium sauce kept on the table;
   - Limiting salty snacks;
   - Checking the nutrition labels to test the amount of sodium in any product prior to buying or consuming it.

   Potassium can neutralize the harmful effects of excess sodium eating on blood pressure. Consuming potassium can be increased by eating more fresh fruits and green vegetables.
4. **Sugars:** Their intake can be lowered by:
   - Restricting the consumption of foods and drinks with high amounts of sugars, like sugary snacks, candies, and sugar-sweetened beverages (like carbonated or even non-carbonated soft drinks, fruit or vegetable juices and drinks, flavored water, liquid or powder concentrates, energy drinks or sports drinks, ready-to-have tea, ready-to-drink coffee, and all flavored milk-based drinks like that).
   - Consuming fresh fruits and raw vegetables as snacks.

12.9 PROMOTING HEALTHY DIETARY PRACTICES

Diet is affected by many social and economic factors. They interact with each other in a complicated way to determine individual dietary patterns. These factors include income, prices of foods (which can affect the affordability and availability of healthy foods), personal preferences and beliefs, culture, and environmental and geographical factors (like climate change). So, promoting a diverse, balanced, and healthy diet, needs the involvement of many sectors like government, private, and public sectors.

Governments have a great role in creating a healthy food environment. This can enable people to follow healthy food habits. Policymakers need to:

1. **Adopt coherent trade, food, and agricultural policies by:**
   i. Enhancing the incentives for producers and retailers so as to grow, use, and sell fresh fruits and veggies.
   ii. Lowering the incentives for the manufacture of processed foods that have high levels of *trans*-fats, saturated lipids, refined sugars, and salt/sodium.
   iii. Encouraging revision of the formula of food products for reducing the amounts of free sugars, saturated fats, *trans*-fats, and salt/sodium.
   iv. Fostering healthy dietary practices by ensuring the availability of healthy, nutritious, and safe foods in Anganwadi centers, schools, and other public institutions.
   v. Exploring regulatory and voluntary ways (like marketing rules and food-labeling methods), and financial incentives as well as disincentives (or taxation and subsidies).
   vi. Encouraging the food services and also the catering joints for improving the quality of nutrition of their foods, thus ensuring the availability of healthy choices that are also affordable.
2. **Encourage consumer demands for health-friendly foods and meals by:**
   i. Promoting awareness of healthy diets among consumers.
   ii. Developing policies and programs in school to encourage children to take up and maintain a healthy diet.
   iii. Educating all about nutrition and healthy dietary practices.
   iv. Encouraging cooking skills, including them in the syllabi of children.
   v. Supporting point-of-sale information, like labeling of food products that makes sure that accurate, standardized, and

comprehensible information on nutrient contents is made available in foods (which is in accordance with the guidelines of the Codex Alimentarius Commission).

vi. Providing counseling for proper nutrition and diet at primary healthcare establishments.

3. **Promote appropriate feeding practices by infants and young children:**
   i. Implementation of the International Code for Marketing of Breast-milk Substitutes;
   ii. Putting to use policies and habits to support working mothers;
   iii. Adopting breastfeeding in healthcare facilities and the community, including Baby-friendly Hospital Initiatives.

12.10 CHANGING THE MEANING OF 'DIET'

Till now, we have been discussing diet as an agglomeration of foodstuff which is consumed for nutrition. The term 'diet' has increasingly been used to imply something restrictive – as a method for losing weight. The Ancient Greeks and Romans knew that food and physical exercise can influence our health and our weight. The Greek word 'diatia' (from which the word 'diet' is derived) means a way of living that is focused on self-control and eating food in moderation.

In 1474, Bartolomeo Sacchi composed the first bestseller book on diet which was read all through Europe. The affluent society was obsessed with the association between gastronomic pleasure (voluptate) and health (valetudine) which was highlighted in the book. However, it was the Venetian oligarch Alvise (Luigi) Cornaro, who, after gauging the risks of his lifestyle, adopted a frugal diet, lived on to become a Centenarian, and wrote the famous book "*How to Live One Hundred Years – Discourses on the Sober Life*" (in AD 1558) to document that a healthy lifestyle starts from a healthy and wholesome diet.

The 20th century witnessed the emergence of a number of diets. In 1918, famous writer Lulu Hunt Peters published the book '*Diet and Health: With Key to the Calories,*' encouraging women readers to contemplate food in terms of calories and not to put in more than 1,200 kilocalories a day. Her recommendations paved the way for other means to lose weight, like the 'Weight Watchers diet' which grants points to foodstuffs according to their calorie and nutritional composition.

After years and years of focus on a simple, minimalistic outlook to dietary fat, energy density, saturated fat, and nutrient density, the emergent complex relationships between different foods and dietary patterns have generated a lot of challenges for comprehending their effects on general health and wellbeing. The growing realization of the true value of overall diet patterns has led to the discovery of many dietary patterns ranging from flexitarian, vegetarian, and vegan, to newer patterns like low carbohydrate, paleo, gluten-free, the famed Atkins diet of the 1970s or later, the Dukan diet of the 2000s.

However, studies have demonstrated that a concentration of total fat, which was a feature of dietary guidelines since 1980, has produced almost no health benefits. Nutrient-based recommendations for particular foods like eggs, red meats, and dairy products (like those based on dietary cholesterol, saturated fats, and Calcium) have proven the observed relationship of specific foods with health outcomes as wrong. Decades of emphasis on low-fat diets for weight control and glycemic control have been challenged by the implications of a series of prospective cohort studies, metabolic feeding studies, and randomized trials. These documents conclusively that foods which are rich in health-friendly fats are beneficial for our health. However, foods rich in starch and sugar can cause some harm. This led to the recognition of the significance of traditional Mediterranean or vegetarian diets having minimally processed foods like fruits, vegetables, nuts, beans, whole grains, and plant oils and low levels of highly processed foods rich in starch, sugar, salt, and additives.

Since the 1990s, researchers have focussed specifically on diets for cardiac health and mental health (Keservani et al., 2015a, 2020). Scientists now promote eating vegetables and whole-grain cereals and advocate reducing the consumption of red meat. The DASH diet evolved in this way. DASH means Dietary Approaches to Stop Hypertension (Keservani et al., 2015b). The DASH diet was planned to help cure or prevent high blood pressure (hypertension). It includes foods that are rich in Calcium, Potassium, and magnesium, restricting food matters which are high in sodium, saturated fat, and added sugar.

The 'DASH' diet is hence very rich in fruits, veggies, and whole grains. It includes fat-free or low-fat dairy products, beans, fish, poultry products, and nuts. It restricts foods that are high in saturated fat, such as fatty meats and full-fat dairy products. Studies have shown that the DASH diet can effectively reduce blood pressure. It can also reduce low-density lipoprotein (LDL).

The MIND diet is a mixture of the DASH diet and the Mediterranean diet. It emphasizes the intake of foods that improve brain health. Green leafy vegetables like spinach and lettuce can reduce the risks of dementia and cognitive decline. This is because they are rich in nutrients linked to better brain health, like folate, vitamin E, carotenoids, and flavonoids. It has been found to slow brain aging. High levels of flavonoids found in berries can improve brain health. The MIND diet also vouches for eating nuts which are abundant in lipid-soluble vitamin E, which has neuroprotective abilities. Brain-healthy eating means low consumption of meat (red meat at a frequency of 4 times a week is the proper MIND diet). It advocates replacing it with lentils, beans, and soybeans, which are rich in protein and fiber, and the B complex group of vitamins, which are vital for a healthy brain. Adults aged 65 and more, who eat fish once or more a week, score better on memory tests and number games, than those who rarely have seafood. While too much alcohol is very harmful to the brain and overall health, studies now suggest that light to moderate drinking may diminish the risks of dementia and allay the onset of full-fledged Alzheimer's disease (AD) by approximately two to three years. Alcohol improves blood flow to the brain and tissues, making them less susceptible to coagulability.

Complex biological effects of dietary patterns are increasingly felt to reflect interactions between carbohydrate quality (or GI, fiber content), fatty acid profiles, types of proteins, micronutrients, phytochemicals, food structure, preparation, and processing methods, and also additives.

During the 1970s to 1990s economic development caused the development of farming-related, food processing, and food formulation techniques, leading to a marked lowering of single nutrient deficiency diseases, but an increase in diet-related chronic diseases, like obesity, type 2 diabetes, and many cancers. This led later to the further development of therapeutic diets. They are also called meal plans, aimed to control the intake of some foods or nutrients as part of the treatment of these diseases. Therapeutic diets are advised to maintain, restore or correct the nutritional status. They can be modulated for: (i) nutrient substances; (ii) texture of food; and/or (iii) food allergy or food intolerance.

12.11 NUTRACEUTICALS AND DIET-RELATED DISEASES

Nutraceuticals provide a new approach to preventing or attenuating diet-related diseases. Nutraceuticals are helpful for heart disease, obesity, cancer, diabetes, osteoporosis, and also some other chronic or degenerative diseases like Parkinson's disease and Alzheimer's dementia. The actions of nutraceuticals involve a wide range of biological processes, like activation of antioxidant activity, signal transduction pathways, cell survival-associated gene expressions, cell multiplication and differentiation, and maintaining mitochondrial integrity (Nasri et al., 2014). These properties help it in protection against many age-related or chronic diseases.

Nutraceutical antioxidants such as curcumin, lycopene, turmerin, lutein, and β-carotene help in combating oxidative stress, lutein, and zeaxanthin can prevent the progress of cataracts and ocular macular degeneration, and beta-carotene and lycopene protect the dermis from UV ray-induced radiation damages. Lutein and lycopene are very for cardiovascular health. Lycopene helps prevent prostate cancer. Nutraceuticals are prescribed in the form of dietary supplements to reduce risk factors like high cholesterol, high blood pressure, and diabetes. Nutraceuticals are also used for postponing the development of dementia.

Functional food includes all whole foods, fortified foods, and enriched or enhanced dietary elements. They lessen the risk of acquiring chronic diseases. Functional foods include foods like cereals, breads, and beverages that are fortified with vitamins, herbs, or other nutraceuticals.

Medicinal food is consumed under the guidance and monitoring of a qualified doctor. It is used for specific management with a diet of a disease or condition for which particular nutritional requirements are established. For instance, to manage inflammatory states, cancers, and other diseases. They also provide protection against several age-related or chronic illnesses. Herbal medicines are also grouped in this category (Rane & Keservani, 2020a–c).

12.12 CONCLUSION

So far, in this chapter we have delineated the importance of nutrition, the different types of nutrients both macro- and micro, and their roles in the upkeep of the good nutrition of the body. The contribution of a healthy, safe, and diverse diet towards nutrition and good health has also been discussed. The changing scenario of nutritional science with an increasing focus on nutraceuticals has also been mentioned. These will be further deliberated in subsequent chapters.

KEYWORDS

- **amino acids**
- **development**
- **dietary protein quality**
- **food fortification**
- **lipids**
- **macronutrients**
- **micronutrients**
- **nutrition**
- **protein energy malnutrition**

REFERENCES

Adolescents, Diets and Nutrition [Internet]. www.unicef.org. (2021). Available from: https://www.unicef.org/india/reports/adolescents-diets-and-nutrition (accessed on 05 June 2023).

All India Institute of Hygiene & Public Health [Internet]. www.aiihph.gov.in., (2016). Available from: http://aiihph.gov.in/annual-report/ (accessed on 05 June 2023).

Child Nutrition – UNICEF DATA [Internet]. UNICEF DATA. [cited 7 September 2021]. Available from: https://data.unicef.org/topic/nutrition/child-nutrition/ (accessed on 05 June 2023).

Country Nutrition Profiles [Internet]. www.globalnutritionreport.org. Available from: https://globalnutritionreport.org/resources/nutrition-profiles/asia/southern-asia/india/ (accessed on 05 June 2023).

Das, L., Bhaumik, E., Raychaudhuri, U., & Chakraborty, R., (2012). Role of nutraceuticals in human health. *J. Food Sci. Technol., 49*(2), 173–183.

Essential Amino Acids: Definition, Benefits and Food Sources [Internet], (2018). Healthline. Available from: https://www.healthline.com/nutrition/essential-amino-acids (accessed on 05 June 2023).

FAO, (2013). *Food and Agriculture Organization of the United Nations* | Feedipedia [Internet]. Feedipedia.org. Available from: https://www.feedipedia.org/node/16584#:~:text=Food%20and%20Agriculture%20Organization%20of%20the%20United%20Nations,-Document%20reference&text=FAOSTAT%20(FAO)%20provides%20time%2D,from%201961%20through%20the%20present (accessed on 05 June 2023).

Globalnutritionreport.org., (2021). Available from: https://globalnutritionreport.org/resources/nutrition-profiles/ (accessed on 05 June 2023).

Guidelines on Food Fortification with Micronutrients [Internet]. Who.int. Available from: https://www.who.int/publications/i/item/9241594012 (accessed on 05 June 2023).

Healthy Diet [Internet]. Who.int. Available from: https://www.who.int/news-room/fact-sheets/detail/healthy-diet (accessed on 05 June 2023).

India Newborn Action Plan (INAP): National Health Mission [Internet], (2012). Nhm.gov.in. Available from: https://nhm.gov.in/index4.php?lang=1&level=0&linkid=153&lid=174 NNMB (accessed on 05 June 2023).

Interventions for Addressing Vitamin and Mineral Inadequacies [Internet]. Usaid.gov. Available from: https://docslib.org/doc/2540965/interventions-for-addressing-vitamin-and-mineral-inadequacies (accessed on 05 June 2023).

Jarouliya, U., & Keservani, R. K., (2019). Pathways leading to child obesity. In: Debasis, B., (ed.), *Global Perspectives on Childhood Obesity: Current Status, Consequences and Prevention* (2nd edn., pp. 137–146). Chapter 12, Academic Press, Elsevier. ISBN: 9780128128404.

Jarouliya, U., & Keservani, R. K., (2019a). Protein function as a cell surface and nuclear receptor in human diseases. In: Sanjay, B., & Debarshikar, M., (eds.), *Medicinal Chemistry with Pharmaceutical Product Development* (pp. 1–32). Apple Academic Press, CRC Press, Taylor & Francis Group. Chapter 1. ISBN: 9781771887106.

Kathleen, M. L., & Escott-Stump, S., (2008). Krause's food & nutrition therapy. *Medical Nutrition Therapy for Psychiatric Conditions,* 12.

Keservani, R. K., & Sharma, A. K., (2014). Flavonoids: Emerging trends and potential health benefits. *Journal of Chinese Pharmaceutical Sciences, 23*(12), 815.

Keservani, R. K., Kesharwani, R. K., & Sharma, A. K., (2015c). Pulmonary and respiratory health: Antioxidants and nutraceuticals. In: Debasis, B., Harry, G. P., & Anand, S., (eds.), *Nutraceutical and Functional Foods in Human Life and Disease Prevention* (pp. 279–295). Chapter 17, Boca Raton, Fl. CRC Press, Taylor and Francis. ISBN: 9781482237214.

Keservani, R. K., Kesharwani, R. K., Sharma, A. K., Vyas, N., & Chadoker, A., (2010b). Nutritional supplements: An overview. *International Journal of Current Pharmaceutical Review and Research, 1*(1), 59–75.

Keservani, R. K., Kesharwani, R. K., Vyas, N., Jain, S., Raghuvanshi, R., & Sharma, A. K., (2010a). Nutraceutical and functional food as future food: A review. *Der Pharmacia Lettre, 2*(1), 106–116.

Keservani, R. K., Sharma, A. K., & Kesharwani, R. K., (2016a). Nutraceutical and functional foods for cardiovascular health. In: *Food Process Engineering* (pp. 257–278, 291–312). Apple Academic Press, CRC Press. ISBN: 9781771884020.

Keservani, R. K., Sharma, A. K., & Kesharwani, R. K., (2016b). Medicinal effect of nutraceutical fruits for the cognition and brain health. *Scientifica,* 10. Article ID 3109254. doi: 10.1155/2016/3109254.

Keservani, R. K., Sharma, A. K., & Kesharwani, R. K., (2020). *Nutraceuticals and Dietary Supplements: Applications in Health Improvement and Disease Management.* CRC Press. ISBN: 9781771888738.

Keservani, R. K., Singh, S., Singh, V., Kesharwani, R. K., & Sharma, A. K., (2015). Nutraceuticals and functional foods in the prevention of mental disorder. In: Debasis, B., Harry, G. P., & Anand, S., (eds.), *Nutraceutical and Functional Foods in Human Life and Disease Prevention* (pp. 255–268). Chapter 15b, Boca Raton, Fl. CRC Press, Taylor and Francis. ISBN: 9781482237214.

Lopez, M. J., & Mohiuddin, S. S., (2021). Biochemistry, essential amino acids. *StatPearls [Internet].*

Micronutrients, (2021). https://www.who.int/health-topics/micronutrients#tab=tab_1 (accessed on 05 June 2023).

Mozaffarian, D., Rosenberg, I., & Uauy, R., (2018). History of modern nutrition science—Implications for current research, dietary guidelines, and food policy. *British Med J., 361*, k2392.

Nasri, H., Baradaran, A., Shirzad, H., & Rafieian-Kopaei, M., (2014). New concepts in nutraceuticals as alternative for pharmaceuticals. *Int. J. Prev. Med., 5*(12), 1487–1499.

National Family Health Survey (NFHS-5), (2021). www.rchiips.org. Available from: http://rchiips.org/nfhs/factsheet_NFHS-5.shtml (accessed on 05 June 2023).

National Institute of Health; Office of dietary supplements, (2016). *Strengthening Knowledge and Understanding of Dietary Supplements.* https://ods.od.nih.gov/factsheets (accessed on 05 June 2023).

Nhm.gov.in. (2021), Available from: https://www.nhm.gov.in/images/pdf/programmes/child-health/IEC-materials/PARTICIPANT-MANUAL_FBCSA-Malnutrition.pdf (accessed on 05 June 2023).

Nutrition, (2021). www.who.int.; https://www.who.int/health-topics/nutrition (accessed on 05 June 2023).

Nutrition. Food and Agriculture Organization of the United Nations. Available from: https://www.fao.org/nutrition/en/. Country Nutrition Profiles [Internet]. (accessed on 05 June 2023).

Protein Energy Malnutrition Classification – Wikidoc, www.wikidoc.org. Available from: https://www.wikidoc.org/index.php/Protein_energy_malnutrition_classification (accessed on 05 June 2023).

Rane, B. R., Bharath, M. S., Patil, R. R., Keservani, R. K., & Jain, A. S., (2020b). Novel approaches in nutraceuticals. In: Rajesh, K. K., Raj, K. K., & Anil, K. S., (eds.), *Enhancing the Therapeutic Efficacy of Herbal Formulations Through Novel Drug Delivery Systems* (pp. 241–266). Chapter 11, IGI Global International Publisher, Pennsylvania, USA. ISBN13: 9781799844532.

Rane, B. R., Patil, A. S., Keservani, R. K., & Jain, A. S., (2020a). Novel approaches in herbal formulation. In: Rajesh, K. K., Raj, K. K., & Anil, K. S., (eds.), *Enhancing the Therapeutic Efficacy of Herbal Formulations Through Novel Drug Delivery Systems* (pp. 43–68). Chapter 2, IGI Global International Publisher, Pennsylvania, USA. ISBN13: 9781799844532.

Rane, B. R., Tadavi, S. A., & Keservani, R. K., (2020c). Naturopathy. In: Anil, K. S., Raj, K. K., & Surya, P. G., (eds.), *Herbal Product Development* (pp. 321–347). Chapter 12, Apple Academic Press, CRC Press, Taylor & Francis Group. ISBN: 9781771888776.

Singh, P., Kesharwani, R. K., & Keservani, R. K., (2017a). Antioxidants and vitamins: Roles in cellular function and metabolism. In: *Sustained Energy for Enhanced Human Functions and Activity* (pp. 385–407). Academic Press. ISBN:9780128054130.

Singh, P., Kesharwani, R. K., & Keservani, R. K., (2017b). Protein, carbohydrates, and fats: Energy metabolism. In: *Sustained Energy for Enhanced Human Functions and Activity* (pp. 103–115). Academic Press. ISBN:9780128054130.

The WHO Child Growth Standards [Internet]. Who.int. Available from: https://www.who.int/tools/child-growth-standards (accessed on 05 June 2023).

Tools and Toolkits. World Health Organization, www.who.int. Available from: https://www.who.int/tools (accessed on 05 June 2023).

WHO EMRO, (2021). *Macronutrients.* Health topics Emro.who.int. Available from: http://www.emro.who.int/health-topics/macronutrients/index.html (accessed on 05 June 2023).

INDEX

B

E

J

K

L

N

O

Q

R

T